M000084404

Seventh Edition

Essential Mathematics with Applications

Student Support Edition

Vernon C. Barker
Palomar College, California

Richard N. Aufmann
Palomar College, California

Joanne S. Lockwood
New Hampshire Community
Technical College

Houghton Mifflin Company
Boston   New York

Publisher: Richard Stratton
Executive Editor: Mary Finch
Senior Marketing Manager: Katherine Greig
Associate Editor: Carl Chudyk
Art and Design Manager: Jill Haber
Cover Design Manager: Anne S. Katzeff
Senior Photo Editor: Jennifer Meyer Dare
Senior Composition Buyer: Chuck Dutton
New Title Project Manager: James Lonergan
Editorial Assistant: Nicole Catavolos
Marketing Assistant: Erin Timm

Cover photo © Bryan Reinhart/Masterfile

Photo Credits

Chapter 1: *p. 1:* Photodisc Blue/Getty Images; *p. 2:* Photodisc Green/Getty Images; *p. 12:* Alan Schein Photography/CORBIS; *p. 15:* Laura Dwight/PhotoEdit, Inc.; *p. 20:* Hulton-Deutsch Collection/CORBIS; *p. 24:* Rachel Epstein/PhotoEdit, Inc.; *p. 25:* CORBIS; *p. 32:* Blaine Harrington III/CORBIS; *p. 39:* Michelle D. Bridwell/PhotoEdit, Inc.; *p. 44:* CORBIS; *p. 54:* CORBIS; *p. 56:* Jonathan Nourak/PhotoEdit, Inc.; *p. 60:* Pete Saloutos/CORBIS; *p. 62:* Ed Young/CORBIS. **Chapter 2:** *p. 63:* AP/Wide World Photos; *p. 64:* Duomo/CORBIS; *p. 68:* Johnny Buzzerio/CORBIS; *p. 84:* Westlight Stock/OZ Productions/ CORBIS; *p. 91:* Reuters/CORBIS; *p. 92:* Timothy A. Clary/Getty Images; *p. 100:* Joe Sohm/The Image Works; *p. 107:* David Young-Wolff/PhotoEdit, Inc.; *p. 108:* Tom McCarthy/PhotoEdit, Inc.; *p. 109:* Bill Aron/PhotoEdit, Inc.; *p. 120:* AP/Wide World Photos; *p. 124:* Kevin Lee/Getty Images. **Chapter 3:** *p. 125:* Michael Newman/PhotoEdit, Inc.; *p. 132:* AFP/Getty Images; *p. 134:* Mug Shots/CORBIS; *p. 148:* Richard Cummins/CORBIS; *p. 149:* CORBIS; *p. 157:* Eric Fowke/PhotoEdit, Inc.; *p. 164:* Tony Freeman/ PhotoEdit, Inc.; *p. 168:* Ariel Skelley/CORBIS; *p. 172:* Dana White/PhotoEdit, Inc. **Chapter 4:** *p. 173:* Joel Simon/STONE/Getty Images; *p. 176:* Charles O'Rear/CORBIS; *p. 178:* Mike Powell/Allsport Concepts/ Getty Images; *p. 180:* John Madere/CORBIS; *p. 181:* AP/Wide World Photos; *p. 182:* Warner Brothers Television/Getty Images; *p. 189:* Michael Newman/PhotoEdit, Inc.; *p. 190:* Digital Image © 1996 CORBIS, Original Image Courtesy of NASA/CORBIS; *p. 191:* Reproduced by Permission of The State Hermitage Museum, St. Petersburg, Russia/CORBIS; *p. 192:* Dallas & John Heaton/CORBIS; *p. 200:* Bob Daemmrich/ PhotoEdit, Inc. **Chapter 5:** *p. 201:* David Sacks/The Image Bank/Getty Images; *p. 206:* David Chasey/ Photodisc/Getty Images; *p. 209:* Galen Rowell/CORBIS; *p. 210:* Joe McBride/STONE /Getty Images; *p. 214:* Ulrike Welsch/PhotoEdit, Inc.; *p. 217:* Ariel Skelley/CORBIS; *p. 218:* Chuck Savage/CORBIS; *p. 224:* Ariel Skelley/CORBIS; *p. 228:* CORBIS. **Chapter 6:** *p. 233:* CORBIS; *p. 234:* Ainaco/CORBIS; *p. 236:* Myrleen Ferguson Cate/PhotoEdit, Inc.; *p. 240:* David Madison/STONE/Getty Images; *p. 247:* Todd A. Gipstein/CORBIS; *p. 248:* Dave Bartruff/CORBIS; *p. 255:* Richard Cummins/CORBIS; *p. 257:* Knut Platon/STONE/Getty Images; *p. 263:* Paul Conklin/PhotoEdit, Inc.; *p. 271:* Jeff Greenberg/ PhotoEdit, Inc.

Copyright © 2009 by Houghton Mifflin Company. All rights reserved.

No part of this work may be reproduced or transmitted in any form or by any means, electronic or mechanical, including photocopying and recording, or by any information storage or retrieval system without the prior written permission of Houghton Mifflin Company unless such copying is expressly permitted by federal copyright law. Address inquiries to College Permissions, Houghton Mifflin Company, 222 Berkeley Street, Boston, MA 02116-3764.

Printed in the U.S.A.

Library of Congress Control Number: 2007932359

ISBN-10: 0-547-01647-6
ISBN-13: 978-0-547-01647-4

23456789-WEB-11 10 09 08

Contents

Copyright © Houghton Mifflin Company. All rights reserved.

3   **Decimals   125**

Copyright © Houghton Mifflin Company. All rights reserved.

Copyright © Houghton Mifflin Company. All rights reserved.

---

6   Applications for Business and Consumers   233

Copyright © Houghton Mifflin Company. All rights reserved.

Copyright © Houghton Mifflin Company. All rights reserved.

Preface

The seventh edition of *Essential Mathematics with Applications* provides mathematically sound and comprehensive coverage of the topics considered essential in a basic college mathematics course. The text has been designed not only to meet the needs of the traditional college student but also to serve the needs of returning students whose mathematical proficiency may have declined during years away from formal education.

In this new edition of *Essential Mathematics with Applications*, we have continued to integrate approaches suggested by AMATYC. Each chapter opens by illustrating and referencing a mathematical application within the chapter. At the end of each section there are Applying the Concepts exercises, that include writing, synthesis, critical thinking, and challenge problems. At the end of each chapter there is a "Focus on Problem Solving" that introduces students to various problem-solving strategies. This is followed by "Projects and Group Activities" that can be used for cooperative learning activities.

One of the main challenges for students is the ability to translate verbal phrases into mathematical expressions. One reason for this difficulty is that students are not exposed to verbal phrases until later in most texts. In *Essential Mathematics with Applications*, we introduce verbal phrases for operations as we introduce the operation. For instance, after addition concepts have been presented, we provide exercises which say "Find the sum of . . ." or "What is 6 more than 7?" In this way, students are constantly confronted with verbal phrases and must make a mathematical connection between the phrase and a mathematical operation.

NEW! Changes to the Seventh Edition

In this textbook, as always, students are provided with ample practice in solving application problems requiring addition, subtraction, multiplication, or division. In this edition, however, we have increased the number of application problems that require a student first to determine which operation is required to solve the problem. See, for example, the Applying the Concept exercises on pages 44 and 45. Here is an opportunity for students to apply their understanding of the four basic operations and the words in an application problem that indicate each. This "mixed practice" will improve their ability to determine what operation to use to solve an application problem, a skill that is vital to their becoming good problem solvers. For further examples, see also the Applying the Concept exercises on pages 108 and 109 of Chapter 2 and pages 158 and 159 of Chapter 3.

We have updated the material in Chapter 1 on estimation. And to emphasize the importance of using estimation to check the result of a calculation performed on a calculator, exercises now ask students first to perform the operation on a calculator and then to use estimation to determine if that answer is reasonable. See, for example, the calculator exercises on pages 14, 23, 31, and 43. In this text, estimation is presented as a problem-solving tool; see, for example, the exercises on page 223 of Chapter 5.

In Chapter 2, we choose to present addition and subtraction of fractions prior to multiplication and division of fractions. For instructors who prefer to teach multiplication first, an alternative sequencing of the sections in the chapter is provided on page 77.

Copyright © Houghton Mifflin Company. All rights reserved.

The exposition in Section 1 of Chapter 3 includes more instruction on the relationship between decimals and fractions, and the corresponding exercise set includes problems asking students to convert fractions to decimals and decimals to fractions.

The in-text examples are now highlighted by a prominent HOW TO bar. Students looking for a worked-out example can easily locate one of these problems.

Throughout the text, data problems have been updated to reflect current data and trends. Also, titles have been added to the application exercises in the exercise sets. These changes emphasize the relevance of mathematics and the variety of problems in real life that require mathematical analysis.

The Chapter Summaries have been remodeled and expanded. Students are provided with definitions, rules, and procedures, along with examples of each. An objective reference and a page reference accompany each entry. We are confident that these will be valuable aids as students review material and study for exams.

Margin notes entitled Integrating Technology provide suggestions for using a scientific calculator. An annotated illustration of a scientific calculator appears on the inside back cover of this text.

NEW! Changes to the Student Support Edition

With the student in mind, we have expanded the **AIM for Success.** Getting Started introduces students to the skills they need to develop to be successful, and to the dedicated organization and study resources available for each chapter. Students can use the Chapter Checklist to track their assignments, solve the Math Word Scramble to cement their understanding of vocabulary, and complete the Concept Review in preparation for a chapter test.

Online homework powered by **WebAssign®** is now available through Houghton Mifflin's course management system in **HM MathSPACE®.** Developed by teachers for teachers, WebAssign allows instructors to focus on what really matters—teaching rather than grading. Instructors can create assignments from a ready-to-use database of algorithmic questions based on end-of-section exercises, or write and customize their own exercises. With WebAssign, instructors can create, post, and review assignments; deliver, collect, grade, and record assignments instantly; offer practice exercises, quizzes, and homework; and assess student performance to keep abreast of individual progress.

An **Online Multimedia eBook** is now available in HM MathSPACE, integrating numerous assets such as video explanations and tutorials to expand upon and reinforce concepts appearing in the text.

Visit **college.hmco.com/pic/barkerSSE7e** to enter **HM MathSPACE.**

NEW! Changes to the Instructor's Annotated Edition

The Instructor's Annotated Edition now contains full-sized pages. Most Instructor Notes, In-Class Exercises, Suggested Assignments, and Quick Quizzes remain at point-of-use for teaching convenience. Additional instructor features are available in **HM MathSPACE®.**

Copyright © Houghton Mifflin Company. All rights reserved.

ACKNOWLEDGMENTS

The authors would like to thank the people who have reviewed this manuscript and provided many valuable suggestions.

Dorothy A. Brown, *Camden County College, NJ*
Kim Doyle, *Monroe Community College, NY*
Said Fariabi, *San Antonio College, TX*
Kimberly A. Gregor, *Delaware Technical and Community College, DE*
Allen Grommet, *East Arkansas Community College, AR*
Anne Haney
Rose M. Kaniper, *Burlington County College, NJ*
Mary Ann Klicka, *Bucks County Community College, PA*
Helen Medley, *Kent State University, OH*
Steve Meidinger, *Merced College, CA*
Dr. James R. Perry, *Owens Community College, OH*
Gowribalan Vamadeva, *University of Cincinnati, OH*
Susan Wessner, *Tallahassee Community College, FL*

The authors also would like to thank the following students for their recommendations and criticisms regarding the material offered in the opening pages of the Student Support Edition.

Matthew Berg, *Vanderbilt University, TN*
Gregory Fulchino, *Middlebury College, VT*
Emma Goehring, *Trinity College, CT*
Gili Malinsky, *Boston University, MA*
Julia Ong, *Boston University, MA*
Anjali Parasnis-Samar, *Mount Holyoke College, MA*
Teresa Reilly, *University of Massachusetts–Amherst, MA*

Copyright © Houghton Mifflin Company. All rights reserved.

Student Success — Aufmann Interactive Method

Essential Mathematics with Applications uses an interactive style that engages students in trying new skills and reinforcing learning through structured exercises.

✓ UPDATED! AIM for Success— Getting Started

Getting Started helps students develop the study skills necessary to achieve success in college mathematics.

It also provides students with an explanation of how to effectively use the features of the text.

AIM for Success—Getting Started can be used as a lesson on the first day of class or as a student project.

Page AIM4

TAKE NOTE
When planning your schedule, give some thought to how much time you realistically have available each week.

For example, if you work 40 hours a week, take 15 units, spend the recommended study time given at the right, and sleep 8 hours a day, you will use over 80% of the available hours in a week. That leaves less than 20% of the hours in a week for family, friends, eating, recreation, and other activities.

Visit

Manage Your Time We know how busy you are outside of school. Do you have a full-time or a part-time job? Do you have children? Visit your family often? Play basketball or write for the school newspaper? It can be stressful to balance all of the important activities and responsibilities in your life. Making a **time management plan** will help you create a schedule that gives you enough time for everything you need to do.

Let's get started! Use the grids on pages AIM6 and AIM7 to fill in your weekly schedule.

First, fill in all of your responsibilities that take up certain set hours during the week. Be sure to include:

✓ each class you are taking

✓ time you spend at work

✓ any other commitments (child care, tutoring, volunteering, etc.)

TAKE NOTE
We realize that your weekly schedule may change. Visit **college.hmco.com/pic/ barkerSSE7e** to print out additional blank schedule forms, if you need them.

Page 184

184 Chapter 4 / Ratio and Proportion

Objective B To solve proportions

Study Tip
An important element of success is practice. We cannot do anything well if we do not practice it repeatedly. Practice is crucial to success in mathematics. In this objective you are learning a new skill: how to solve a proportion. You will need to practice this skill over and over again in order to be successful at it.

Sometimes one of the numbers in a proportion is unknown. In this case, it is necessary to *solve* the proportion.

To **solve a proportion**, find a number to replace the unknown so that the proportion is true.

HOW TO Solve: $\frac{9}{6} = \frac{3}{n}$

$$\frac{9}{6} = \frac{3}{n}$$

$$9 \times n = 6 \times 3$$

• Find the cross products.

• Think of $9 \times n = 18$ as $9\overline{)18}$

$6 \times 3 = 18$
$9 \times 2 = 18$

Study Tip

An important element of success is practice. We cannot do anything well if we do not practice it repeatedly. Practice is crucial to success in mathematics. In this objective you are learning a new skill: how to solve a proportion. You will need to practice this skill over and over again in order to be successful at it.

Page 184

✓ Interactive Approach

Each section is divided into objectives, and every objective contains one or more **HOW TO** examples. Annotations explain what is happening in key steps of each complete worked-out solution.

Each objective continues with one or more matched-pair examples. The first example in each set is worked out, much like the **HOW TO** examples. The second example, called "**You Try It**," is for the student.

Complete worked-out solutions to these examples appear in the appendix for students to check their work.

✓ Study Tips

These margin notes provide reminders of study skills and habits presented in the **AIM for Success**.

Page 259

Section 6.4 / Real Estate Expenses **259**

6.4 Real Estate Expenses

Objective A To calculate the initial expenses of buying a home

One of the largest investments most people ever make is the purchase of a home. The major initial expense in the purchase is the **down payment**, which is normally a percent of the purchase price. This percent varies among banks, but it usually ranges from 5% to 25%.

The **mortgage** is the amount that is borrowed to buy real estate. The mortgage amount is the difference between the purchase price and the down payment.

HOW TO A home is purchased for $140,000, and a down payment of $21,000 is made. Find the mortgage.

Purchase price	−	down payment	=	mortgage
140,000	−	21,000	=	119,000

The mortgage is $119,000.

TAKE NOTE
Because *points* means percent, a loan origination fee of
$2\frac{1}{2}$ points $= 2\frac{1}{2}\%$ =
2.5% = 0.025.

Another initial expense in buying a home is the **loan origination fee**, which is a fee that the bank charges for processing the mortgage papers. The loan origination fee is usually a percent of the mortgage and is expressed in **points**, which is the term banks use to mean percent. For example, "5 points" means "5 percent."

Points	×	mortgage	=	loan origination fee

Example 1
A house is purchased for $125,000, and a down payment, which is 20% of the purchase price, is made. Find the mortgage.

Strategy
To find the mortgage:
• Find the down payment by solving the basic percent equation for *amount*.
• Subtract the down payment from the purchase price.

Solution

Percent	×	base	=	amount
Percent	×	purchase price	=	down payment
0.20	×	125,000	=	n
		25,000	=	n

Purchase price	−	down payment	=	mortgage
125,000	−	25,000	=	100,000

The mortgage is $100,000.

You Try It 1
An office building is purchased for $216,000, and a down payment, which is 25% of the purchase price, is made. Find the mortgage.

Your strategy

SECTION 6.4

You Try It 1
Strategy To find the mortgage:
• Find the down payment by solving the basic percent equation for *amount*.
• Subtract the down payment from the purchase price.

Solution

Percent	×	base	=	amount
Percent	×	purchase price	=	down payment
0.25	×	216,000	=	n
		54,000	=	n

Purchase price	−	down payment	=	mortgage
216,000	−	54,000	=	162,000

The mortgage is $162,000.

Page S17

Copyright © Houghton Mifflin Company. All rights reserved.

Copyright © Houghton Mifflin Company. All rights reserved.

Student Success — Objective-Based Approach

Essential Mathematics with Applications is designed to foster student success through an integrated text and media program.

OBJECTIVES

Section 3.1
A To write decimals in standard form and in words
B To round a decimal to a given place value

Section 3.2
A To add decimals
B To solve application problems

Section 3.3
A To subtract decimals
B To solve application problems

Section 3.4
A To multiply decimals
B To solve application problems

Section 3.5
A To divide decimals
B To solve application problems

Section 3.6
A To convert fractions to decimals
B To convert decimals to fractions
C To identify the order relation between two decimals... fraction

...is the owner and operator of a wholesale groceries ...ch, she is a self-employed person. Anyone who ...trade, business, or profession is self-employed. There ... million self-employed people in the United States. ...m are authors, musicians, computer technicians, ...landscape designers, lawyers, psychologists, sales ...dentists. The table associated with **Exercises 28 to 30** ...lists the numbers of self-employed persons in the ...es by annual earnings, from less than $5000 per year to ...ore.

Page 125

✓ Objective-Based Approach

Each chapter's objectives are listed on the chapter opener page, which serves as a guide for student learning. All lessons, exercise sets, tests, and supplements are organized around this carefully constructed hierarchy of objectives.

A numbered objective describes the topic of each lesson.

3.2 Addition of Decimals

Objective A To add decimals

To add decimals, write the numbers so that the decimal points are on a vertical line. Add as for whole numbers, and write the decimal point in the sum directly below the decimal points in the addends.

Page 133

All exercise sets correspond directly to objectives.

3.2 Exercises

Objective A To add decimals

For Exercises 1 to 17, add.

1. $16.008 + 2.0385 + 132.06$
2. $17.32 + 1.0579 + 16.5$
3. $1.792 + 67 + 27.0526$
4. $8.772 + 1.09 + 26.5027$
5. $3.02 + 62.7 + 3.924$
7. $82.006 + 9.95 + 0.927$
8. $0.826 + 8.76 + 79.005$

10.
$$\begin{array}{r} 0.3 \\ + 0.07 \end{array}$$
11.
$$\begin{array}{r} 0.29 \\ + 0.4 \end{array}$$
12.
$$\begin{array}{r} 1.007 \\ + 2.1 \end{array}$$

14.
$$\begin{array}{r} 4.9257 \\ 27.05 \\ + 9.0063 \end{array}$$
15.
$$\begin{array}{r} 8.72 \\ 99.073 \\ + 2.9736 \end{array}$$
16.
$$\begin{array}{r} 62.4\ldots \\ 9.8\ldots \\ + 692.4\ldots \end{array}$$

Page 135

Answers to the Prep Tests, Chapter Review Exercises, Chapter Tests, and Cumulative Review Exercises refer students back to the original objectives for further study.

A8 Chapter 4

11. 0.778 [3.6A]   12. $\frac{2}{3}$ [3.6B]   13. 22.8635 [3.3A]   14. 7.94 [3.1B]   15. 8.932 [3.4A]
16. Three hundred forty-two and thirty-seven hundredths [3.1A]   17. 3.06753 [3.1A]   18. 25.7446 [3.4A]
19. 6.594 [3.5A]   20. 4.8785 [3.3A]   21. The new balance in your account is $661.51. [3.3B]   22. There are 53.466 million children in grades K–12. [3.2B]   23. There are 40.49 million more children in public school than in private school. [3.3B]   24. During a 5-day school week, 9.5 million gallons of milk are served. [3.4B]   25. The number who drove was 6.4 times greater than the number who flew. [3.5B]

CHAPTER 3 TEST

1. $0.66 < 0.666$ [3.6C]   2. 4.087 [3.3A]   3. Forty-five and three hundred two ten-thousandths [3.1A]
4. 0.692 [3.6A]   5. $\frac{33}{40}$ [3.6B]   6. 0.0740 [3.1B]   7. 1.538 [3.5A]   8. 27.76626 [3.3A]   9. 7.095 [3.1B]
10. 232 [3.5A]   11. 458.581 [3.2A]   12. The missing dimension is 1.37 inches. [3.3B]   13. 0.00548 [3.4A]
14. 255.957 [3.2A]   15. 209.07086 [3.1A]   16. Each payment is $395.40. [3.5B]   17. Your total income is $3087.14. [3.2B]   18. The cost of the call is $4.63. [3.4B]   19. The yearly average computer use by a 10th-grade student is 348.4 hours. [3.4B]   20. On average a 2nd-grade student uses the computer 36.4 hours more per year than a 5th-grade student. [3.4B]

CUMULATIVE REVIEW EXERCISES

1. 235 r17 [1.5C]   2. 128 [1.6A]   3. 3 [1.6B]   4. 72 [2.1A]   5. $4\frac{2}{5}$ [2.2B]   6. $\frac{37}{8}$ [2.2B]
7. $\frac{25}{60}$ [2.3A]   8. $1\frac{17}{48}$ [2.4B]   9. $8\frac{35}{36}$ [2.4C]   10. $5\frac{23}{36}$ [2.5C]   11. $\frac{1}{12}$ [2.6A]   12. $9\frac{1}{8}$ [2.6B]
13. $1\frac{2}{9}$ [2.7A]   14. $\frac{19}{20}$ [2.7B]   15. $\frac{3}{16}$ [2.8B]   16. $2\frac{5}{18}$ [2.8C]   17. Sixty-five and three hundred nine ten-thousandths [3.1A]   18. 504.6991 [3.2A]   19. 21.0764 [3.3A]   20. 55.26066 [3.4A]   21. 2.154 [3.5A]
22. 0.733 [3.6A]   23. $\frac{1}{6}$ [3.6B]   24. $\frac{8}{9} < 0.98$ [3.6C]   25. Sweden mandates 14 days more vacation than Germany. [1.3C]   26. The patient must lose $7\frac{3}{4}$ pounds the third month to achieve the goal. [2.5D]

Page A8

Student Success — Assessment and Review

Essential Mathematics with Applications appeals to students' different study styles with a variety of review methods.

Expanded! AIM for Success

For every chapter, there are four corresponding pages in the **AIM for Success** that enable students to track their progress during the chapter and to review critical material after completing the chapter.

Chapter Checklist

The **Chapter Checklist** allows students to track homework assignments and check off their progress as they master the material. The list can be used in an ad hoc manner or as a semester-long study plan.

Page AIM20

AIM20   AIM for Success

Chapter 3: Chapter Checklist

Keep track of your progress. Record the due dates of your assignments and check them off as you complete them.

Section 1 Introduction to Decimals

Tutorial Online	/ /	Video Explanation Online	/ /	Homework Online	/ /	ACE Practice Test Online	/ /

VOCABULARY	OBJECTIVES	Odd Text Exercises	Even Text Exercises	HM Assess Online
❑ decimal notation	A To write decimals in standard form and in words	/ /	/ /	/ /
❑ decimal				
❑ whole-number part	B To round a decimal to a given place value	/ /	/ /	/ /
❑ decimal point				
❑ decimal part				
❑ standard form				
❑ place value				
❑ rounding				

Section 2 Addition of Decimals

Tutorial Online	/ /	Video Explanation Online	/ /	Homework Online	/ /	ACE Practice Test Online	/ /

VOCABULARY	OBJECTIVES	Odd Text Exercises	Even Text Exercises	HM Assess Online
...dend	A To add decimals	/ /	/ /	/ /
...m	B To solve application problems	/ /	/ /	/ /

AIM22   AIM for Success

Chapter 3: Math Word Scramble

Answer each question. Unscramble the circled letters of the answers to solve the riddle.

1. In decimal notation, that part of the number that appears to the left of the decimal is the _____ part.

2. To _____ a fraction to a decimal, divide the numerator of the fraction by the denominator.

3. The part of the number that appears to the right of the decimal point is the _____ part.

4. To write a decimal in _____ form when it is written in words, write the whole-number part, replace the word *and* with a decimal point, and write the decimal part so that the last digit is in the given place-value position.

5. A number written in decimal _____ has three parts.

6. To _____ a decimal to a given place value, use the same rules used with whole numbers, except drop the digits to the right of the given place value instead of replacing them with zeros.

7. A number written in decimal notation is often called simply a _____.

8. For a number written in decimal notation, the _____ is read as "and."

Riddle What do you call a musical rhythm that has lost its life?

A ⬚⬚⬚⬚ ⬚⬚⬚⬚

Page AIM22

Math Word Scramble

The **Math Word Scramble** is a fun way for students to show off what they've learned in each chapter. The answer to each question can be found somewhere in the corresponding chapter. The circled letters of each answer can be unscrambled to find the solution to the riddle at the bottom of the page!

Page AIM23

AIM for Success   **AIM23**

Chapter 3: Concept Review

Review the basics and be prepared! Keep track of what you know and what you are expected to know.

	NOTES
1. How do you round a decimal to the nearest tenth? **p. 129**	_____
2. Write the decimal 0.37 as a fraction. **p. 128**	_____
3. Write the fraction $\frac{173}{10,000}$ as a decimal. **p. 128**	_____

Concept Review

Students can use the **Concept Review** as a quick study guide after the chapter is complete. The page number describing the corresponding concept follows each question. To the right is space for personal study notes, such as reminder pneumonics or expectations about the next test.

Copyright © Houghton Mifflin Company. All rights reserved.

Student Success — Conceptual Understanding

Essential Mathematics with Applications helps students understand the course concepts through the textbook exposition and feature set.

Page 207

✔ Key Terms and Concepts

Key Terms, in bold, emphasize important terms. Key terms can also be found in the **Glossary** at the back of the text.

Key Concepts are presented in orange boxes for easy reference.

The broker receives a payment of $11,400.

The solution was found by solving the **basic percent equation** for amount.

The Basic Percent Equation

$$\text{Percent} \times \text{base} = \text{amount}$$

In most cases, the percent is written as a decimal before the basic percent equation is solved. However, some percents are more easily written as a fraction than as a decimal. For example,

---

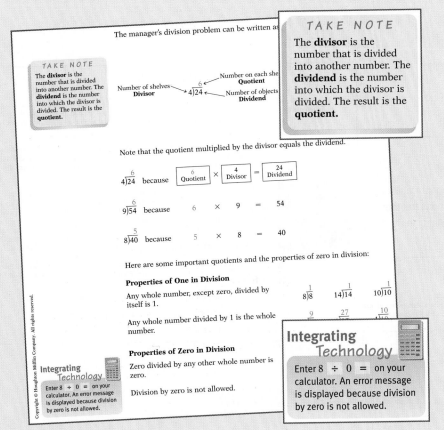

The manager's division problem can be written as

TAKE NOTE

The **divisor** is the number that is divided into another number. The **dividend** is the number into which the divisor is divided. The result is the **quotient.**

Number of shelves → Divisor

$4\overline{)24}$ quotient 6

Number on each shelf → Quotient

Number of objects → Dividend

Note that the quotient multiplied by the divisor equals the dividend.

$4\overline{)24}^{6}$ because $\boxed{6 \atop \text{Quotient}} \times \boxed{4 \atop \text{Divisor}} = \boxed{24 \atop \text{Dividend}}$

$9\overline{)54}^{6}$ because $6 \times 9 = 54$

$8\overline{)40}^{5}$ because $5 \times 8 = 40$

Here are some important quotients and the properties of zero in division:

Properties of One in Division

Any whole number, except zero, divided by itself is 1.

$8\overline{)8}^{1} \qquad 14\overline{)14}^{1} \qquad 10\overline{)10}^{1}$

Any whole number divided by 1 is the whole number.

$1\overline{)9}^{9} \qquad 1\overline{)27}^{27} \qquad 1\overline{)10}^{10}$

Properties of Zero in Division

Zero divided by any other whole number is zero.

Division by zero is not allowed.

Integrating Technology

Enter 8 ÷ 0 = on your calculator. An error message is displayed because division by zero is not allowed.

✔ Take Note

TAKE NOTE

The **divisor** is the number that is divided into another number. The **dividend** is the number into which the divisor is divided. The result is the **quotient.**

Some concepts require special attention!

✔ Integrating Technology

Integrating Technology

Enter 8 ÷ 0 = on your calculator. An error message is displayed because division by zero is not allowed.

Proper scientific calculator usage for accomplishing specific tasks is introduced as needed.

Page 23

Page 33

✔ Exercises

Icons designate when writing or a calculator 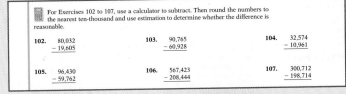 is needed to complete certain exercises.

✔ Applying the Concepts

The final exercises of each set offer deeper challenges.

For Exercises 102 to 107, use a calculator to subtract. Then round the numbers to the nearest ten-thousand and use estimation to determine whether the difference is reasonable.

| 102. | 80,032 | 103. | 90,765 | 104. | 32,574 |
| | − 19,605 | | − 60,928 | | − 10,961 |

| 105. | 96,430 | 106. | 567,423 | 107. | 300,712 |
| | − 59,762 | | − 208,444 | | − 198,714 |

APPLYING THE CONCEPTS

119. Answer true or false.
  a. The phrases "the difference between 9 and 5" and "5 less than 9" mean the same thing.
  b. $9 - (5 - 3) = (9 - 5) - 3$.
  c. Subtraction is an associative operation. *Hint:* See part b of this exercise.

120. Make up a word problem for which the difference between 15 and 8 is the answer.

Page 24

Copyright © Houghton Mifflin Company. All rights reserved.

Student Success Problem Solving

Essential Mathematics with Applications emphasizes applications, problem solving, and critical thinking.

Page 88

Example 6

A $2\frac{2}{3}$-inch piece is cut from a $6\frac{5}{8}$-inch board. How much of the board is left?

Strategy
To find the length remaining, subtract the length of the piece cut from the total length of the board.

Solution

$$6\frac{5}{8} = 6\frac{15}{24} = 5\frac{39}{24}$$
$$-2\frac{2}{3} = 2\frac{16}{24} = 2\frac{16}{24}$$
$$\phantom{-2\frac{2}{3} = 2\frac{16}{24} = }3\frac{23}{24}$$

$3\frac{23}{24}$ inches of the board are left.

You Try It 6

A flight from New York to Los Angeles takes $5\frac{1}{2}$ hours. After the plane has been in the air for $2\frac{3}{4}$ hours, how much flight time remains?

Your strategy

Your solution

✔ Problem-Solving Strategies

Encourage students to develop their own problem-solving strategies such as drawing diagrams and writing out the solution steps in words. Refer to these model **Strategies** as examples.

Page 164

Focus on Problem Solving

Relevant Information

Problems in mathematics or real life involve a question or a need and information or circumstances related to that question or need. Solving problems in the sciences usually involves a question, an observation, and measurements of some kind.

One of the challenges of problem solving in the sciences is to separate the information that is relevant to the problem from other information. Following is an example from the physical sciences in which some relevant information was omitted.

Hooke's Law states that the distance that a weight will stretch a spring is directly proportional to the weight on the spring. That is, $d = kF$, where d is the distance the spring is stretched and F is the force. In an experiment to verify this law, some physics students were continually getting inconsistent results. Finally, the instructor discovered that the heat produced when the lights were turned on was affecting the experiment. In this case, relevant information was omitted—namely, that the temperature of the spring can affect the distance it will stretch.

A lawyer drove 8 miles to the train station. After a 35-minute ride of 18 miles, the lawyer walked 10 minutes to the office. Find the total time it took the lawyer to get to work.

From this situation, answer the following before reading on.

a. What is asked for?
b. Is there enough information to answer the question?
c. Is information given that is not needed?

✔ Focus on Problem Solving

Foster further discovery of new problem-solving strategies at the end of each chapter, such as applying solutions to other problems, working backwards, inductive reasoning, and trial and error.

Projects and Group Activities

The Golden Ratio

There are certain designs that have been repeated over and over in both art and architecture. One of these involves the **golden rectangle.**

A golden rectangle is drawn at the right. Begin with a square that measures, say, 2 inches on a side. Let A be the midpoint of a side (halfway between two corners). Now measure the distance from A to B. Place this length along the bottom of the square, starting at A. The resulting rectangle is a golden rectangle.

The **golden ratio** is the ratio of the length of the golden rectangle to its width. If you have drawn the rectangle following the procedure above, you will find that the golden ratio is approximately 1.6 to 1.

The golden ratio appears in many different situations. Some historians claim that some of the great pyramids of Egypt are based on the golden ratio. The drawing at the right shows the Pyramid of Giza, which dates from approximately 2600 B.C. The ratio of the height to a side of the base is approximately 1.6 to 1.

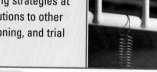

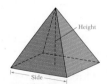

1. There are instances of the golden rectangle in the Mona Lisa painted by Leonardo da Vinci. Do some research on this painting and write a few paragraphs summarizing your findings.
2. What do 3 × 5 and 5 × 8 index cards have to do with the golden rectangle?
3. What does the United Nations Building in New York City have to do with the golden rectangle?
4. When was the Parthenon in Athens, Greece, built? What does the front of that building have to do with the golden rectangle?

✔ Projects and Group Activities

Tackle calculator usage, the Internet, data analysis, applications, and more in group settings or as longer assignments.

Copyright © Houghton Mifflin Company. All rights reserved.

Copyright © Houghton Mifflin Company. All rights reserved.

Page 192

✓ Applications

Problem-solving strategies and skills are tested in depth in the last objective of many sections: *To solve application problems.*

Applications are taken from agriculture, business, carpentry, chemistry, construction, education, finance, nutrition, real estate, sports, weather, and more.

12   Chapter 1 / Whole Numbers

Objective B   To solve application problems

To solve an application problem, first read the problem carefully. The **strategy** involves identifying the quantity to be found and planning the steps that are necessary to find that quantity. The **solution of an application problem** involves performing each operation stated in the strategy and writing the answer.

The table below displays the Wal-Mart store counts as reported in the Wal-Mart 2003 Annual Report.

	Discount Stores	Supercenters	SAM'S CLUBS	Neighborhood Markets
Within the United States	1568	1258	525	49
Outside the United States	942	238	71	37

HOW TO   Find the total number of Wal-Mart stores in the United States.

Strategy   To find the total number of Wal-Mart stores in the United States, read the table to find the number of each type of store in the United States. Then add these numbers.

Solution
$$\begin{array}{r}1568\\1258\\525\\+\ \ 49\\\hline3400\end{array}$$

Wal-Mart has a total of 3400 stores in the United States.

You Try It 4

...ove, how many ...–Mart have

Use the table above to determine the total number of Wal-Mart stores that are outside the United States.

...r of discount stores ...e to find the number ...the United States ...unt stores outside ...add the two

Your strategy

Your solution

...unt stores

Solution on p. S1

---

222   Chapter 5 / Percents

27.   **Lodging**   The graph at the right shows the breakdown of the locations of the 53,500 hotels throughout the United States. How many hotels in the United States are located along highways?

28.   **Poultry**   In a recent year, North Carolina produced 1,300,000,000 pounds of turkey. This was 18.6% of the U.S. total in that year. Calculate the U.S. total turkey production for that year. Round to the nearest billion.

29.   **Mining**   During 1 year, approximately 2,240,000 ounces of gold went into the manufacturing of electronic equipment in the United States. This is 16% of all the gold mined in the United States that year. How many ounces of gold were mined in the United States that year?

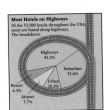

Most Hotels on Highways
Of the 53,500 hotels throughout the USA, most are found along highways. The breakdown:

Highways 42.2%
Suburban 33.6%
Resort 6.3%
Urban 10.2%
Airport 7.7%

Source: American Hotel and Lodging Association

30.   **Demography**   The table at the right shows the predicted increase in population from 2000 to 2040 for each of four counties in the Central Valley of California.
   a. What percent of the 2000 population of Sacramento County is the increase in population?
   b. What percent of the 2000 population of Kern County is the increase in population? Round to the nearest tenth of a percent.

County	2000 Population	Projected Increase
Sacramento	1,200,000	900,000
Kern	651,700	948,300
Fresno	794,200	705,800
San Joaquin	562,000	737,400

Source: California Department of Finance

31.   **Police Officers**   The graph at the right shows the causes of death for all police officers killed in the line of duty during a recent year. What percent of the deaths were due to traffic accidents? Round to the nearest tenth of a percent.

Job-related illness 19
Other 6
Violent attacks 58
Traffic accidents 73

Causes of Death for Police Officers Killed in the Line of Duty

Source: International Union of Police Associations

32.   **Demography**   According to a 25-city survey of the status of hunger and homelessness by the U.S. Conference of Mayors, 41% of the homeless in the United States are single men, 41% are families with children, 13% are single women, and 5% are unaccompanied minors. How many homeless people in the United States are single men?

APPLYING THE CONCEPTS

33.   **The Federal Government**   In the 108th Senate, there were 51 Republicans, 48 Democrats, and 1 Independent. In the 108th House of Representatives, there were 229 Republicans, 205 Democrats, and 1 Independent. Which had the larger percent of Republicans, the 108th Senate or the 108th House of Representatives?

✓ Real Data

Many of the problems based on real data require students to work with tables, graphs, and charts.

Copyright © Houghton Mifflin Company. All rights reserved.

ADDITIONAL RESOURCES   Get More from Your Textbook!

INSTRUCTOR RESOURCES	STUDENT RESOURCES
Instructor's Annotated Edition (IAE) Now with full-size pages, the IAE contains additional instructor support material, including notes, examples, quizzes, and suggested assignments.	**Student Solutions Manual** Contains complete solutions to all odd-numbered exercises, and all of the solutions to the end-of-unit material.
HM Testing™ (Powered by Diploma™) Provides instructors with a wide array of new algorithmic exercises along with improved functionality and ease of use. Instructors can create, author/edit algorithmic questions, customize, and deliver multiple types of tests.	*Math Study Skills Workbook*, **Third Edition, by Paul Nolting** Helps students identify their strengths, weaknesses, and personal learning styles in math. Offers study tips, proven test-taking strategies, a homework system, and recommendations for reducing anxiety and improving grades.

Instructional DVDs   Hosted by Dana Mosely, these text-specific DVDs cover all sections of the text and provide explanations of key concepts, examples, exercises, and applications in a lecture-based format. DVDs are now close-captioned for the hearing-impaired.

HM [math] SPACE. HM MathSPACE® encompasses the interactive online products and services integrated with Houghton Mifflin textbook programs. HM MathSPACE is available through text-specific student and instructor websites and via Houghton Mifflin's online course management system. **HM MathSPACE** now includes homework powered by **WebAssign®;** a new **Multimedia eBook;** self-assessment and remediation tools; videos, tutorials, and **SMARTHINKING®.**

- **NEW! WebAssign®** Developed by teachers, for teachers, WebAssign allows instructors to create assignments from an abundant ready-to-use database of algorithmic questions, or write and customize their own exercises. With WebAssign, instructors can create, post, and review assignments 24 hours a day, 7 days a week; deliver, collect, grade, and record assignments instantly; offer more practice exercises, quizzes, and homework; assess student performance to keep abreast of individual progress; and capture the attention of online or distance learning students.

- **NEW! Online Multimedia eBook** Integrates numerous assets such as video explanations and tutorials to expand upon and reinforce concepts as they appear in the text.

- **SMARTHINKING® Live, Online Tutoring** Provides an easy-to-use and effective online, text-specific tutoring service. A dynamic **Whiteboard** and a **Graphing Calculator** function enable students and e-structors to collaborate easily.

- **Student Website** Students can continue their learning here with a new Multimedia eBook, ACE practice tests, glossary flashcards, and more.

- **Instructor Website** Instructors can download AIM for Success practice sheets, chapter tests, instructor's solutions, course sequences, a printed test bank, a lesson plan (PowerPoint®) to use with the AIM for Success preface, and digital art and figures.

Online Course Management Content for Blackboard®, WebCT®, and eCollege®   Deliver program- or text-specific Houghton Mifflin content online using your institution's local course management system. Houghton Mifflin offers homework, tutorials, videos, and other resources formatted for Blackboard, WebCT, eCollege, and other course management systems. Add to an existing online course or create a new one by selecting from a wide range of powerful learning and instructional materials.

For more information, visit college.hmco.com/pic/barkerSSE7e or contact your local Houghton Mifflin sales representative.

Copyright © Houghton Mifflin Company. All rights reserved.

AIM for Success: Getting Started

Welcome to *Essential Mathematics with Applications!* Students come to this course with varied backgrounds and different experiences learning math. We are committed to your success in learning mathematics and have developed many tools and resources to support you along the way. Want to excel in this course? Read on to learn the skills you'll need, and how best to use this book to get the results you want.

Motivate Yourself

You'll find many real-life problems in this book, relating to sports, money, cars, music, and more. We hope that these topics will help you understand how you will use mathematics in your real life. However, to learn all of the necessary skills, and how you can apply them to your life outside this course, you need to stay motivated.

TAKE NOTE

Motivation alone won't lead to success. For example, suppose a person who cannot swim is rowed out to the middle of a lake and thrown overboard. That person has a lot of motivation to swim, but will most likely drown without some help. You'll need motivation **and** learning in order to succeed.

> **THINK ABOUT WHY YOU WANT TO SUCCEED IN THIS COURSE. LIST THE REASONS HERE (NOT IN YOUR HEAD . . . ON THE PAPER!)**

We also know that this course may be a requirement for you to graduate or complete your major. That's OK. If you have a goal for the future, such as becoming a nurse or a teacher, you will need to succeed in mathematics first. Picture yourself where you want to be, and use this image to stay on track.

Make the Commitment

Stay committed to success! With practice, you will improve your math skills. Skeptical? Think about when you first learned to ride a bike or drive a car. You probably felt self-conscious and worried that you might fail. But with time and practice, it became second nature to you.

You will also need to put in the time and practice to do well in mathematics. Think of us as your "driving" instructors. We'll lead you along the path to success, but we need you to stay focused and energized along the way.

> **LIST A SITUATION IN WHICH YOU ACCOMPLISHED YOUR GOAL BY SPENDING TIME PRACTICING AND PERFECTING YOUR SKILLS (SUCH AS LEARNING TO PLAY THE PIANO OR PLAYING BASKETBALL):**

Copyright © Houghton Mifflin Company. All rights reserved.

If you spend time learning and practicing the skills in this book, you will also succeed in math.

Think You Can't Do Math? Think Again!

You can do math! When you first learned the skills you just listed, you may have not done them well. With practice, you got better. With practice, you will be better at math. Stay focused, motivated, and committed to success.

It is difficult for us to emphasize how important it is to overcome the "I Can't Do Math Syndrome." If you listen to interviews of very successful athletes after a particularly bad performance, you will note that they focus on the positive aspect of what they did, not the negative. Sports psychologists encourage athletes to always be positive—to have a "Can Do" attitude. Develop this attitude toward math and you will succeed.

Skills for Success

Get the Big Picture If this were an English class, we wouldn't encourage you to look ahead in the book. But this is mathematics—go right ahead! Take a few minutes to read the table of contents. Then, look through the entire book. Move quickly: scan titles, look at pictures, notice diagrams.

Getting this big picture view will help you see where this course is going. To reach your goal, it's important to get an idea of the steps you will need to take along the way.

As you look through the book, find topics that interest you. What's your preference? Horse racing? Sailing? TV? Amusement parks? Find the Index of Applications at the back of the book and pull out three subjects that interest you. Then, flip to the pages in the book where the topics are featured, and read the exercises or problems where they appear. Write these topics here:

WRITE THE TOPIC HERE	WRITE THE CORRESPONDING EXERCISE/PROBLEM HERE

You'll find it's easier to work at learning the material if you are interested in how it can be used in your everyday life.

Use the following activities to think about more ways you might use mathematics in your daily life. Flip open your book to the following exercises to answer the questions.

- (see p. 84, #93) I just started a new job and will be paid hourly, but my hours change every week. I need to use mathematics to . . .

  _____

  _____

  _____

- (see p. 228, #24) I'd like to buy a new video camera, but it's very expensive. I need math to . . .

  _____

  _____

  _____

Copyright © Houghton Mifflin Company. All rights reserved.

- (see p. 258, #39) I want my child to be able to attend college in the future. I need math to . . .

_____

_____

_____

You know that the activities you just completed are from daily life, but do you notice anything else they have in common? That's right—they are **word problems.** Try not to be intimidated by word problems. You just need a strategy. It's true that word problems can be challenging because we need to use multiple steps to solve them:

✓ Read the problem.

✓ Determine the quantity we must find.

✓ Think of a method to find it.

✓ Solve the problem.

✓ Check the answer.

In short, we must come up with a *strategy* and then use that strategy to find the *solution*.

We'll teach you about strategies for tackling word problems that will make you feel more confident in branching out to these problems from daily life. After all, even though no one will ever come up to you on the street and ask you to solve a multiplication problem, you will need to use math every day to balance your checkbook, evaluate credit card offers, etc.

Take a look at the following example. You'll see that solving a word problem includes finding a *strategy* and using that strategy to find a *solution*. If you find yourself struggling with a word problem, try writing down the information you know about the problem. Be as specific as you can. Write out a phrase or a sentence that states what you are trying to find. Ask yourself whether there is a formula that expresses the known and unknown quantities. Then, try again!

Example 8

Jason Ng earns a salary of $440 for a 40-hour workweek. This week he worked 12 hours of overtime at a rate of $16.50 for each hour of overtime worked. Find his total income for the week.

Strategy

To find Jason's total income for the week:

- Find the overtime pay by multiplying the hourly overtime rate (16.50) by the number of hours of overtime worked (12).
- Add the overtime pay to the weekly salary (440).

Solution

$$\begin{array}{r} 16.50 \\ \times\ \ \ 12 \\ \hline 33\ 00 \\ 165\ 0 \\ \hline 198.00 \end{array}$$ Overtime pay

$$\begin{array}{r} 440.00 \\ +\ 198.00 \\ \hline 638.00 \end{array}$$

Jason's total income for this week is $638.00.

You Try It 8

You make a down payment of $175 on a stereo and agree to make payments of $37.18 a month for the next 18 months to repay the remaining balance. Find the total cost of the stereo.

Your strategy

Your solution

Solutions on p. S10

Copyright © Houghton Mifflin Company. All rights reserved.

Copyright © Houghton Mifflin Company. All rights reserved.

TAKE NOTE

Take a look at your syllabus to see if your instructor has an **attendance policy** that is part of your overall grade in the course.

The attendance policy will tell you:

- How many classes you can miss without a penalty
- What to do if you miss an exam or quiz
- If you can get the lecture notes from the professor if you miss a class

Get the Basics   On the first day of class, your instructor will hand out a **syllabus** listing the requirements of your course. Think of this syllabus as your personal roadmap to success. It shows you the *destinations* (topics you need to learn) and the *dates* you need to arrive at those destinations (when you need to learn the topics by). Learning mathematics is a journey. But, to get the most out of this course, you'll need to know what the important stops are, and what skills you'll need to learn for your arrival at those stops.

You've quickly scanned the table of contents, but now we want you to take a closer look. Flip open to the table of contents and look at it next to your syllabus. Identify when your major exams are, and what material you'll need to learn by those dates. For example, if you know you have an exam in the second month of the semester, how many chapters of this text will you need to learn by then? Write these down using the chart on the inside cover of the book. What homework do you have to do during this time? Use the Chapter Checklists provided later in the AIM for Success to record your assignments and their due dates. Managing this important information will help keep you on track for success.

Manage Your Time   We know how busy you are outside of school. Do you have a full-time or a part-time job? Do you have children? Visit your family often? Play basketball or write for the school newspaper? It can be stressful to balance all of the important activities and responsibilities in your life. Making a **time management plan** will help you create a schedule that gives you enough time for everything you need to do.

Let's get started! Use the grids on pages AIM6 and AIM7 to fill in your weekly schedule.

First, fill in all of your responsibilities that take up certain set hours during the week. Be sure to include:

✔ each class you are taking

✔ time you spend at work

✔ any other commitments (child care, tutoring, volunteering, etc.)

Then, fill in all of your responsibilities that are more flexible. Remember to make time for:

✔ **Studying** You'll need to study to succeed, but luckily you get to choose what times work best for you. Keep in mind:

- Most instructors ask students to spend twice as much time studying as they do in class. (3 hours of class = 6 hours of study)

- Try studying in chunks. We've found it works better to study an hour each day, rather than studying for 6 hours on one day.

- Studying can be even more helpful if you're able to do it right after your class meets, when the material is fresh in your mind.

TAKE NOTE

When planning your schedule, give some thought to how much time you realistically have available each week.

For example, if you work 40 hours a week, take 15 units, spend the recommended study time given at the right, and sleep 8 hours a day, you will use over 80% of the available hours in a week. That leaves less than 20% of the hours in a week for family, friends, eating, recreation, and other activities.

Visit http://college.hmco.com/masterstudent/shared/content/time_chart/chart.html and use the Interactive Time Chart to see how you're spending your time—you may be surprised.

TAKE NOTE

We realize that your weekly schedule may change. Visit **college.hmco.com/pic/barkerSSE7e** to print out additional blank schedule forms, if you need them.

✔ **Meals** Eating well gives you energy and stamina for attending classes and studying.

✔ **Entertainment** It's impossible to stay focused on your responsibilities 100% of the time. Giving yourself a break for entertainment will reduce your stress and help keep you on track.

✔ **Exercise** Exercise contributes to overall health. You'll find you're at your most productive when you have both a healthy mind **and** a healthy body.

Here is a sample of what part of your schedule might look like:

	8–9	9–10	10–11	11–12	12–1	1–2	2–3	3–4	4–5	5–6
Monday	History class Jenkins Hall 8–9:15	Eat 9:15–10	Study/Homework for History 10–12		Lunch and Nap! 12–1:30		Work 2–6			
Tuesday	Sleep!	Math Class Douglas Hall 9–9:45	Study/Homework for Math 10–12		Eat 12–1	English Class Scott Hall 1–1:45	Study/Homework for English 2–4		Hang out with Alli and Mike 4 until whenever!	

Copyright © Houghton Mifflin Company. All rights reserved.

	MORNING					AFTERNOON						EVENING					
	8–9	9–10	10–11	11–12	12–1	1–2	2–3	3–4	4–5	5–6	6–7	7–8	8–9	9–10	10–11
Monday															
Tuesday															
Wednesday															
Thursday															
Friday															
Saturday															
Sunday															

Copyright © Houghton Mifflin Company. All rights reserved.

Copyright © Houghton Mifflin Company. All rights reserved.

	MORNING				AFTERNOON						EVENING				
	8–9	9–10	10–11	11–12	12–1	1–2	2–3	3–4	4–5	5–6	6–7	7–8	8–9	9–10	10–11
Monday															
Tuesday															
Wednesday															
Thursday															
Friday															
Saturday															
Sunday															

Features for Success in This Text

Organization Let's look again at the Table of Contents. There are 6 chapters in this book. You'll see that every chapter is divided into **sections,** and each section contains a number of **learning objectives.** Each learning objective is labeled with a letter from A to D. Knowing how this book is organized will help you locate important topics and concepts as you're studying.

Preparation Ready to start a new chapter? Take a few minutes to be sure you're ready, using some of the tools in this book.

✓ **Cumulative Review Exercises:** You'll find these exercises after every chapter, starting with Chapter 2. The questions in the Cumulative Review Exercises are taken from the previous chapters. For example, the Cumulative Review for Chapter 3 will test all of the skills you have learned in Chapters 1, 2, and 3. Use this to refresh yourself before moving on to the next chapter, or to test what you know before a big exam.

Here's an example of how to use the Cumulative Review:

- Turn to page 171 and look at the questions for the Chapter 3 Cumulative Review, which are taken from the current chapter and the previous chapters.

- We have the answers to all of the Cumulative Review Exercises in the back of the book. Flip to page A8 to see the answers for this chapter.

- Got the answer wrong? We can tell you where to go in the book for help! For example, scroll down page A8 to find the answer for the first exercise, which is 235 r17. You'll see that after this answer, there is an **objective reference** [1.5C]. This means that the question was taken from Chapter 1, Section 5, Objective C. Go here to restudy the objective.

✓ **Prep Tests:** These tests are found at the beginning of every chapter and will help you see if you've mastered all of the skills needed for the new chapter.

Here's an example of how to use the Prep Test:

- Turn to page 174 and look at the Prep Test for Chapter 4.

- All of the answers to the Prep Tests are in the back of the book. You'll find them in the first set of answers in each answer section for a chapter. Turn to page A8 to see the answers for this Prep Test.

- Restudy the objectives if you need some extra help.

✓ Before you start a new section, take a few minutes to read the **Objective Statement** for that section. Then, browse through the objective material. Especially note the words or phrases in bold type—these are important concepts that you'll need as you're moving along in the course.

✓ As you start moving through the chapter, pay special attention to the **rule boxes.** These rules give you the reasons certain types of problems are solved the way they are. When you see a rule, try to rewrite the rule in your own words.

The Order of Operations Agreement

Step 1. Do all the operations inside parentheses.
Step 2. Simplify any number expressions containing exponents.
Step 3. Do multiplications and divisions as they occur from left to right.
Step 4. Do additions and subtractions as they occur from left to right.

Copyright © Houghton Mifflin Company. All rights reserved.

Knowing what to pay attention to as you move through a chapter will help you study and prepare.

Interaction We want you to be actively involved in learning mathematics and have given you many ways to get hands-on with this book.

✓ **HOW TO Examples** Take a look at page 70 shown here. See the HOW TO example? This contains an explanation by each step of the solution to a sample problem.

HOW TO Write $\frac{13}{5}$ as a mixed number.

Divide the numerator by the denominator.	To write the fractional part of the mixed number, write the remainder over the divisor.	Write the answer.
$\begin{array}{r} 2 \\ 5\overline{)13} \\ -10 \\ \hline 3 \end{array}$	$\begin{array}{r} 2\frac{3}{5} \\ 5\overline{)13} \\ -10 \\ \hline 3 \end{array}$	$\frac{13}{5} = 2\frac{3}{5}$

Page 70

Grab a paper and pencil and work along as you're reading through each example. When you're done, get a clean sheet of paper. Write down the problem and try to complete the solution without looking at your notes or at the book. When you're done, check your answer. If you got it right, you're ready to move on.

✓ **Example/You Try It Pairs** You'll need hands-on practice to succeed in mathematics. When we show you an example, work it out beside our solution. Use the Example/You Try It Pairs to get the practice you need.

Take a look at page 70, Example 5 and You Try It 5 shown here:

Example 5 Write $21\frac{3}{4}$ as an improper fraction.	**You Try It 5** Write $14\frac{5}{8}$ as an improper fraction.
Solution $21\frac{3}{4} = \frac{84 + 3}{4} = \frac{87}{4}$	**Your solution**

Solutions on pp. S4–S5

Page 70

You'll see that each Example is fully worked-out. Study this Example carefully by working through each step. Then, try your hand at it by completing the You Try It. If you get stuck, the solutions to the You Try Its are provided in the back of the book. There is a page number following the You Try It, which shows you where you can find the completely worked out solution. Use the solution to get a hint for the step on which you are stuck. Then, try again!

When you've finished the solution, check your work against the solution in the back of the book. Turn to page S5 to see the solution for You Try It 5.

Remember that sometimes there can be more than one way to solve a problem. But, your answer should always match the answers we've given in the back of the book. If you have any questions about whether your method will always work, check with your instructor.

Copyright © Houghton Mifflin Company. All rights reserved.

Review We have provided many opportunities for you to practice and review the skills you have learned in each chapter.

✓ **Section Exercises** After you're done studying a section, flip to the end of the section and complete the exercises. If you immediately practice what you've learned, you'll find it easier to master the core skills. Want to know if you answered the questions correctly? The answers to the odd-numbered exercises are given in the back of the book.

✓ **Chapter Summary** Once you've completed a chapter, look at the Chapter Summary. This is divided into two sections: *Key Words* and *Essential Rules and Procedures*. Flip to pages 165 and 166 to see the Chapter Summary for Chapter 3. This summary shows all of the important topics covered in the chapter. See the page number following each topic? This shows you the objective reference and the page in the text where you can find more information on the concept.

✓ **Math Word Scramble** When you feel you have a handle on the vocabulary introduced in a chapter, try your hand at the Math Word Scramble for that chapter. All the Math Word Scrambles can be found later in the AIM for Success, in between the Chapter Checklists. Flip to page AIM22 to see the Math Word Scramble for Chapter 3. As you fill in the correct vocabulary, you discover the identity of the circled letters. You can unscramble these circled letters to form the answer to the riddle on the bottom of the page.

✓ **Concept Review** Following the Math Word Scramble for each chapter is the Concept Review. Flip to page AIM23 to see the Concept Review for Chapter 3. When you read each question, jot down a reminder note on the right about whatever you feel will be most helpful to remember if you need to apply that concept during an exam. You can also use the space on the right to mark what concepts your instructor expects you to know for the next test. If you are unsure of the answer to a concept review question, flip to the page listed after the question to review that particular concept.

✓ **Chapter Review Exercises** You'll find the Chapter Review Exercises after the Chapter Summary. Flip to pages 167 and 168 to see the Chapter Review Exercises for Chapter 3. When you do the review exercises, you're giving yourself an important opportunity to test your understanding of the chapter. The answer to each review exercise is given at the back of the book, along with the objective the question relates to. When you're done with the Chapter Review Exercises, check your answers. If you had trouble with any of the questions, you can restudy the objectives and retry some of the exercises in those objectives for extra help.

✓ **Chapter Tests** The Chapter Tests can be found after the Chapter Review Exercises and can be used to prepare for your exams. Think of these tests as "practice runs" for your in-class tests. Take the test in a quiet place and try to work through it in the same amount of time you will be allowed for your exam.

Here are some strategies for success when you're taking your exams:

• Scan the entire test to get a feel for the questions (get the big picture).

• Read the directions carefully.

• Work the problems that are easiest for you first.

• Stay calm, and remember that you will have lots of opportunities for success in this class!

Copyright © Houghton Mifflin Company. All rights reserved.

Excel Go online to HM MathSPACE® by visiting **college.hmco.com/pic/ barkerSSE7e,** and take advantage of more study tools!

✓ **WebAssign®** online practice exercises and homework problems match the textbook exercises.

✓ Live tutoring is available with **SMARTHINKING®.**

✓ **DVD clips,** hosted by Dana Mosely, are short segments you can view to get a better handle on topics that are giving you trouble. If you find the DVD clips helpful and want even more coverage, the comprehensive set of DVDs for the entire course can be ordered.

✓ **Glossary Flashcards** will help refresh you on key terms.

Get Involved

Have a question? Ask! Your professor and your classmates are there to help. Here are some tips to help you jump in to the action:

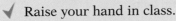

✓ Raise your hand in class.

✓ If your instructor prefers, email or call your instructor with your question. If your professor has a website where you can post your question, also look there for answers to previous questions from other students. Take advantage of these ways to get your questions answered.

✓ Visit a **math center.** Ask your instructor for more information about the math center services available on your campus.

✓ Your instructor will have **office hours** where he or she will be available to help you. Take note of where and when your instructor holds office hours. Use this time for one-on-one help, if you need it. Write down your instructor's office hours in the space provided on the inside front cover of the book.

✓ Form a **study group** with students from your class. This is a great way to prepare for tests, catch up on topics you may have missed, or get extra help on problems you're struggling with. Here are a few suggestions to make the most of your study group:

- **Test each other by asking questions.** Have each person bring a few sample questions when you get together.
- **Practice teaching each other.** We've found that you can learn a lot about what you know when you have to explain it to someone else.
- **Compare class notes.** Couldn't understand the last five minutes of class? Missed class because you were sick? Chances are someone in your group has the notes for the topics you missed.
- **Brainstorm test questions.**
- **Make a plan for your meeting.** Agree on what topics you'll talk about and how long you'll be meeting for. When you make a plan, you'll be sure that you make the most of your meeting.

Ready, Set, Succeed!

It takes hard work and commitment to succeed, but we know you can do it! Doing well in mathematics is just one step you'll take along the path to success.

We want you to check the following box, once you have accomplished your goals for this course.

Copyright © Houghton Mifflin Company. All rights reserved.

TAKE NOTE

Visit http://college.hmco .com/collegesurvival/ downing/on_course/4e/ students/affirmation/ affirmation.swf to create your very own customized certificate celebrating your accomplishments!

☐ I succeeded in Essential Mathematics!

We are confident that if you follow our suggestions, you will succeed. Good luck!

Chapter 1: Chapter Checklist

Keep track of your progress. Record the due dates of your assignments and check them off as you complete them.

Section 1 Introduction to Whole Numbers

Tutorial Online	/ /	Video Explanation Online	/ /	Homework Online	/ /	ACE Practice Test Online	/ /

VOCABULARY

- ❑ whole numbers
- ❑ number line
- ❑ graph of a whole number
- ❑ is less than (<)
- ❑ is greater than (>)
- ❑ standard form
- ❑ place value
- ❑ place-value chart
- ❑ period
- ❑ expanded form
- ❑ rounding

OBJECTIVES

		Odd Text Exercises	Even Text Exercises	HM Assess Online
A	To identify the order relation between two numbers	/ /	/ /	/ /
B	To write whole numbers in words and in standard form	/ /	/ /	/ /
C	To write whole numbers in expanded form	/ /	/ /	/ /
D	To round a whole number to a given place value	/ /	/ /	/ /

Section 2 Addition of Whole Numbers

Tutorial Online	/ /	Video Explanation Online	/ /	Homework Online	/ /	ACE Practice Test Online	/ /

VOCABULARY

- ❑ addition
- ❑ addend
- ❑ sum
- ❑ carrying
- ❑ approximately equal to (≈)
- ❑ strategy
- ❑ solution of an application problem
- ❑ Addition Property of Zero
- ❑ Commutative Property of Addition
- ❑ Associative Property of Addition

OBJECTIVES

		Odd Text Exercises	Even Text Exercises	HM Assess Online
A	To add whole numbers	/ /	/ /	/ /
B	To solve application problems	/ /	/ /	/ /

Section 3 Subtraction of Whole Numbers

Tutorial Online	/ /	Video Explanation Online	/ /	Homework Online	/ /	ACE Practice Test Online	/ /

VOCABULARY

- ❑ subtraction
- ❑ minuend
- ❑ subtrahend
- ❑ difference
- ❑ borrowing

OBJECTIVES

		Odd Text Exercises	Even Text Exercises	HM Assess Online
A	To subtract whole numbers without borrowing	/ /	/ /	/ /
B	To subtract whole numbers with borrowing	/ /	/ /	/ /
C	To solve application problems	/ /	/ /	/ /

Copyright © Houghton Mifflin Company. All rights reserved.

Section 4 Multiplication of Whole Numbers

Tutorial Online	/ /	Video Explanation Online	/ /	Homework Online	/ /	ACE Practice Test Online	/ /

VOCABULARY	OBJECTIVES	Odd Text Exercises	Even Text Exercises	HM Assess Online
❏ multiplication	**A** To multiply a number by a single digit	/ /	/ /	/ /
❏ factors				
❏ product	**B** To multiply larger whole numbers	/ /	/ /	/ /
❏ Multiplication Property of Zero				
❏ Multiplication Property of One	**C** To solve application problems	/ /	/ /	/ /
❏ Commutative Property of Multiplication				
❏ Associative Property of Multiplication				

Section 5 Division of Whole Numbers

Tutorial Online	/ /	Video Explanation Online	/ /	Homework Online	/ /	ACE Practice Test Online	/ /

VOCABULARY	OBJECTIVES	Odd Text Exercises	Even Text Exercises	HM Assess Online
❏ division	**A** To divide by a single digit with no remainder in the quotient	/ /	/ /	/ /
❏ divisor				
❏ quotient	**B** To divide by a single digit with a remainder in the quotient	/ /	/ /	/ /
❏ dividend				
❏ remainder	**C** To divide by larger whole numbers	/ /	/ /	/ /
❏ average	**D** To solve application problems	/ /	/ /	/ /
❏ properties of zero and one in division				

Section 6 Exponential Notation and the Order of Operations Agreement

Tutorial Online	/ /	Video Explanation Online	/ /	Homework Online	/ /	ACE Practice Test Online	/ /

VOCABULARY	OBJECTIVES	Odd Text Exercises	Even Text Exercises	HM Assess Online
❏ exponent	**A** To simplify expressions that contain exponents	/ /	/ /	/ /
❏ exponential notation				
❏ power	**B** To use the Order of Operations Agreement to simplify expressions	/ /	/ /	/ /
❏ Order of Operations Agreement				

Section 7 Prime Numbers and Factoring

Tutorial Online	/ /	Video Explanation Online	/ /	Homework Online	/ /	ACE Practice Test Online	/ /

VOCABULARY	OBJECTIVES	Odd Text Exercises	Even Text Exercises	HM Assess Online
❏ factors of a number	**A** To factor numbers	/ /	/ /	/ /
❏ prime number	**B** To find the prime factorization of a number	/ /	/ /	/ /
❏ composite number				
❏ prime factorization				
❏ rules for finding factors of a number				

Test Preparation

❏ KEY TERMS
❏ ESSENTIAL RULES AND PROCEDURES
❏ CHAPTER REVIEW EXERCISES
❏ CHAPTER TEST

Copyright © Houghton Mifflin Company. All rights reserved.

Chapter 1: Math Word Scramble

Answer each question. Unscramble the circled letters of the answers to solve the riddle.

1. Multiplying a number by _____ does not change the number.

2. Adding _____ to a number does not change the number.

3. 0, 1, 2, 3, 4, and so on are called _____ numbers.

4. Numbers that are multiplied are called _____.

5. In the expression 4^6, the 6 is called the _____.

6. When you divide one number by another number, the result is called the _____.

7. To effectively check an answer from a calculator, take time to quickly _____ the answer.

8. The place value of the digit 7 in the number 4,057,862 is _____.

9. If a number is even and greater than 2, it must be a _____ number.

10. When you subtract one number from another number, the result is called the _____.

Riddle What do you call someone who always knows the way out?

An ☐☐☐☐☐ ☐☐☐☐☐☐

Copyright © Houghton Mifflin Company. All rights reserved.

Chapter 1: Concept Review

Review the basics and be prepared! Keep track of what you know and what you are expected to know.

NOTES

1. What is the difference between the symbols < and >? **p. 3**
    _____

2. How do you round a four-digit number to the nearest hundred? **p. 6**
    _____

3. What is the difference between the Commutative Property of Addition and the Associative Property of Addition? **p. 9**
    _____

4. How do you estimate the sum of two numbers? **p. 1**
    _____

5. When is it necessary to borrow when performing subtraction? **p. 18**
    _____

6. What is the difference between the Multiplication Property of Zero and the Multiplication Property of One? **p. 25**
    _____

7. How do you multiply a number by 100? **p. 26**
    _____

8. How do you estimate the product of two numbers? **p. 27**
    _____

9. What is the difference between $0 \div 9$ and $9 \div 0$? **p. 33**
    _____

10. How do you check the answer to a division problem that has a remainder? **p. 36**
    _____

11. What are the steps in the Order of Operations Agreement? **p. 47**
    _____

12. How do you know if a number is a factor of another number? **p. 50**
    _____

13. What is a quick way to determine if 3 is a factor of a number? **p. 50**
    _____

Copyright © Houghton Mifflin Company. All rights reserved.

Chapter 2: Chapter Checklist

Keep track of your progress. Record the due dates of your assignments and check them off as you complete them.

Section 1 The Least Common Multiple and Greatest Common Factor

| Tutorial Online | / / | Video Explanation Online | / / | Homework Online | / / | ACE Practice Test Online | / / |

VOCABULARY

- ❑ multiples of number
- ❑ common multiple
- ❑ least common multiple (LCM)
- ❑ common factor
- ❑ greatest common factor (GCF)
- ❑ factors of a number

OBJECTIVES

	Odd Text Exercises	Even Text Exercises	HM Assess Online
A To find the least common multiple (LCM)	/ /	/ /	/ /
B To find the greatest common factor (GCF)	/ /	/ /	/ /

Section 2 Introduction to Fractions

| Tutorial Online | / / | Video Explanation Online | / / | Homework Online | / / | ACE Practice Test Online | / / |

VOCABULARY

- ❑ fraction
- ❑ fraction bar
- ❑ numerator
- ❑ denominator
- ❑ proper fraction
- ❑ mixed number
- ❑ improper fraction

OBJECTIVES

	Odd Text Exercises	Even Text Exercises	HM Assess Online
A To write a fraction that represents part of a whole	/ /	/ /	/ /
B To write an improper fraction as a mixed number or a whole number, and a mixed number as an improper fraction	/ /	/ /	/ /

Section 3 Writing Equivalent Fractions

| Tutorial Online | / / | Video Explanation Online | / / | Homework Online | / / | ACE Practice Test Online | / / |

VOCABULARY

- ❑ equivalent fractions
- ❑ simplest form of a fraction
- ❑ common factors
- ❑ Multiplication Property of One

OBJECTIVES

	Odd Text Exercises	Even Text Exercises	HM Assess Online
A To find equivalent fractions by raising to higher terms	/ /	/ /	/ /
B To write a fraction in simplest form	/ /	/ /	/ /

Section 4 Addition of Fractions and Mixed Numbers

| Tutorial Online | / / | Video Explanation Online | / / | Homework Online | / / | ACE Practice Test Online | / / |

VOCABULARY

- ❑ numerator
- ❑ denominator
- ❑ least common multiple (LCM)
- ❑ least common denominator (LCD)

OBJECTIVES

	Odd Text Exercises	Even Text Exercises	HM Assess Online
A To add fractions with the same denominator	/ /	/ /	/ /
B To add fractions with different denominators	/ /	/ /	/ /
C To add whole numbers, mixed numbers, and fractions	/ /	/ /	/ /
D To solve application problems	/ /	/ /	/ /

Copyright © Houghton Mifflin Company. All rights reserved.

Section 5 Subtraction of Fractions and Mixed Numbers

Tutorial Online	/ /	Video Explanation Online	/ /	Homework Online	/ /	ACE Practice Test Online	/ /

VOCABULARY

OBJECTIVES

		Odd Text Exercises	Even Text Exercises	HM Assess Online
A To subtract fractions with the same denominator		/ /	/ /	/ /
B To subtract fractions with different denominators		/ /	/ /	/ /
C To subtract whole numbers, mixed numbers, and fractions		/ /	/ /	/ /
D To solve application problems		/ /	/ /	/ /

Section 6 Multiplication of Fractions and Mixed Numbers

Tutorial Online	/ /	Video Explanation Online	/ /	Homework Online	/ /	ACE Practice Test Online	/ /

VOCABULARY

❑ product

OBJECTIVES

		Odd Text Exercises	Even Text Exercises	HM Assess Online
A To multiply fractions		/ /	/ /	/ /
B To multiply whole numbers, mixed numbers, and fractions		/ /	/ /	/ /
C To solve application problems		/ /	/ /	/ /

Section 7 Division of Fractions and Mixed Numbers

Tutorial Online	/ /	Video Explanation Online	/ /	Homework Online	/ /	ACE Practice Test Online	/ /

VOCABULARY

❑ reciprocal of a fraction
❑ inverting a fraction

OBJECTIVES

		Odd Text Exercises	Even Text Exercises	HM Assess Online
A To divide fractions		/ /	/ /	/ /
B To divide whole numbers, mixed numbers, and fractions		/ /	/ /	/ /
C To solve application problems		/ /	/ /	/ /

Section 8 Order, Exponents, and the Order of Operations Agreement

Tutorial Online	/ /	Video Explanation Online	/ /	Homework Online	/ /	ACE Practice Test Online	/ /

VOCABULARY

❑ number line
❑ graphing of a whole number
❑ exponent
❑ exponential notation
❑ Order of Operations Agreement

OBJECTIVES

		Odd Text Exercises	Even Text Exercises	HM Assess Online
A To identify the order relation between two fractions		/ /	/ /	/ /
B To simplify expressions containing exponents		/ /	/ /	/ /
C To use the Order of Operations Agreement to simplify expressions		/ /	/ /	/ /

Test Preparation

❑ KEY TERMS
❑ ESSENTIAL RULES AND PROCEDURES
❑ CHAPTER REVIEW EXERCISES
❑ CHAPTER TEST
❑ CUMULATIVE REVIEW EXERCISES

Copyright © Houghton Mifflin Company. All rights reserved.

Chapter 2: Math Word Scramble

Answer each question. Unscramble the circled letters of the answers to solve the riddle.

1. The _____ common multiple is the smallest common multiple of two or more numbers.

2. The _____ common factor is the largest common factor of two or more numbers.

3. The _____ is the part of the fraction that appears above the fraction bar.

4. The _____ is the part of the fraction that appears below the fraction bar.

5. A _____ number is a number greater than 1 with a whole-number part and a fractional part.

6. A fraction is in _____ form when the numerator and denominator have no common factors other than 1.

7. In an _____ fraction, the numerator is greater than or equal to the denominator.

8. The _____ of a fraction is the fraction with the numerator and denominator interchanged.

9. In a _____ fraction, the numerator is smaller than the denominator.

10. A number that is a multiple of two or more numbers is a _____ multiple.

11. Equal fractions with different denominators are called _____ fractions.

Riddle   If you walk through a scene of a movie, you get this at the end.

Copyright © Houghton Mifflin Company. All rights reserved.

Chapter 2: Concept Review

Review the basics and be prepared! Keep track of what you know and what you are expected to know.

NOTES

1. How do you find the LCM of 75, 30, and 50? **p. 65**   _____

2. How do you find the GCF of 6, 14, and 21? **p. 66**   _____

3. How do you write an improper fraction as a mixed number? **p. 70**   _____

4. Explain when a fraction is in simplest form. **p. 74**   _____

5. When adding fractions, why do you have to convert to equivalent fractions   _____
   with a common denominator? **p. 77**

6. How do you add mixed numbers? **p. 78**   _____

7. If you are subtracting a mixed number from a whole number, why do you   _____
   need to borrow? **p. 86**

8. When multiplying two fractions, why is it better to eliminate the common   _____
   factors before multiplying the remaining factors in the numerator and
   denominator? **p. 93**

9. When multiplying two positive fractions that are less than one, will the   _____
   product be greater than 1, less than the smaller number, or between the
   smaller number and the bigger number? **p. 93**

10. How are reciprocals used when dividing fractions? **p. 101**   _____

11. When a fraction is divided by a whole number, why is it important to write   _____
    the whole number as a fraction before dividing? **p. 102**

12. When comparing two fractions, why is it important to look at both the   _____
    numerators and denominators to determine which is larger? **p. 110**

13. In the expression $\left(\frac{5}{7}\right)^2 - \left(\frac{3}{4} - \frac{1}{3}\right) \div \frac{1}{2}$, in what order should the operations   _____
    be performed? **p. 111**

Copyright © Houghton Mifflin Company. All rights reserved.

Chapter 3: Chapter Checklist

Keep track of your progress. Record the due dates of your assignments and check them off as you complete them.

Section 1 Introduction to Decimals

Tutorial Online	/ /	Video Explanation Online	/ /	Homework Online	/ /	ACE Practice Test Online	/ /

VOCABULARY	OBJECTIVES	Odd Text Exercises	Even Text Exercises	HM Assess Online
❑ decimal notation	**A** To write decimals in standard form and in words	/ /	/ /	/ /
❑ decimal				
❑ whole-number part	**B** To round a decimal to a given place value	/ /	/ /	/ /
❑ decimal point				
❑ decimal part				
❑ standard form				
❑ place value				
❑ rounding				

Section 2 Addition of Decimals

Tutorial Online	/ /	Video Explanation Online	/ /	Homework Online	/ /	ACE Practice Test Online	/ /

VOCABULARY	OBJECTIVES	Odd Text Exercises	Even Text Exercises	HM Assess Online
❑ addend	**A** To add decimals	/ /	/ /	/ /
❑ sum	**B** To solve application problems	/ /	/ /	/ /

Section 3 Subtraction of Decimals

Tutorial Online	/ /	Video Explanation Online	/ /	Homework Online	/ /	ACE Practice Test Online	/ /

VOCABULARY	OBJECTIVES	Odd Text Exercises	Even Text Exercises	HM Assess Online
❑ minuend	**A** To subtract decimals	/ /	/ /	/ /
❑ subtrahend	**B** To solve application problems	/ /	/ /	/ /
❑ difference				

Copyright © Houghton Mifflin Company. All rights reserved.

Section 4 Multiplication of Decimals

Tutorial Online	/ /	Video Explanation Online	/ /	Homework Online	/ /	ACE Practice Test Online	/ /

VOCABULARY

❑ product
❑ power of 10

OBJECTIVES

		Odd Text Exercises	Even Text Exercises	HM Assess Online
A	To multiply decimals	/ /	/ /	/ /
B	To solve application problems	/ /	/ /	/ /

Section 5 Division of Decimals

Tutorial Online	/ /	Video Explanation Online	/ /	Homework Online	/ /	ACE Practice Test Online	/ /

VOCABULARY

❑ divisor
❑ dividend
❑ quotient
❑ power of 10

OBJECTIVES

		Odd Text Exercises	Even Text Exercises	HM Assess Online
A	To divide decimals	/ /	/ /	/ /
B	To solve application problems	/ /	/ /	/ /

Section 6 Comparing and Converting Fractions and Decimals

Tutorial Online	/ /	Video Explanation Online	/ /	Homework Online	/ /	ACE Practice Test Online	/ /

VOCABULARY

❑ numerator
❑ denominator
❑ place value

OBJECTIVES

		Odd Text Exercises	Even Text Exercises	HM Assess Online
A	To convert fractions to decimals	/ /	/ /	/ /
B	To convert decimals to fractions	/ /	/ /	/ /
C	To identify the order relation between two decimals or between a decimal and a fraction	/ /	/ /	/ /

Test Preparation

❑ KEY TERMS
❑ ESSENTIAL RULES AND PROCEDURES
❑ CHAPTER REVIEW EXERCISES
❑ CHAPTER TEST
❑ CUMULATIVE REVIEW EXERCISES

Copyright © Houghton Mifflin Company. All rights reserved.

Chapter 3: Math Word Scramble

Answer each question. Unscramble the circled letters of the answers to solve the riddle.

1. In decimal notation, that part of the number that appears to the left of the decimal is the _____ part.

   ☐☐☐☐☐ – ☐☐☐◯☐

2. To _____ a fraction to a decimal, divide the numerator of the fraction by the denominator.

   ☐☐☐☐☐◯

3. The part of the number that appears to the right of the decimal point is the _____ part.

   ☐☐☐◯☐

4. To write a decimal in _____ form when it is written in words, write the whole-number part, replace the word *and* with a decimal point, and write the decimal part so that the last digit is in the given place-value position.

   ☐☐☐◯☐☐

5. A number written in decimal _____ has three parts.

   ☐☐☐◯☐☐☐

6. To _____ a decimal to a given place value, use the same rules used with whole numbers, except drop the digits to the right of the given place value instead of replacing them with zeros.

   ☐☐☐☐◯

7. A number written in decimal notation is often called simply a _____.

   ☐◯☐☐☐

8. For a number written in decimal notation, the _____ is read as "and."

   ☐◯☐☐☐☐ ☐☐☐☐

Riddle What do you call a musical rhythm that has lost its life?

A ☐☐☐☐ ☐☐☐☐

Copyright © Houghton Mifflin Company. All rights reserved.

Chapter 3: Concept Review

Review the basics and be prepared! Keep track of what you know and what you are expected to know.

NOTES

1. How do you round a decimal to the nearest tenth? **p. 129**

2. Write the decimal 0.37 as a fraction. **p. 128**

3. Write the fraction $\frac{173}{10,000}$ as a decimal. **p. 128**

4. When adding decimals of different place value, what do you do with the decimal points? **p. 133**

5. Where do you put the decimal point in the product of two decimals? **p. 142**

6. How do you estimate the product of two decimals? **p. 143**

7. What do you do with the decimal point when dividing decimals? **p. 143**

8. Which is greater, the decimal 0.63 or the fraction $\frac{5}{8}$? **p. 161**

9. How many zeros must be inserted when dividing 0.763 by 0.6 and rounding to the nearest hundredth? **p. 151**

10. How do you subtract a decimal from a whole number that has no decimal point? **p. 137**

Copyright © Houghton Mifflin Company. All rights reserved.

Chapter 4: Chapter Checklist

Keep track of your progress. Record the due dates of your assignments and check them off as you complete them.

Section 1 Ratio

Tutorial Online	/ /	Video Explanation Online	/ /	Homework Online	/ /	ACE Practice Test Online	/ /

VOCABULARY

❑ ratio
❑ simplest form of a ratio

OBJECTIVES

		Odd Text Exercises	Even Text Exercises	HM Assess Online
A	To write the ratio of two quantities in simplest form	/ /	/ /	/ /
B	To solve application problems	/ /	/ /	/ /

Section 2 Rates

Tutorial Online	/ /	Video Explanation Online	/ /	Homework Online	/ /	ACE Practice Test Online	

VOCABULARY

❑ rate
❑ simplest form of a rate
❑ unit rate

OBJECTIVES

		Odd Text Exercises	Even Text Exercises	HM Assess Online
A	To write rates	/ /	/ /	/ /
B	To write unit rates	/ /	/ /	/ /
C	To solve application problems	/ /	/ /	/ /

Copyright © Houghton Mifflin Company. All rights reserved.

Section 3 Proportions

Tutorial Online	/ /	Video Explanation Online	/ /	Homework Online	/ /	ACE Practice Test Online	/ /

VOCABULARY

❑ proportion
❑ cross products
❑ solve a proportion
❑ ratio
❑ rate

OBJECTIVES

		Odd Text Exercises	Even Text Exercises	HM Assess Online
A	To determine whether a proportion is true	/ /	/ /	/ /
B	To solve proportions	/ /	/ /	/ /
C	To solve application problems	/ /	/ /	/ /

Test Preparation

❑ KEY TERMS
❑ ESSENTIAL RULES AND PROCEDURES
❑ CHAPTER REVIEW EXERCISES
❑ CHAPTER TEST
❑ CUMULATIVE REVIEW EXERCISES

Copyright © Houghton Mifflin Company. All rights reserved.

Chapter 4: Math Word Scramble

Answer each question. Unscramble the circled letters of the answers to solve the riddle.

1. A _____ is the comparison of two quantities with the same units.

⬜⬜⬜⬜

2. A ratio is in _____ form when the numbers that form the ratio do not have a common factor.

⬜⬜⬜⬜⬜⬜⬜

3. A _____ is the comparison of two quantities with different units.

⬜⬜⬜⬜

4. A _____ rate is a rate in which the number in the denominator is 1.

⬜⬜⬜⬜

5. A _____ is an expression of the equality of two ratios or rates.

⬜⬜⬜⬜⬜⬜⬜

6. In any true proportion, the _____ products are equal.

⬜⬜⬜⬜

7. To _____ a proportion, find a number to replace the unknown so that the proportion is true.

⬜⬜⬜⬜⬜

8. A ratio can be written as a _____, as two numbers separated by a _____, or as two numbers separated by the word _____.

⬜⬜⬜⬜⬜⬜⬜ , ⬜⬜⬜⬜⬜ , ⬜⬜⬜

Riddle   What do you call sleep in a no-war zone?

⬜⬜⬜⬜ ⬜⬜⬜ ⬜⬜⬜⬜⬜

Copyright © Houghton Mifflin Company. All rights reserved.

Chapter 4: Concept Review

Review the basics and be prepared! Keep track of what you know and what you are expected to know.

NOTES

1. If the units in a comparison are different, is it a ratio or a rate? **p. 179** _____

2. How do you find a unit rate? **p. 179** _____

3. How do you write the ratio $\frac{6}{7}$ using a colon? **p. 175** _____

4. How do you write the ratio 12:15 in simplest form? **p. 175** _____

5. How do you write the rate $\frac{347 \text{ miles}}{9.5 \text{ gallons}}$ as a unit rate? **p. 179** _____

6. When is a proportion true? **p. 183** _____

7. How do you solve a proportion? **p. 184** _____

8. How do the units help you to set up a proportion? **p. 185** _____

9. How do you check the solution of a proportion? **p. 185** _____

10. How do you write the ratio 19:6 as a fraction? **p. 177** _____

Copyright © Houghton Mifflin Company. All rights reserved.

Chapter 5: Chapter Checklist

Keep track of your progress. Record the due dates of your assignments and check them off as you complete them.

Section 1 Introduction to Percents

Tutorial Online / /   Video Explanation Online / /   Homework Online / /   ACE Practice Test Online / /

VOCABULARY
- ❏ percent
- ❏ percent sign

OBJECTIVES

		Odd Text Exercises	Even Text Exercises	HM Assess Online
A To write a percent as a fraction or a decimal		/ /	/ /	/ /
B To write a fraction or a decimal as a percent		/ /	/ /	/ /

Section 2 Percent Equations: Part I

Tutorial Online / /   Video Explanation Online / /   Homework Online / /   ACE Practice Test Online / /

VOCABULARY
- ❏ the basic percent equation
- ❏ base
- ❏ amount

OBJECTIVES

		Odd Text Exercises	Even Text Exercises	HM Assess Online
A To find the amount when the percent and base are given		/ /	/ /	/ /
B To solve application problems		/ /	/ /	/ /

Section 3 Percent Equations: Part II

Tutorial Online / /   Video Explanation Online / /   Homework Online / /   ACE Practice Test Online / /

VOCABULARY
- ❏ the basic percent equation
- ❏ base
- ❏ amount

OBJECTIVES

		Odd Text Exercises	Even Text Exercises	HM Assess Online
A To find the percent when the base and amount are given		/ /	/ /	/ /
B To solve application problems		/ /	/ /	/ /

Copyright © Houghton Mifflin Company. All rights reserved.

Section 4 Percent Equations: Part III

Tutorial Online	/ /	Video Explanation Online	/ /	Homework Online	/ /	ACE Practice Test Online	/ /

VOCABULARY

❏ the basic percent equation
❏ base
❏ amount

OBJECTIVES

	Odd Text Exercises	Even Text Exercises	HM Assess Online
A To find the base when the percent and amount are given	/ /	/ /	/ /
B To solve application problems	/ /	/ /	/ /

Section 5 Percent Problems: Proportion Method

Tutorial Online	/ /	Video Explanation Online	/ /	Homework Online	/ /	ACE Practice Test Online	/ /

VOCABULARY

❏ ratio
❏ proportion
❏ base
❏ amount

OBJECTIVES

	Odd Text Exercises	Even Text Exercises	HM Assess Online
A To solve percent problems using proportions	/ /	/ /	/ /
B To solve application problems	/ /	/ /	/ /

Test Preparation

❏ KEY TERMS
❏ ESSENTIAL RULES AND PROCEDURES
❏ CHAPTER REVIEW EXERCISES
❏ CHAPTER TEST
❏ CUMULATIVE REVIEW EXERCISES

Copyright © Houghton Mifflin Company. All rights reserved.

Chapter 5: Math Word Scramble

Answer each question. Unscramble the circled letters of the answers to solve the riddle.

1. To write a percent as a _____, drop the percent sign and multiply by $\frac{1}{100}$.

2. A _____ means "parts of 100."

3. To write a percent as a _____, drop the percent sign and multiply by 0.01.

4. The symbol % is the _____.

5. To write a decimal as a _____, multiply by 100.

6. The Basic Percent Equation is Percent × _____ = amount.

7. To write a fraction as a _____, multiply by 100.

8. For the proportion method, we use the equation $\frac{percent}{100} = \frac{}{base}$.

Riddle   What do you call a tree's revenge on fingers?

A

Copyright © Houghton Mifflin Company. All rights reserved.

Chapter 5: Concept Review

Review the basics and be prepared! Keep track of what you know and what you are expected to know.

NOTES

1. How do you write 197% as a fraction? **p. 203**

_____

2. How do you write 6.7% as a decimal? **p. 204**

_____

3. How do you write $\frac{9}{5}$ as a percent? **p. 204**

_____

4. How do you write 56.3 as a percent? **p. 204**

_____

5. What is the Basic Percent Equation? **p. 207**

_____

6. Find what percent of 42 is 30. Do you multiply or divide? **p. 211**

_____

7. Find 11.7% of 532. Do you multiply or divide? **p. 207**

_____

8. 36 is 240% of what number? Do you multiply or divide? **p. 215**

_____

9. How do you use the proportion method to solve a percent problem? **p. 219**

_____

10. What percent of 1400 is 763? Use the proportion method to solve. **p. 220**

_____

Copyright © Houghton Mifflin Company. All rights reserved.

Chapter 6: Chapter Checklist

Keep track of your progress. Record the due dates of your assignments and check them off as you complete them.

Section 1 Applications to Purchasing

Tutorial Online	/ /	Video Explanation Online	/ /	Homework Online	/ /	ACE Practice Test Online	/ /

VOCABULARY

❏ unit cost
❏ total cost

OBJECTIVES

	Odd Text Exercises	Even Text Exercises	HM Assess Online
A To find unit cost	/ /	/ /	/ /
B To find the most economical purchase	/ /	/ /	/ /
C To find total cost	/ /	/ /	/ /

Section 2 Percent Increase and Percent Decrease

Tutorial Online	/ /	Video Explanation Online	/ /	Homework Online	/ /	ACE Practice Test Online	/ /

VOCABULARY

❏ percent increase
❏ cost
❏ selling price
❏ markup
❏ markup rate
❏ percent decrease
❏ sale price
❏ discount
❏ discount rate

OBJECTIVES

	Odd Text Exercises	Even Text Exercises	HM Assess Online
A To find percent increase	/ /	/ /	
B To apply percent increase to business—markup	/ /	/ /	
C To find percent decrease	/ /	/ /	
D To apply percent decrease to business—discount	/ /	/ /	

Section 3 Interest

Tutorial Online	/ /	Video Explanation Online	/ /	Homework Online	/ /	ACE Practice Test Online	/ /

VOCABULARY

❏ interest
❏ principal
❏ interest rate
❏ simple interest
❏ maturity value of a loan
❏ finance charges
❏ compound interest
❏ Simple Interest Formula for Annual Interest Rates: Principal × annual interest rate × time (in years) = interest
❏ Maturity Value Formula for Simple Interest Loans: Principle + interest = maturity value
❏ Monthly Payment on a Simple Interest Loan: Maturity value ÷ length of the loan in months = monthly payment

OBJECTIVES

	Odd Text Exercises	Even Text Exercises	HM Assess Online
A To calculate simple interest	/ /	/ /	/ /
B To calculate finance charges on a credit card bill	/ /	/ /	/ /
C To calculate compound interest	/ /	/ /	/ /

Copyright © Houghton Mifflin Company. All rights reserved.

Section 4 Real Estate Expenses

Tutorial Online	/ /	Video Explanation Online	/ /	Homework Online	/ /	ACE Practice Test Online	/ /

		Odd Text Exercises	Even Text Exercises	HM Assess Online

VOCABULARY

❏ down payment
❏ mortgage
❏ loan origination fee
❏ points
❏ monthly mortgage payment
❏ property tax
❏ fixed-rate mortgage

OBJECTIVES

A To calculate the initial expenses of buying a home   / /   / /   / /

B To calculate ongoing expenses of owning a home   / /   / /   / /

Section 5 Car Expenses

Tutorial Online	/ /	Video Explanation Online	/ /	Homework Online	/ /	ACE Practice Test Online	/ /

		Odd Text Exercises	Even Text Exercises	HM Assess Online

VOCABULARY

❏ license fee
❏ sales tax

OBJECTIVES

A To calculate the initial expenses of buying a car   / /   / /   / /

B To calculate ongoing expenses of owning a car   / /   / /   / /

Section 6 Wages

Tutorial Online	/ /	Video Explanation Online	/ /	Homework Online	/ /	ACE Practice Test Online	/ /

		Odd Text Exercises	Even Text Exercises	HM Assess Online

VOCABULARY

❏ commission
❏ hourly wage
❏ salary

OBJECTIVES

A To calculate commissions, total hourly wages, and salaries   / /   / /   / /

Section 7 Bank Statements

Tutorial Online	/ /	Video Explanation Online	/ /	Homework Online	/ /	ACE Practice Test Online	/ /

		Odd Text Exercises	Even Text Exercises	HM Assess Online

VOCABULARY

❏ checking account
❏ check
❏ deposit slip
❏ bank statement
❏ balancing a checkbook
❏ service charges

OBJECTIVES

A To calculate checkbook balances   / /   / /   / /

B To balance a checkbook   / /   / /   / /

Test Preparation

❏ KEY TERMS
❏ ESSENTIAL RULES AND PROCEDURES
❏ CHAPTER REVIEW EXERCISES
❏ CHAPTER TEST
❏ CUMULATIVE REVIEW EXERCISES

Copyright © Houghton Mifflin Company. All rights reserved.

Chapter 6: Math Word Scramble

Answer each question. Unscramble the circled letters of the answers to solve the riddle.

1. _____ is the difference between the selling price and the cost.

2. The interest charges on purchases made with a credit card are called _____.

3. _____ interest is computed not only on the original principal but also on the interest already earned.

4. The _____ cost is the cost of one item.

5. _____ is the amount paid for the privilege of using someone else's money.

6. _____ is the amount of money originally deposited or borrowed.

7. _____ is the price a business pays for a product.

8. The principal plus the interest owed on a loan is called the _____ value.

9. A _____ is usually paid to a salesperson and is calculated as a percent of sales.

10. An employee who is paid a _____ receives payment based on a weekly, biweekly, monthly, or annual time schedule.

Riddle What do you call a vertically challenged pile of breakfast carbohydrates?

A

Copyright © Houghton Mifflin Company. All rights reserved.

Chapter 6: Concept Review

Review the basics and be prepared! Keep track of what you know and what you are expected to know.

NOTES

1. Find the unit cost if 4 cans cost $2.96. **p. 235**

2. Find the total cost of 3.4 pounds if apples are $.85 per pound. **p. 236**

3. How do you find the selling price if you know the cost and the markup? **p. 240**

4. How do you use the markup rate to find the markup? **p. 240**

5. How do you find the amount of decrease if you know the percent decrease? **p. 242**

6. How do you find the discount amount if you know the regular price and the sale price? **p. 243**

7. How do you find the discount rate? **p. 244**

8. How do you find simple interest? **p. 249**

9. How do you find the maturity value for simple interest loans? **p. 249**

10. What is the principal? **p. 249**

11. How do you find the monthly payment for a loan of 18 months if you know the maturity value of the loan? **p. 250**

12. What is compound interest? **p. 252**

13. What is a fixed-rate mortgage? **p. 260**

14. What expenses are involved in owning a car? **p. 266**

15. How do you balance a checkbook? **p. 274**

Copyright © Houghton Mifflin Company. All rights reserved.

1

Whole Numbers

How much would it cost to take a vacation that looks as relaxing as this one? What would you have to pay for travel, accommodations, and food? Do you have enough money in the bank, or would you need to earn and save more money? If you are considering taking a vacation, you will need to know the answers to these and other questions. The **Focus on Problem Solving on page 54** illustrates how we use operations on whole numbers to determine costs associated with planning a vacation.

Copyright © Houghton Mifflin Company. All rights reserved.

Need help? For online student resources, such as section quizzes, visit this textbook's website at **math.college.hmco.com/students.**

OBJECTIVES

Section 1.1

A To identify the order relation between two numbers

B To write whole numbers in words and in standard form

C To write whole numbers in expanded form

D To round a whole number to a given place value

Section 1.2

A To add whole numbers

B To solve application problems

Section 1.3

A To subtract whole numbers without borrowing

B To subtract whole numbers with borrowing

C To solve application problems

Section 1.4

A To multiply a number by a single digit

B To multiply larger whole numbers

C To solve application problems

Section 1.5

A To divide by a single digit with no remainder in the quotient

B To divide by a single digit with a remainder in the quotient

C To divide by larger whole numbers

D To solve application problems

Section 1.6

A To simplify expressions that contain exponents

B To use the Order of Operations Agreement to simplify expressions

Section 1.7

A To factor numbers

B To find the prime factorization of a number

Do these exercises to prepare for Chapter 1.

1. Name the number of ♦s shown below.

♦ ♦ ♦ ♦ ♦ ♦ ♦

2. Write the numbers from 1 to 10.

1 ___ ___ ___ ___ ___ ___ ___ ___ 10

3. Match the number with its word form.

a. 4   **A.** five
b. 2   **B.** one
c. 5   **C.** zero
d. 1   **D.** four
e. 3   **E.** two
f. 0   **F.** three

GO FIGURE ● ● ●

Five adults and two children want to cross a river in a canoe. A trip across the river in the canoe can be made by one adult or by two children or by one child. Everyone is able to paddle the canoe. What is the minimum number of trips that will be necessary for everyone to get to the other side?

Copyright © Houghton Mifflin Company. All rights reserved.

1.1 Introduction to Whole Numbers

Objective A

To identify the order relation between two numbers

The **whole numbers** are 0, 1, 2, 3, 4, 5, 6, 7, 8, 9, 10, 11, 12, 13, 14, ...

The three dots mean that the list continues on and on and that there is no largest whole number.

Just as distances are associated with the markings on the edge of a ruler, the whole numbers can be associated with points on a line. This line is called the **number line.** The arrow on the number line below indicates that there is no largest whole number.

The **graph of a whole number** is shown by placing a heavy dot directly above that number on the number line. Here is the graph of 7 on the number line:

The number line can be used to show the order of whole numbers. A number that appears to the left of a given number **is less than (<)** the given number. A number that appears to the right of a given number **is greater than (>)** the given number.

Four is less than seven.
4 < 7

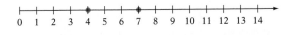

Twelve is greater than seven.
12 > 7

Example 1 Graph 11 on the number line.

Solution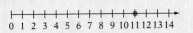

You Try It 1 Graph 6 on the number line.

Your solution

Example 2 Place the correct symbol, < or >, between the two numbers.

a. 39 > 24
b. 0 < 51

Solution a. 39 > 24
b. 0 < 51

You Try It 2 Place the correct symbol, < or >, between the two numbers.

a. 45   29
b. 27   0

Your solution a.
b.

Solutions on p. S1

Copyright © Houghton Mifflin Company. All rights reserved.

Point of Interest

The Babylonians had a place-value system based on 60. Its influence is still with us in angle measurement and time: 60 seconds in 1 minute, 60 minutes in 1 hour. It appears that the earliest record of a base-10 place-value system for natural numbers dates from the 8th century.

Objective B   To write whole numbers in words and in standard form

When a whole number is written using the digits 0, 1, 2, 3, 4, 5, 6, 7, 8, and 9, it is said to be in **standard form.** The position of each digit in the number determines the digit's **place value.** The diagram below shows a **place-value chart** naming the first 12 place values. The number 37,462 is in standard form and has been entered in the chart.

In the number 37,462, the position of the digit 3 determines that its place value is ten-thousands.

When a number is written in standard form, each group of digits separated from the digits by a comma (or commas) is called a **period.** The number 3,786,451,294 has four periods. The period names are shown in red in the place-value chart above.

To write a number in words, start from the left. Name the number in each period. Then write the period name in place of the comma.

3,786,451,294 is read "three billion seven hundred eighty-six million four hundred fifty-one thousand two hundred ninety-four."

To write a whole number in standard form, write the number named in each period, and replace each period name with a comma.

Four million sixty-two thousand five hundred eighty-four is written 4,062,584. The zero is used as a place holder for the hundred-thousands place.

Example 3   Write 25,478,083 in words.

Solution   Twenty-five million four hundred seventy-eight thousand eighty-three

You Try It 3   Write 36,462,075 in words.

Your solution

Example 4   Write three hundred three thousand three in standard form.

Solution   303,003

You Try It 4   Write four hundred fifty-two thousand seven in standard form.

Your solution

Solutions on p. S1

Objective C   To write whole numbers in expanded form

The whole number 26,429 can be written in **expanded form** as

$20,000 + 6000 + 400 + 20 + 9.$

The place-value chart can be used to find the expanded form of a number.

Copyright © Houghton Mifflin Company. All rights reserved.

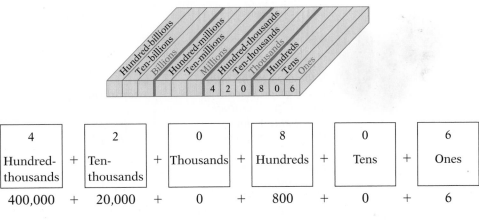

2 Ten- thousands	+	6 Thousands	+	4 Hundreds	+	2 Tens	+	9 Ones
20,000	+	6000	+	400	+	20	+	9

The number 420,806 is written in expanded form below. Note the effect of having zeros in the number.

4 Hundred- thousands	+	2 Ten- thousands	+	0 Thousands	+	8 Hundreds	+	0 Tens	+	6 Ones
400,000	+	20,000	+	0	+	800	+	0	+	6

or simply 400,000 + 20,000 + 800 + 6.

Example 5   Write 23,859 in expanded form.

Solution   20,000 + 3000 + 800 + 50 + 9

You Try It 5   Write 68,281 in expanded form.

Your solution

Example 6   Write 709,542 in expanded form.

Solution   700,000 + 9000 + 500 + 40 + 2

You Try It 6   Write 109,207 in expanded form.

Your solution

Solutions on p. S1

Objective D   **To round a whole number to a given place value**

When the distance to the moon is given as 240,000 miles, the number represents an approximation to the true distance. Taking an approximate value for an exact number is called **rounding.** A rounded number is always rounded to a given place value.

Copyright © Houghton Mifflin Company. All rights reserved.

37 is closer to 40 than it is to 30. 37 rounded to the nearest ten is 40.

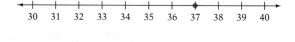

673 rounded to the nearest ten is 670. 673 rounded to the nearest hundred is 700.

A whole number is rounded to a given place value without using the number line by looking at the first digit to the right of the given place value.

If the digit to the right of the given place value is less than 5, that digit and all digits to the right are replaced by zeros.

HOW TO Round 13,834 to the nearest hundred.

┌──── Given place value
13,834
    └──── 3 < 5

13,834 rounded to the nearest hundred is 13,800.

If the digit to the right of the given place value is greater than or equal to 5, increase the digit in the given place value by 1, and replace all other digits to the right by zeros.

HOW TO Round 386,217 to the nearest ten-thousand.

┌──── Given place value
386,217
     └──── 6 > 5

386,217 rounded to the nearest ten-thousand is 390,000.

Example 7 Round 525,453 to the nearest ten-thousand.

Solution

┌──── Given place value
525,453
    └──── 5 = 5

525,453 rounded to the nearest ten-thousand is 530,000.

You Try It 7 Round 368,492 to the nearest ten-thousand.

Your solution

Example 8 Round 1972 to the nearest hundred.

Solution

┌──── Given place value
1972
  └──── 7 > 5

1972 rounded to the nearest hundred is 2000.

You Try It 8 Round 3962 to the nearest hundred.

Your solution

Solutions on p. S1

Copyright © Houghton Mifflin Company. All rights reserved.

1.1 Exercises

Objective A   **To identify the order relation between two numbers**

For Exercises 1 to 4, graph the number on the number line.

1. 3

2. 5

3. 9

4. 0

For Exercises 5 to 13, place the correct symbol, < or >, between the two numbers.

5. 37   49

6. 58   21

7. 101   87

8. 16   5

9. 245   158

10. 2701   2071

11. 0   45

12. 107   0

13. 815   928

Objective B   **To write whole numbers in words and in standard form**

For Exercises 14 to 17, name the place value of the digit 3.

14. 83,479

15. 3,491,507

16. 2,634,958

17. 76,319,204

For Exercises 18 to 25, write the number in words.

18. 2675

19. 3790

20. 42,928

21. 58,473

22. 356,943

23. 498,512

24. 3,697,483

25. 6,842,715

For Exercises 26 to 31, write the number in standard form.

26. Eighty-five

27. Three hundred fifty-seven

28. Three thousand four hundred fifty-six

29. Sixty-three thousand seven hundred eighty

Copyright © Houghton Mifflin Company. All rights reserved.

30. Six hundred nine thousand nine hundred forty-eight

31. Seven million twenty-four thousand seven hundred nine

Objective C   **To write whole numbers in expanded form**

For Exercises 32 to 39, write the number in expanded form.

32. 5287   **33.** 6295   **34.** 58,943   **35.** 453,921

36. 200,583   **37.** 301,809   **38.** 403,705   **39.** 3,000,642

Objective D   **To round a whole number to a given place value**

For Exercises 40 to 51, round the number to the given place value.

40. 926   Tens   **41.** 845   Tens   **42.** 1439   Hundreds

43. 3973   Hundreds   **44.** 43,607   Thousands   **45.** 52,715   Thousands

46. 389,702   Thousands   **47.** 629,513   Thousands   **48.** 647,989   Ten-thousands

49. 253,678   Ten-thousands **50.** 36,702,599 Millions   **51.** 71,834,250 Millions

---

APPLYING THE CONCEPTS

Answer true or false for Exercise 52a and 52b. If the answer is false, give an example to show that it is false.

52. **a.** If you are given two distinct whole numbers, then one of the numbers is always greater than the other number.
   b. A rounded-off number is always less than its exact value.

53. What is the largest three-digit whole number? What is the smallest five-digit whole number?

54. If 3846 is rounded to the nearest ten and then that number is rounded to the nearest hundred, is the result the same as what you get when you round 3846 to the nearest hundred? If not, which of the two methods is correct for rounding to the nearest hundred?

Copyright © Houghton Mifflin Company. All rights reserved.

1.2 Addition of Whole Numbers

Objective A **To add whole numbers**

Addition is the process of finding the total of two or more numbers.

TAKE NOTE

The numbers being added are called **addends.** The result is the **sum.**

By counting, we see that the total of $3 and $4 is $7.

$3 + $4 = $7

Addend   Addend   Sum

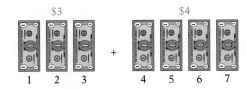

Addition can be illustrated on the number line by using arrows to represent the addends. The size, or magnitude, of a number can be represented on the number line by an arrow.

The number 3 can be represented anywhere on the number line by an arrow that is 3 units in length.

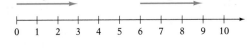

Point of Interest

The first use of the plus sign appeared in 1489 in *Mercantile Arithmetic.* It was used to indicate a surplus, not as the symbol for addition. That use did not appear until about 1515.

To add on the number line, place the arrows representing the addends head to tail, with the first arrow starting at zero. The sum is represented by an arrow starting at zero and stopping at the tip of the last arrow.

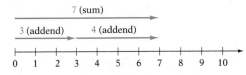

$$3 + 4 = 7$$

More than two numbers can be added on the number line.

$$3 + 2 + 4 = 9$$

Some special properties of addition that are used frequently are given below.

Addition Property of Zero

Zero added to a number does not change the number.

$$4 + 0 = 4$$
$$0 + 7 = 7$$

Commutative Property of Addition

Two numbers can be added in either order; the sum will be the same.

$$4 + 8 = 8 + 4$$
$$12 = 12$$

Associative Property of Addition

TAKE NOTE

This is the same addition problem shown on the number line above.

Grouping the addition in any order gives the same result. The parentheses are grouping symbols and have the meaning "Do the operations inside the parentheses first."

$$(3 + 2) + 4 = 3 + (2 + 4)$$
$$5 + 4 = 3 + 6$$
$$9 = 9$$

Copyright © Houghton Mifflin Company. All rights reserved.

The number line is not useful for adding large numbers. The basic addition facts for adding one digit to one digit should be memorized. Addition of larger numbers requires the repeated use of the basic addition facts.

To add large numbers, begin by arranging the numbers vertically, keeping the digits of the same place value in the same column.

HOW TO   Add: 321 + 6472

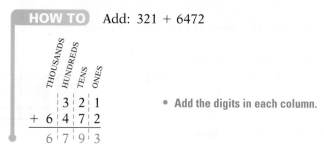

$$\begin{array}{r} 3\,2\,1 \\ +\ 6\,4\,7\,2 \\ \hline 6\,7\,9\,3 \end{array}$$

• Add the digits in each column.

There are several words or phrases in English that indicate the operation of addition. Here are some examples:

added to	3 added to 5	5 + 3
more than	7 more than 5	5 + 7
the sum of	the sum of 3 and 9	3 + 9
increased by	4 increased by 6	4 + 6
the total of	the total of 8 and 3	8 + 3
plus	5 plus 10	5 + 10

Integrating Technology

Most scientific calculators use *algebraic logic:* the add (+), subtract (−), multiply (×), and divide (÷) keys perform the indicated operation on the number in the display and the next number keyed in. For instance, for the example at the right, enter 24 + 71 = . The display reads 95.

HOW TO   What is the sum of 24 and 71?

$$\begin{array}{r} 24 \\ +\ 71 \\ \hline 95 \end{array}$$

• The phrase *the sum of* means to add.

The sum of 24 and 71 is 95.

When the sum of the digits in a column exceeds 9, the addition will involve **carrying.**

HOW TO   Add: 487 + 369

$$\begin{array}{r} {}^{\ \ 1} \\ 4\,8\,7 \\ +\ 3\,6\,9 \\ \hline 6 \end{array}$$

• Add the ones column.
  7 + 9 = 16 (1 ten + 6 ones).
  Write the 6 in the ones column and carry the 1 ten to the tens column.

$$\begin{array}{r} {}^{1\ 1} \\ 4\,8\,7 \\ +\ 3\,6\,9 \\ \hline 5\,6 \end{array}$$

• Add the tens column.
  1 + 8 + 6 = 15 (1 hundred + 5 tens).
  Write the 5 in the tens column and carry the 1 hundred to the hundreds column.

$$\begin{array}{r} {}^{1\ 1} \\ 4\,8\,7 \\ +\ 3\,6\,9 \\ \hline 8\,5\,6 \end{array}$$

• Add the hundreds column.
  1 + 4 + 3 = 8 (8 hundreds).
  Write the 8 in the hundreds column.

Copyright © Houghton Mifflin Company. All rights reserved.

Example 1 Find the total of 17, 103, and 8.

Solution

$$\begin{array}{r}\overset{1}{17}\\103\\+\ \ 8\\\hline128\end{array}$$

• 7 + 3 + 8 = 18
Write the 8 in the ones column. Carry the 1 to the tens column.

You Try It 1 What is 347 increased by 12,453?

Your solution

Example 2 Add: 89 + 36 + 98

Solution

$$\begin{array}{r}\overset{2}{89}\\36\\+\ 98\\\hline223\end{array}$$

• 9 + 6 + 8 = 23
Write the 3 in the ones column. Carry the 2 to the tens column.

You Try It 2 Add: 95 + 88 + 67

Your solution

Example 3 Add:

$$\begin{array}{r}41,395\\4,327\\497,625\\+\ 32,991\end{array}$$

Solution

$$\begin{array}{r}{}^{1\ 1\ 2\ \ 2\ 1}\\41,395\\4,327\\497,625\\+\ 32,991\\\hline576,338\end{array}$$

You Try It 3 Add:

$$\begin{array}{r}392\\4,079\\89,035\\+\ 4,992\end{array}$$

Your solution

Solutions on p. S1

● E S T I M A T I O N ●

Estimation and Calculators

At some places in the text, you will be asked to use your calculator. Effective use of a calculator requires that you estimate the answer to the problem. This helps ensure that you have entered the numbers correctly and pressed the correct keys.

For example, if you use your calculator to find 22,347 + 5896 and the answer in the calculator's display is 131,757,912, you should realize that you have entered some part of the calculation incorrectly. In this case, you pressed ✕ instead of + . By estimating the answer to a problem, you can help ensure the accuracy of your calculations. We have a special symbol for **approximately equal to (≈).**

For example, to estimate the answer to 22,347 + 5896, round each number to the same place value. In this case, we will round to the nearest thousand. Then add.

$$\begin{array}{r}22{,}347 \approx\ \ \ \ 22{,}000\\+\ \ \ 5{,}896 \approx\ +\ \ 6{,}000\\\hline28{,}000\end{array}$$

The sum 22,347 + 5896 is approximately 28,000. Knowing this, you would know that 131,757,912 is much too large and is therefore incorrect.

To estimate the sum of two numbers, first round each whole number to the same place value and then add. Compare this answer with the calculator's answer.

Integrating Technology

This example illustrates that estimation is important when one is using a calculator.

Copyright © Houghton Mifflin Company. All rights reserved.

Copyright © Houghton Mifflin Company. All rights reserved.

Objective B   **To solve application problems**

To solve an application problem, first read the problem carefully. The **strategy** involves identifying the quantity to be found and planning the steps that are necessary to find that quantity. The **solution of an application problem** involves performing each operation stated in the strategy and writing the answer.

 The table below displays the Wal-Mart store counts as reported in the Wal-Mart 2003 Annual Report.

	Discount Stores	Supercenters	SAM'S CLUBS	Neighborhood Markets
Within the United States	1568	1258	525	49
Outside the United States	942	238	71	37

HOW TO   Find the total number of Wal-Mart stores in the United States.

Strategy   To find the total number of Wal-Mart stores in the United States, read the table to find the number of each type of store in the United States. Then add these numbers.

Solution
$$
\begin{array}{r}
1568 \\
1258 \\
525 \\
+\ \ \ 49 \\
\hline
3400
\end{array}
$$

Wal-Mart has a total of 3400 stores in the United States.

Example 4

According to the table above, how many discount stores does Wal-Mart have worldwide?

Strategy
To determine the number of discount stores worldwide, read the table to find the number of discount stores within the United States and the number of discount stores outside the United States. Then add the two numbers.

Solution
$$
\begin{array}{r}
1568 \\
+\ 942 \\
\hline
2510
\end{array}
$$

Wal-Mart has 2510 discount stores worldwide.

You Try It 4

Use the table above to determine the total number of Wal-Mart stores that are outside the United States.

Your strategy

Your solution

Solution on p. S1

1.2 Exercises

Objective A   **To add whole numbers**

For Exercises 1 to 32, add.

1.   17
       + 11

2.   25
       + 63

3.   83
       + 42

4.   63
       + 94

5.   77
       + 25

6.   63
       + 49

7.   56
       + 98

8.   86
       + 68

9.   658
       + 831

10.   842
        + 936

11.   735
        + 93

12.   189
        + 50

13.   859
        + 725

14.   637
        + 829

15.   470
        + 749

16.   427
        + 690

17.   36,925
        + 65,392

18.   56,772
        + 51,239

19.   50,873
        + 28,453

20.   34,872
        + 46,079

21.   878
        737
       + 189

22.   768
        461
       + 669

23.   319
        348
       + 912

24.   292
        579
       + 315

25.   9409
        3253
       + 7078

26.   8188
        8020
       + 7104

27.   2038
        2243
       + 3139

28.   4252
        6882
       + 5235

29.   67,428
        32,171
       + 20,971

30.   52,801
        11,664
       + 89,638

31.   76,290
        43,761
       + 87,402

32.   43,901
        98,301
       + 67,943

Copyright © Houghton Mifflin Company. All rights reserved.

For Exercises 33 to 40, add.

33. 20,958 + 3218 + 42

34. 80,973 + 5168 + 29

35. 392 + 37 + 10,924 + 621

36. 694 + 62 + 70,129 + 217

37. 294 + 1029 + 7935 + 65

38. 692 + 2107 + 3196 + 92

39. 97 + 7234 + 69,532 + 276

40. 87 + 1698 + 27,317 + 727

41. What is 9874 plus 4509?

42. What is 7988 plus 5678?

43. What is 3487 increased by 5986?

44. What is 99,567 added to 126,863?

45. What is 23,569 more than 9678?

46. What is 7894 more than 45,872?

47. What is 479 added to 4579?

48. What is 23,902 added to 23,885?

49. Find the total of 659, 55, and 1278.

50. Find the total of 4561, 56, and 2309.

51. Find the sum of 34, 329, 8, and 67,892.

52. Find the sum of 45, 1289, 7, and 32,876.

For Exercises 53 to 56, use a calculator to add. Then round the numbers to the nearest hundred, and use estimation to determine whether the sum is reasonable.

53. 1234 + 9780 + 6740

54. 919 + 3642 + 8796

55. 241 + 569 + 390 + 1672

56. 107 + 984 + 1035 + 2904

For Exercises 57 to 60, use a calculator to add. Then round the numbers to the nearest thousand, and use estimation to determine whether the sum is reasonable.

57. 32,461
9,844
+ 59,407

58. 29,036
22,904
+ 7,903

59. 25,432
62,941
+ 70,390

60. 66,541
29,365
+ 98,742

Copyright © Houghton Mifflin Company. All rights reserved.

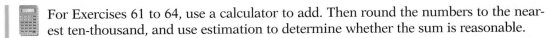

For Exercises 61 to 64, use a calculator to add. Then round the numbers to the nearest ten-thousand, and use estimation to determine whether the sum is reasonable.

61.	**62.**	**63.**	**64.**
67,421	21,896	281,421	542,698
82,984	4,235	9,874	97,327
66,361	62,544	34,394	7,235
10,792	21,892	526,398	73,667
+ 34,037	+ 1,334	+ 94,631	+ 173,201

For Exercises 65 to 68, use a calculator to add. Then round the numbers to the nearest million, and use estimation to determine whether the sum is reasonable.

65.	**66.**	**67.**	**68.**
28,627,052	1,792,085	12,377,491	46,751,070
983,073	29,919,301	3,409,723	6,095,832
+ 3,081,496	+ 3,406,882	7,928,026	280,011
		+ 10,705,682	+ 1,563,897

Objective B | **To solve application problems**

69. **Demographics** In a recent year, according to the U.S. Department of Health and Human Services, there were 110,670 twin births in this country, 6919 triplet births, 627 quadruplet deliveries, and 79 quintuplet and other higher-order multiple births. Find the total number of multiple births during the year.

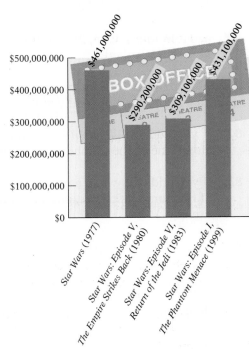

70. **Demographics** The Census Bureau estimates that the U.S. population will grow by 296 million people from 2000 to 2100. Given that the U.S. population in 2000 was 281 million, find the Census Bureau's estimate of the U.S. population in 2100.

The Film Industry The graph at the right shows the domestic box-office income from the first four *Star Wars* movies. Use this information for Exercises 71 to 73.

71. Estimate the total income from the first four *Star Wars* movies.

72. Find the total income from the first four *Star Wars* movies.

73. **a.** Find the total income from the two movies with the lowest box-office incomes.
   b. Does the total income from the two movies with the lowest box-office incomes exceed the income from the 1977 *Star Wars* production?

Source: **www.worldwideboxoffice.com**

Copyright © Houghton Mifflin Company. All rights reserved.

74. **Geometry**   The perimeter of a triangle is the sum of the lengths of the three sides of the triangle. Find the perimeter of a triangle that has sides that measure 12 inches, 14 inches, and 17 inches.

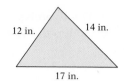

12 in.   14 in.

17 in.

75. **Travel**   The odometer on a moving van reads 68,692. The driver plans to drive 515 miles the first day, 492 miles the second day, and 278 miles the third day.
   a. How many miles will be driven during the three days?
   b. What will the odometer reading be at the end of the trip?

Investments   The table below shows the average amounts of money that all Americans have invested in selected assets and the average amounts invested for Americans between the ages of 16 and 34. Use this table for Exercises 76 to 79.

76. What is the total average amount for all Americans in checking accounts, savings accounts, and U.S. Savings Bonds?

	All Americans	Ages 16 to 34
Checking accounts	$487	$375
Savings accounts	$3,494	$1,155
U.S. Savings Bonds	$546	$266
Money market	$10,911	$4,427
Stock/mutual funds	$4,510	$1,615
Home equity	$43,070	$17,184
Retirement	$9,016	$4,298

77. What is the total average amount for Americans aged 16 to 34 in checking accounts, savings accounts, and U.S. Savings Bonds?

78. What is the total average amount for Americans aged 16 to 34 in all categories except home equity and retirement?

79. Is the sum of the average amounts invested in home equity and retirement for all Americans greater than or less than the sum of all categories for Americans between the ages of 16 and 34?

APPLYING THE CONCEPTS

80. **a.** How many two-digit numbers are there? **b.** How many three-digit numbers are there?

81. If you roll two ordinary six-sided dice and add the two numbers that appear on top, how many different sums are possible?

82. If you add two *different* whole numbers, is the sum always greater than either one of the numbers? If not, give an example.

83. If you add two whole numbers, is the sum always greater than either one of the numbers? If not, give an example. (Compare this with the previous exercise.)

84. ✏   Make up a word problem for which the answer is the sum of 34 and 28.

85. Call a number "lucky" if it ends in a 7. How many lucky numbers are less than 100?

Copyright © Houghton Mifflin Company. All rights reserved.

1.3 Subtraction of Whole Numbers

Objective A | **To subtract whole numbers without borrowing**

Subtraction is the process of finding the difference between two numbers.

TAKE NOTE

The **minuend** is the number from which another number is subtracted. The **subtrahend** is the number that is subtracted from another number. The result is the **difference.**

By counting, we see that the difference between $8 and $5 is $3.

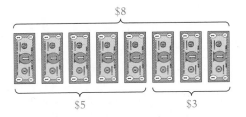

$8	–	$5	=	$3
Minuend		**Subtrahend**		**Difference**

The difference $8 - 5$ can be shown on the number line.

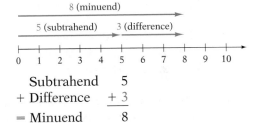

Note from the number line that addition and subtraction are related.

$$
\begin{array}{rr}
\text{Subtrahend} & 5 \\
+ \text{ Difference} & +\ 3 \\
\hline
= \text{Minuend} & 8
\end{array}
$$

Point of Interest

The use of the minus sign dates from the same period as the plus sign, around 1515.

The fact that the sum of the subtrahend and the difference equals the minuend can be used to check subtraction.

To subtract large numbers, begin by arranging the numbers vertically, keeping the digits that have the same place value in the same column. Then subtract the digits in each column.

HOW TO Subtract $8955 - 2432$ and check.

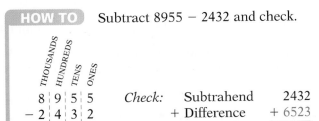

$$
\begin{array}{r}
8\ 9\ 5\ 5 \\
-\ 2\ 4\ 3\ 2 \\
\hline
6\ 5\ 2\ 3
\end{array}
$$

Check:
$$
\begin{array}{lr}
\text{Subtrahend} & 2432 \\
+\ \text{Difference} & +\ 6523 \\
\hline
=\ \text{Minuend} & 8955
\end{array}
$$

Example 1 Subtract $6594 - 3271$ and check.

Solution
$$
\begin{array}{r}
6594 \\
-\ 3271 \\
\hline
3323
\end{array}
\qquad
Check:
\begin{array}{r}
3271 \\
+\ 3323 \\
\hline
6594
\end{array}
$$

You Try It 1 Subtract $8925 - 6413$ and check.

Your solution

Example 2 Subtract $15{,}762 - 7541$ and check.

Solution
$$
\begin{array}{r}
15{,}762 \\
-\ 7{,}541 \\
\hline
8{,}221
\end{array}
\qquad
Check:
\begin{array}{r}
7{,}541 \\
+\ 8{,}221 \\
\hline
15{,}762
\end{array}
$$

You Try It 2 Subtract $17{,}504 - 9302$ and check.

Your solution

Solutions on p. S1

Copyright © Houghton Mifflin Company. All rights reserved.

Objective B **To subtract whole numbers with borrowing**

In all the subtraction problems in the previous objective, for each place value the lower digit was not larger than the upper digit. When the lower digit *is* larger than the upper digit, subtraction will involve **borrowing.**

HOW TO Subtract: 692 − 378

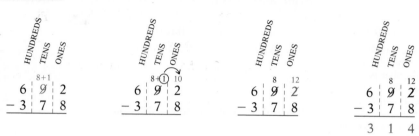

Because 8 > 2, borrowing is necessary. 9 tens = 8 tens + 1 ten.

Borrow 1 ten from the tens column and write 10 in the ones column.

Add the borrowed 10 to 2.

Subtract the digits in each column.

The phrases below are used to indicate the operation of subtraction. An example is shown at the right of each phrase.

minus	8 minus 5	8 − 5
less	9 less 3	9 − 3
less than	2 less than 7	7 − 2
the difference between	the difference between 8 and 2	8 − 2
decreased by	5 decreased by 1	5 − 1

HOW TO Find the difference between 1234 and 485, and check.

"The difference between 1234 and 485" means 1234 − 485.

$$
\begin{array}{r}
{}^{}{}^{2}{}^{14} \\
1\ 2\ \cancel{3}\ \cancel{4} \\
-\ \ \ 4\ 8\ 5 \\
\hline
9
\end{array}
\qquad
\begin{array}{r}
{}^{1}{}^{12}{}^{14} \\
1\ \cancel{2}\ \cancel{3}\ \cancel{4} \\
-\ \ \ 4\ 8\ 5 \\
\hline
4\ 9
\end{array}
\qquad
\begin{array}{r}
{}^{0}{}^{11}{}^{12}{}^{14} \\
\cancel{1}\ \cancel{2}\ \cancel{3}\ \cancel{4} \\
-\ \ \ 4\ 8\ 5 \\
\hline
7\ 4\ 9
\end{array}
\qquad
\textit{Check:}
\begin{array}{r}
{}^{1\ 1} \\
485 \\
+\ 749 \\
\hline
1234
\end{array}
$$

Subtraction with a zero in the minuend involves repeated borrowing.

Study Tip

The HOW TO feature indicates an example with explanatory remarks. Using paper and pencil, you should work through the example. See *AIM for Success* at the front of the book.

HOW TO Subtract: 3904 − 1775

$$
\begin{array}{r}
{}^{}{}^{8}{}^{10} \\
3\ \cancel{9}\ \cancel{0}\ 4 \\
-\ 1\ 7\ 7\ 5 \\
\hline
\end{array}
\qquad
\begin{array}{r}
{}^{}{}^{9} \\
{}^{8}{}^{10}{}^{14} \\
3\ \cancel{9}\ \cancel{0}\ \cancel{4} \\
-\ 1\ 7\ 7\ 5 \\
\hline
\end{array}
\qquad
\begin{array}{r}
{}^{}{}^{9} \\
{}^{8}{}^{10}{}^{14} \\
3\ \cancel{9}\ \cancel{0}\ \cancel{4} \\
-\ 1\ 7\ 7\ 5 \\
\hline
2\ 1\ 2\ 9
\end{array}
$$

5 > 4
There is a 0 in the tens column. Borrow 1 hundred (= 10 tens) from the hundreds column and write 10 in the tens column.

Borrow 1 ten from the tens column and add 10 to the 4 in the ones column.

Subtract the digits in each column.

Copyright © Houghton Mifflin Company. All rights reserved.

Example 3 Subtract 4392 − 678 and check.

Solution

$$\begin{array}{r} \overset{3}{\cancel{4}}\ \overset{13}{\cancel{3}}\ \overset{8}{\cancel{9}}\ \overset{12}{\cancel{2}} \\ -\ \ 6\ \ 7\ \ 8 \\ \hline 3\ \ 7\ \ 1\ \ 4 \end{array}$$
Check:
$$\begin{array}{r} 678 \\ +\ 3714 \\ \hline 4392 \end{array}$$

You Try It 3 Subtract 3481 − 865 and check.

Your solution

Example 4 Find 23,954 less than 63,221 and check.

Solution

$$\begin{array}{r} \overset{5}{\cancel{6}}\ \overset{12}{\cancel{3}},\overset{11}{\cancel{2}}\ \overset{11}{\cancel{2}}\ \overset{11}{\cancel{1}} \\ -\ 2\ \ 3,9\ \ 5\ \ 4 \\ \hline 3\ \ 9,2\ \ 6\ \ 7 \end{array}$$
Check:
$$\begin{array}{r} 23,954 \\ +39,267 \\ \hline 63,221 \end{array}$$

You Try It 4 Find 54,562 decreased by 14,485 and check.

Your solution

Example 5 Subtract 46,005 − 32,167 and check.

Solution

$$\begin{array}{r} \overset{5}{\cancel{6}}\ \overset{10}{\cancel{0}} \\ 4\ \ 6,\cancel{0}\ \ 0\ \ 5 \\ -3\ \ 2,1\ \ 6\ \ 7 \end{array}$$

• There are two zeros in the minuend. Borrow 1 thousand from the thousands column and write 10 in the hundreds column.

$$\begin{array}{r} \overset{9}{} \\ \overset{5}{} \overset{10}{\cancel{10}}\ \overset{10}{} \\ 4\ \ \cancel{6},\cancel{0}\ \ \cancel{0}\ \ 5 \\ -3\ \ 2,1\ \ 6\ \ 7 \end{array}$$

• Borrow 1 hundred from the hundreds column and write 10 in the tens column.

$$\begin{array}{r} \overset{9}{}\ \overset{9}{} \\ \overset{5}{} \overset{10}{\cancel{10}}\ \overset{10}{\cancel{10}}\ \overset{15}{} \\ 4\ \ \cancel{6},\cancel{0}\ \ \cancel{0}\ \ \cancel{5} \\ -3\ \ 2,1\ \ 6\ \ 7 \\ \hline 1\ \ 3,8\ \ 3\ \ 8 \end{array}$$

• Borrow 1 ten from the tens column and add 10 to the 5 in the ones column.

Check:
$$\begin{array}{r} 32,167 \\ +\ 13,838 \\ \hline 46,005 \end{array}$$

You Try It 5 Subtract 64,003 − 54,936 and check.

Your solution

Solutions on pp. S1–S2

● E S T I M A T I O N ●

Estimating the Difference Between Two Whole Numbers

Calculate 323,502 − 28,912. Then use estimation to determine whether the difference is reasonable.

Subtract to find the exact difference. To estimate the difference, round each number to the same place value. Here we have rounded to the nearest ten-thousand. Then subtract. The estimated answer is 290,000, which is very close to the exact difference 294,590.

$$\begin{array}{r} 323,502 \approx\ \ \ 320,000 \\ -\ 28,912 \approx\ -\ 30,000 \\ \hline 294,590\ \ \ \ \ \ 290,000 \end{array}$$

Copyright © Houghton Mifflin Company. All rights reserved.

Objective C   **To solve application problems**

The table at the right shows the number of personnel on active duty in the branches of the U.S. military in 1940 and 1945. Use this table for Example 6 and You Try It 6.

Branch	1940	1945
U.S. Army	267,767	8,266,373
U.S. Navy	160,997	3,380,817
U.S. Air Force	51,165	2,282,259
U.S. Marine Corps	28,345	474,680

Source: Dept. of the Army, Dept. of the Navy, Air Force Dept., Dept. of the Marines, U.S. Dept. of Defense

Example 6

Find the difference between the number of U.S. Army personnel on active duty in 1945 and the number in 1940.

Strategy
To find the difference, subtract the number of U.S. Army personnel on active duty in 1940 (267,767) from the number on active duty in 1945 (8,266,373).

Solution
$$\begin{array}{r} 8{,}266{,}373 \\ -267{,}767 \\ \hline 7{,}998{,}606 \end{array}$$

There were 7,998,606 more personnel on active duty in the U.S. Army in 1945 than in 1940.

You Try It 6

Find the difference between the number of personnel on active duty in the Navy and the number in the Air Force in 1945.

Your strategy

Your solution

Example 7

You had a balance of $415 on your student debit card. You then used the card, deducting $197 for books, $48 for art supplies, and $24 for theater tickets. What is your new student debit card balance?

Strategy
To find your new debit card balance:
• Add to find the total of the three deductions (197 + 48 + 24).
• Subtract the total of the three deductions from the old balance (415).

Solution
$$\begin{array}{r} 197 \\ 48 \\ +24 \\ \hline 269 \end{array} \text{ total deductions}$$

$$\begin{array}{r} 415 \\ -269 \\ \hline 146 \end{array}$$

Your new debit card balance is $146.

You Try It 7

Your total weekly salary is $638. Deductions of $127 for taxes, $18 for insurance, and $35 for savings are taken from your pay. Find your weekly take-home pay.

Your strategy

Your solution

Solutions on p. S2

Copyright © Houghton Mifflin Company. All rights reserved.

1.3 Exercises

Objective A   **To subtract whole numbers without borrowing**

For Exercises 1 to 35, subtract.

1. $\begin{array}{r} 9 \\ -\ 5 \\ \hline \end{array}$ **2.** $\begin{array}{r} 8 \\ -\ 7 \\ \hline \end{array}$ **3.** $\begin{array}{r} 8 \\ -\ 4 \\ \hline \end{array}$ **4.** $\begin{array}{r} 7 \\ -\ 3 \\ \hline \end{array}$ **5.** $\begin{array}{r} 10 \\ -\ 0 \\ \hline \end{array}$

6. $\begin{array}{r} 11 \\ -\ 4 \\ \hline \end{array}$ **7.** $\begin{array}{r} 12 \\ -\ 8 \\ \hline \end{array}$ **8.** $\begin{array}{r} 19 \\ -\ 8 \\ \hline \end{array}$ **9.** $\begin{array}{r} 15 \\ -\ 6 \\ \hline \end{array}$ **10.** $\begin{array}{r} 16 \\ -\ 7 \\ \hline \end{array}$

11. $\begin{array}{r} 25 \\ -\ 3 \\ \hline \end{array}$ **12.** $\begin{array}{r} 55 \\ -\ 4 \\ \hline \end{array}$ **13.** $\begin{array}{r} 68 \\ -\ 8 \\ \hline \end{array}$ **14.** $\begin{array}{r} 77 \\ -\ 3 \\ \hline \end{array}$ **15.** $\begin{array}{r} 89 \\ -\ 23 \\ \hline \end{array}$

16. $\begin{array}{r} 54 \\ -\ 21 \\ \hline \end{array}$ **17.** $\begin{array}{r} 88 \\ -\ 57 \\ \hline \end{array}$ **18.** $\begin{array}{r} 1202 \\ -\ 701 \\ \hline \end{array}$ **19.** $\begin{array}{r} 1305 \\ -\ 404 \\ \hline \end{array}$ **20.** $\begin{array}{r} 1763 \\ -\ 801 \\ \hline \end{array}$

21. $\begin{array}{r} 1497 \\ -\ 706 \\ \hline \end{array}$ **22.** $\begin{array}{r} 8974 \\ -\ 3972 \\ \hline \end{array}$ **23.** $\begin{array}{r} 2836 \\ -\ 1711 \\ \hline \end{array}$ **24.** $\begin{array}{r} 8976 \\ -\ 7463 \\ \hline \end{array}$ **25.** $\begin{array}{r} 9273 \\ -\ 6142 \\ \hline \end{array}$

26. $77 - 36$ **27.** $129 - 82$ **28.** $132 - 61$ **29.** $969 - 44$ **30.** $1347 - 103$

31. $4865 - 304$ **32.** $1525 - 702$ **33.** $9999 - 6794$ **34.** $7806 - 3405$ **35.** $8843 - 7621$

36. What is 3795 minus 1092?   **37.** What is 9071 minus 6050?

38. Find the difference between 9763 and 541.   **39.** Find the difference between 6094 and 3072.

40. What is 3701 less than 6932?   **41.** What is 2031 less than 5071?

42. Find 6509 decreased by 3102.   **43.** Find 7994 decreased by 7782.

44. Find 23,907 less 12,705.   **45.** Find 65,986 less 5741.

Copyright © Houghton Mifflin Company. All rights reserved.

Objective B   **To subtract whole numbers with borrowing**

For Exercises 46 to 89, subtract.

46.   71
      -18

47.   93
      -28

48.   47
      -18

49.   44
      -27

50.   37
      -29

51.   50
      -27

52.   70
      -33

53.   993
      -537

54.   250
      -192

55.   840
      -783

56.   768
      -194

57.   770
      -395

58. $674 - 337$

59. $3526 - 387$

60. $1712 - 289$

61. $4350 - 729$

62. $1702 - 948$

63. $1607 - 869$

64. $5933 - 3754$

65. $7293 - 3748$

66. $9407 - 2918$

67. $3706 - 2957$

68. $8605 - 7716$

69. $8052 - 2709$

70. $80,305 - 9176$

71. $70,702 - 4239$

72. $10,004 - 9306$

73. $80,009 - 63,419$

74. $70,618 - 41,213$

75. $80,053 - 27,649$

76. $70,700 - 21,076$

77. $80,800 - 42,023$

78.   2600
      -1972

79.   8400
      -3762

80.   9003
      -2471

81.   6004
      -2392

82.   8202
      -3916

83.   7050
      -4137

84.   7015
      -2973

85.   4207
      -1624

86.   7005
      -1796

87.   8003
      -2735

88.   20,005
      $-9,627$

89.   80,004
      $-8,237$

Copyright © Houghton Mifflin Company. All rights reserved.

90. Find 10,051 less 9027.

91. Find 17,031 less 5792.

92. Find the difference between 1003 and 447.

93. What is 29,874 minus 21,392?

94. What is 29,797 less than 68,005?

95. What is 69,379 less than 70,004?

96. What is 25,432 decreased by 7994?

97. What is 86,701 decreased by 9976?

For Exercises 98 to 101, use the relationship between addition and subtraction to complete the statement.

98. ___ + 39 = 104   **99.** 67 + ___ = 90   **100.** ___ + 497 = 862   **101.** 253 + ___ = 4901

For Exercises 102 to 107, use a calculator to subtract. Then round the numbers to the nearest ten-thousand and use estimation to determine whether the difference is reasonable.

102.   80,032
           − 19,605

103.   90,765
           − 60,928

104.   32,574
           − 10,961

105.   96,430
           − 59,762

106.   567,423
           − 208,444

107.   300,712
           − 198,714

Objective C   **To solve application problems**

108. Banking   You have $304 in your checking account. If you write a check for $139, how much is left in your checking account?

109. Identity Theft   Use the graph at the right to determine the increase in the number of identity theft complaints from 2001 to 2002.

110. Consumerism   The tennis coach at a high school purchased a video camera that costs $1079 and made a down payment of $180. Find the amount that remains to be paid.

111. Consumerism   Rod Guerra, an engineer, purchased a used car that cost $11,225 and made a down payment of $950. Find the amount that remains to be paid.

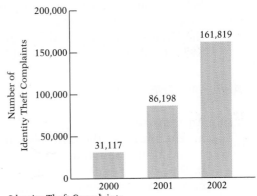

Identity Theft Complaints
Source: Federal Trade Commission

Copyright © Houghton Mifflin Company. All rights reserved.

112. 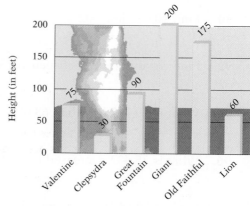 **Earth Science** Use the graph at the right to find the difference between the maximum height to which Great Fountain geyser erupts and the maximum height to which Valentine erupts.

113. **Earth Science** According to the graph at the right, how much higher is the eruption of the Giant than that of Old Faithful?

114. **Education** The National Center for Education Statistics estimates that 850,000 women and 587,000 men will earn a bachelor's degree in 2012. How many more women than men are expected to earn a bachelor's degree in 2012?

The Maximum Heights of the Eruptions of Six Geysers at Yellowstone National Park

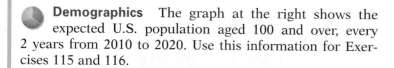 **Demographics** The graph at the right shows the expected U.S. population aged 100 and over, every 2 years from 2010 to 2020. Use this information for Exercises 115 and 116.

115. What is the expected growth in the population aged 100 and over during the 10-year period?

116. **a.** Which 2-year period has the smallest expected increase in the number of people aged 100 and over?
   b. Which 2-year period has the greatest expected increase?

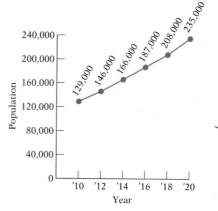

Expected U.S. Population Aged 100 and Over
Source: Census Bureau

117. **Expense Accounts** A sales executive has a monthly expense account of $1500. This month the executive has already spent $479 for transportation, $268 for food, and $317 for lodging. Find the amount remaining in this month's expense account.

118. **Finances** You had a credit card balance of $409 before you used the card to purchase books for $168, CDs for $36, and a pair of shoes for $97. You then made a payment to the credit card company of $350. Find your new credit card balance.

APPLYING THE CONCEPTS

119. Answer true or false.
   a. The phrases "the difference between 9 and 5" and "5 less than 9" mean the same thing.
   b. $9 - (5 - 3) = (9 - 5) - 3$.
   c. Subtraction is an associative operation. *Hint:* See part b of this exercise.

120.  Make up a word problem for which the difference between 15 and 8 is the answer.

Copyright © Houghton Mifflin Company. All rights reserved.

1.4 Multiplication of Whole Numbers

Objective A   **To multiply a number by a single digit**

Six boxes of toasters are ordered. Each box contains eight toasters. How many toasters are ordered?

This problem can be worked by adding 6 eights.

$8 + 8 + 8 + 8 + 8 + 8 = 48$

This problem involves repeated addition of the same number and can be worked by a shorter process called **multiplication.** Multiplication is the repeated addition of the same number.

$$8 + 8 + 8 + 8 + 8 + 8 = 48$$

The numbers that are multiplied are called **factors.** The result is called the **product.**

or

$$6 \times 8 = 48$$

Factor   Factor   Product

The product of 6×8 can be represented on the number line. The arrow representing the whole number 8 is repeated 6 times. The result is the arrow representing 48.

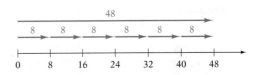

The times sign "$\times$" is only one symbol that is used to indicate multiplication. Each of the expressions that follow represent multiplication.

$$7 \times 8 \qquad 7 \cdot 8 \qquad 7(8) \qquad (7)(8) \qquad (7)8$$

As with addition, there are some useful properties of multiplication.

Multiplication Property of Zero
The product of a number and zero is zero.

$0 \times 4 = 0$
$7 \times 0 = 0$

Multiplication Property of One
The product of a number and one is the number.

$1 \times 6 = 6$
$8 \times 1 = 8$

Commutative Property of Multiplication
Two numbers can be multiplied in either order. The product will be the same.

$4 \times 3 = 3 \times 4$
$12 = 12$

Associative Property of Multiplication
Grouping the numbers to be multiplied in any order gives the same result. Do the multiplication inside the parentheses first.

$(4 \times 2) \times 3 = 4 \times (2 \times 3)$
$8 \quad \times 3 = 4 \times \quad 6$
$24 = 24$

Copyright © Houghton Mifflin Company. All rights reserved.

Study Tip

Some students think that they can "coast" at the beginning of this course because the topic of Chapter 1 is whole numbers. However, this chapter lays the foundation for the entire course. Be sure you know and understand all the concepts presented. For example, study the properties of multiplication presented in this lesson.

The basic facts for multiplying one-digit numbers should be memorized. Multiplication of larger numbers requires the repeated use of the basic multiplication facts.

HOW TO   Multiply: 37×4

$$\begin{array}{r} \overset{2}{3}\,7 \\ \times\quad 4 \\ \hline 8 \end{array}$$

- $4 \times 7 = 28$ (2 tens + 8 ones).
  Write the 8 in the ones column and carry the 2 to the tens column.

$$\begin{array}{r} \overset{2}{3}\,7 \\ \times\quad 4 \\ \hline 14\,8 \end{array}$$

- **The 3 in 37 is 3 tens.**
  4×3 tens = **12 tens**
  Add the carry digit.   + 2 tens
  _____
  14 tens

- **Write the 14.**

The phrases below are used to indicate the operation of multiplication. An example is shown at the right of each phrase.

times	7 times 3	$7 \cdot 3$
the product of	the product of 6 and 9	$6 \cdot 9$
multiplied by	8 multiplied by 2	$2 \cdot 8$

Example 1   Multiply: 735×9

Solution

$$\begin{array}{r} \overset{3\,4}{735} \\ \times\quad 9 \\ \hline 6615 \end{array}$$

- $9 \times 5 = 45$
  Write the 5 in the ones column. Carry the 4 to the tens column.
  $9 \times 3 = 27, 27 + 4 = 31$
  $9 \times 7 = 63, 63 + 3 = 66$

You Try It 1   Multiply: 648×7

Your solution

Solution on p. S2

Objective B   **To multiply larger whole numbers**

Note the pattern when the following numbers are multiplied.

Multiply the nonzero part of the factors.

Now attach the same number of zeros to the product as the total number of zeros in the factors.

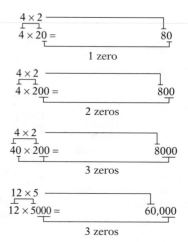

Copyright © Houghton Mifflin Company. All rights reserved.

HOW TO Find the product of 47 and 23.

Multiply by the ones digit.	Multiply by the tens digit.	Add.

$$\begin{array}{r} 47 \\ \times\ 23 \\ \hline 141 \end{array} (= 47 \times 3)$$

$$\begin{array}{r} 47 \\ \times\ 23 \\ \hline 141 \\ 940 \end{array} (= 47 \times 20)$$

$$\begin{array}{r} 47 \\ \times\ 23 \\ \hline 141 \\ 940 \\ \hline 1081 \end{array}$$

Writing the 0 is optional.

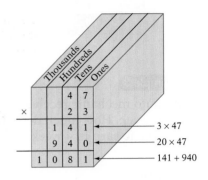

The place-value chart illustrates the placement of the products.

Note the placement of the products when we are multiplying by a factor that contains a zero.

HOW TO Multiply: 439×206

$$\begin{array}{r} 439 \\ \times\ 206 \\ \hline 2634 \\ 000 \\ 878 \\ \hline 90{,}434 \end{array}$$

0×439

When working the problem, we usually write only one zero. Writing this zero ensures the proper placement of the products.

$$\begin{array}{r} 439 \\ \times\ 206 \\ \hline 2634 \\ 8780 \\ \hline 90{,}434 \end{array}$$

Example 2 Find 829 multiplied by 603.

Solution

$$\begin{array}{r} 829 \\ \times\ 603 \\ \hline 2487 \\ 49740 \\ \hline 499{,}887 \end{array}$$

- $3 \times 829 = 2487$
- Write a zero in the tens column for 0×829.
- $6 \times 829 = 4974$

You Try It 2 Multiply: 756×305

Your solution

Solution on p. S2

● E S T I M A T I O N ●

Estimating the Product of Two Whole Numbers

Calculate 3267×389. Then use estimation to determine whether the product is reasonable.

Multiply to find the exact product. $3267\ \times\ 389\ =\ 1{,}270{,}863$

To estimate the product, round each number so there is one nonzero digit. Then multiply. The estimated answer is 1,200,000, which is very close to the exact product 1,270,863.

$$\begin{array}{r} 3267 \approx \quad 3000 \\ \times\ 389 \approx \quad \times\ 400 \\ \hline 1{,}200{,}000 \end{array}$$

Copyright © Houghton Mifflin Company. All rights reserved.

Objective C   **To solve application problems**

Example 3

An auto mechanic receives a salary of $1050 each week. How much does the auto mechanic earn in 4 weeks?

Strategy

To find the mechanic's earnings for 4 weeks, multiply the weekly salary (1050) by the number of weeks (4).

Solution

```
  1050
×    4
  4200
```

The mechanic earns $4200 in 4 weeks.

You Try It 3

A new-car dealer receives a shipment of 37 cars each month. Find the number of cars the dealer will receive in 12 months.

Your strategy

Your solution

Example 4

A press operator earns $640 for working a 40-hour week. This week the press operator also worked 7 hours of overtime at $26 an hour. Find the press operator's total pay for the week.

Strategy

To find the press operator's total pay for the week:

• Find the overtime pay by multiplying the hours of overtime (7) by the overtime rate of pay (26).

• Add the weekly salary (640) to the overtime pay.

Solution

```
    26                 640
×    7               +182
  182 overtime pay    822
```

The press operator earned $822 this week.

You Try It 4

The buyer for Ross Department Store can buy 80 men's suits for $4800. Each sports jacket will cost the store $23. The manager orders 80 men's suits and 25 sports jackets. What is the total cost of the order?

Your strategy

Your solution

Solutions on p. S2

Copyright © Houghton Mifflin Company. All rights reserved.

1.4 Exercises

Objective A   **To multiply a number by a single digit**

For Exercises 1 to 4, write the expression as a product.

1. $2 + 2 + 2 + 2 + 2 + 2$ **2.** $4 + 4 + 4 + 4 + 4$ **3.** $7 + 7 + 7 + 7$ **4.** $18 + 18 + 18$

For Exercises 5 to 39, multiply.

5. $\begin{array}{r} 3 \\ \times\, 4 \\ \hline \end{array}$ **6.** $\begin{array}{r} 2 \\ \times\, 8 \\ \hline \end{array}$ **7.** $\begin{array}{r} 5 \\ \times\, 7 \\ \hline \end{array}$ **8.** $\begin{array}{r} 6 \\ \times\, 4 \\ \hline \end{array}$ **9.** $\begin{array}{r} 5 \\ \times\, 5 \\ \hline \end{array}$

10. $\begin{array}{r} 7 \\ \times\, 7 \\ \hline \end{array}$ **11.** $\begin{array}{r} 0 \\ \times\, 7 \\ \hline \end{array}$ **12.** $\begin{array}{r} 8 \\ \times\, 0 \\ \hline \end{array}$ **13.** $\begin{array}{r} 8 \\ \times\, 9 \\ \hline \end{array}$ **14.** $\begin{array}{r} 7 \\ \times\, 6 \\ \hline \end{array}$

15. $\begin{array}{r} 66 \\ \times\, 3 \\ \hline \end{array}$ **16.** $\begin{array}{r} 70 \\ \times\, 4 \\ \hline \end{array}$ **17.** $\begin{array}{r} 67 \\ \times\, 5 \\ \hline \end{array}$ **18.** $\begin{array}{r} 127 \\ \times\, 9 \\ \hline \end{array}$ **19.** $\begin{array}{r} 623 \\ \times\, 4 \\ \hline \end{array}$

20. $\begin{array}{r} 802 \\ \times\, 5 \\ \hline \end{array}$ **21.** $\begin{array}{r} 607 \\ \times\, 9 \\ \hline \end{array}$ **22.** $\begin{array}{r} 300 \\ \times\, 5 \\ \hline \end{array}$ **23.** $\begin{array}{r} 600 \\ \times\, 7 \\ \hline \end{array}$ **24.** $\begin{array}{r} 906 \\ \times\, 8 \\ \hline \end{array}$

25. $\begin{array}{r} 703 \\ \times\, 9 \\ \hline \end{array}$ **26.** $\begin{array}{r} 127 \\ \times\, 5 \\ \hline \end{array}$ **27.** $\begin{array}{r} 632 \\ \times\, 3 \\ \hline \end{array}$ **28.** $\begin{array}{r} 559 \\ \times\, 4 \\ \hline \end{array}$ **29.** $\begin{array}{r} 632 \\ \times\, 8 \\ \hline \end{array}$

30. $\begin{array}{r} 524 \\ \times\, 4 \\ \hline \end{array}$ **31.** $\begin{array}{r} 337 \\ \times\, 5 \\ \hline \end{array}$ **32.** $\begin{array}{r} 841 \\ \times\, 6 \\ \hline \end{array}$ **33.** $\begin{array}{r} 6709 \\ \times\, 7 \\ \hline \end{array}$ **34.** $\begin{array}{r} 3608 \\ \times\, 5 \\ \hline \end{array}$

35. $\begin{array}{r} 8568 \\ \times\, 7 \\ \hline \end{array}$ **36.** $\begin{array}{r} 5495 \\ \times\, 4 \\ \hline \end{array}$ **37.** $\begin{array}{r} 4780 \\ \times\, 4 \\ \hline \end{array}$ **38.** $\begin{array}{r} 3690 \\ \times\, 5 \\ \hline \end{array}$ **39.** $\begin{array}{r} 9895 \\ \times\, 2 \\ \hline \end{array}$

40. Find the product of 5, 7, and 4.   **41.** Find the product of 6, 2, and 9.

42. Find the product of 5304 and 9.   **43.** Find the product of 458 and 8.

Copyright © Houghton Mifflin Company. All rights reserved.

44. What is 3208 multiplied by 7?

45. What is 5009 multiplied by 4?

46. What is 3105 times 6?

47. What is 8957 times 8?

Objective B **To multiply larger whole numbers**

For Exercises 48 to 79, multiply.

48. 16 $\times$ 21

49. 18 $\times$ 24

50. 35 $\times$ 26

51. 27 $\times$ 72

52. 693 $\times$ 91

53. 581 $\times$ 72

54. 419 $\times$ 80

55. 727 $\times$ 60

56. 8279 $\times$ 46

57. 9577 $\times$ 35

58. 6938 $\times$ 78

59. 8875 $\times$ 67

60. 7035 $\times$ 57

61. 6702 $\times$ 48

62. 3009 $\times$ 35

63. 6003 $\times$ 57

64. 809 $\times$ 530

65. 607 $\times$ 460

66. 800 $\times$ 325

67. 700 $\times$ 274

68. 987 $\times$ 349

69. 688 $\times$ 674

70. 312 $\times$ 134

71. 423 $\times$ 427

72. 379 $\times$ 500

73. 684 $\times$ 700

74. 985 $\times$ 408

75. 758 $\times$ 209

76. 3407 $\times$ 309

77. 5207 $\times$ 902

78. 4258 $\times$ 986

79. 6327 $\times$ 876

80. What is 5763 times 45?

81. What is 7349 times 27?

82. Find the product of 2, 19, and 34.

83. Find the product of 6, 73, and 43.

Copyright © Houghton Mifflin Company. All rights reserved.

84. What is 376 multiplied by 402?

85. What is 842 multiplied by 309?

86. Find the product of 3005 and 233,489.

87. Find the product of 9007 and 34,985.

For Exercises 88 to 95, use a calculator to multiply. Then use estimation to determine whether the product is reasonable.

88.
$$\begin{array}{r} 8745 \\ \times\ \ 63 \\ \hline \end{array}$$

89.
$$\begin{array}{r} 4732 \\ \times\ \ 93 \\ \hline \end{array}$$

90.
$$\begin{array}{r} 2937 \\ \times\ 206 \\ \hline \end{array}$$

91.
$$\begin{array}{r} 8941 \\ \times\ 726 \\ \hline \end{array}$$

92.
$$\begin{array}{r} 3097 \\ \times\ 1025 \\ \hline \end{array}$$

93.
$$\begin{array}{r} 6379 \\ \times\ 2936 \\ \hline \end{array}$$

94.
$$\begin{array}{r} 32{,}508 \\ \times\ \ \ 591 \\ \hline \end{array}$$

95.
$$\begin{array}{r} 62{,}504 \\ \times\ \ 923 \\ \hline \end{array}$$

Objective C   **To solve application problems**

96. **Fuel Efficiency**   Rob Hill owns a compact car that averages 43 miles on 1 gallon of gas. How many miles could the car travel on 12 gallons of gas?

97. **Fuel Efficiency**   A plane flying from Los Angeles to Boston uses 865 gallons of jet fuel each hour. How many gallons of jet fuel were used on a 6-hour flight?

98. **Geometry**   The perimeter of a square is equal to four times the length of a side of the square. Find the perimeter of a square whose side measures 16 miles.

16 mi

99. **Geometry**   To find the area of a square, multiply the length of one side of the square times itself. What is the area of a square whose side measures 16 miles? The area of the square will be in square miles.

100. **Geometry**   The area of a rectangle is equal to the product of the length of the rectangle times its width. Find the area of a rectangle that has a length of 24 meters and a width of 15 meters. The area will be in square meters.

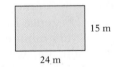

15 m

24 m

Interior Design   A lighting consultant to a bank suggests that the bank lobby contain 43 can lights, 15 high-intensity lights, 20 fire safety lights, and one chandelier. The table at the right gives the costs for each type of light from two companies. Use this table for Exercises 101 and 102.

	Company A	Company B
Can lights	$2 each	$3 each
High-intensity	$6 each	$4 each
Fire safety	$12 each	$11 each
Chandelier	$998 each	$1089 each

101. Which company offers the lights for the lower total price?

102. How much can the lighting consultant save by purchasing the lights from the company that offers the lower total price?

Copyright © Houghton Mifflin Company. All rights reserved.

Construction   The table at the right shows the hourly wages of four different job classifications at a small construction company. Use this table for Exercises 103 to 105.

Type of Work	Wage per Hour
Electrician	$34
Plumber	$30
Clerk	$16
Bookkeeper	$20

103. The owner of this company wants to provide the electrical installation for a new house. On the basis of the architectural plans for the house, it is estimated that it will require 3 electricians, each working 50 hours, to complete the job. What is the estimated cost for the electricians' labor?

104. Carlos Vasquez, a plumbing contractor, hires 4 plumbers from this company at the hourly wage given in the table. If each plumber works 23 hours, what are the total wages paid by Carlos?

105. The owner of this company estimates that remodeling a kitchen will require 1 electrician working 30 hours and 1 plumber working 33 hours. This project also requires 3 hours of clerical work and 4 hours of bookkeeping. What is the total cost for these four components of this remodeling?

APPLYING THE CONCEPTS

106. Determine whether each of the following statements is always true, sometimes true, or never true.
a. A whole number times zero is zero.
b. A whole number times one is the whole number.
c. The product of two whole numbers is greater than either one of the whole numbers.

107. **Safety**   According to the National Safety Council, in a recent year a death resulting from an accident occurred at the rate of 1 every 5 minutes. At this rate, how many accidental deaths occurred each hour? Each day? Throughout the year? Explain how you arrived at your answers.

108. **Demographics**   According to the Population Reference Bureau, in the world today, 261 people are born every minute and 101 people die every minute. Using this statistic, what is the increase in the world's population every hour? Every day? Every week? Every year? Use a 365-day year. Explain how you arrived at your answers.

109. Pick your favorite number between 1 and 9. Multiply the number by 3. Multiply that product by 37,037. How is the product related to your favorite number? Explain why this works. (*Suggestion:* Multiply 3 and 37,037 first.)

110. There are quite a few tricks based on whole numbers. Here's one about birthdays. Write down the month in which you were born. Multiply by 5. Add 7. Multiply by 20. Subtract 100. Add the day of the month on which you were born. Multiply by 4. Subtract 100. Multiply by 25. Add the year you were born. Subtract 3400. The answer is the month/day/year of your birthday.

Copyright © Houghton Mifflin Company. All rights reserved.

1.5 Division of Whole Numbers

Objective A   **To divide by a single digit with no remainder in the quotient**

Division is used to separate objects into equal groups.

A store manager wants to display 24 new objects equally on 4 shelves. From the diagram, we see that the manager would place 6 objects on each shelf.

The manager's division problem can be written as follows:

TAKE NOTE

The **divisor** is the number that is divided into another number. The **dividend** is the number into which the divisor is divided. The result is the **quotient.**

Number of shelves
Divisor

Number on each shelf
Quotient

$$6$$
$$4\overline{)24}$$

Number of objects
Dividend

Note that the quotient multiplied by the divisor equals the dividend.

$$4\overline{)24}^{\,6}$$ because | 6 (Quotient) | × | 4 (Divisor) | = | 24 (Dividend) |

$$9\overline{)54}^{\,6}$$ because   6   ×   9   =   54

$$8\overline{)40}^{\,5}$$ because   5   ×   8   =   40

Here are some important quotients and the properties of zero in division:

Properties of One in Division

Any whole number, except zero, divided by itself is 1.

$$8\overline{)8}^{\,1} \qquad 14\overline{)14}^{\,1} \qquad 10\overline{)10}^{\,1}$$

Any whole number divided by 1 is the whole number.

$$1\overline{)9}^{\,9} \qquad 1\overline{)27}^{\,27} \qquad 1\overline{)10}^{\,10}$$

Properties of Zero in Division

Zero divided by any other whole number is zero.

$$7\overline{)0}^{\,0} \qquad 13\overline{)0}^{\,0} \qquad 10\overline{)0}^{\,0}$$

Division by zero is not allowed.

$$0\overline{)8}^{\,?}$$ There is no number whose product with 0 is 8.

Integrating Technology

Enter 8 ÷ 0 = on your calculator. An error message is displayed because division by zero is not allowed.

Copyright © Houghton Mifflin Company. All rights reserved.

When the dividend is a larger whole number, the digits in the quotient are found in steps.

HOW TO Divide $4\overline{)3192}$ and check.

$$\begin{array}{r} 7 \\ 4\overline{)3192} \\ -28 \\ \hline 39 \end{array}$$

- Think $4\overline{)31}$.
- Subtract 7×4.
- Bring down the 9.

$$\begin{array}{r} 79 \\ 4\overline{)3192} \\ -28 \\ \hline 39 \\ -36 \\ \hline 32 \end{array}$$

- Think $4\overline{)39}$.
- Subtract 9×4.
- Bring down the 2.

$$\begin{array}{r} 798 \\ 4\overline{)3192} \\ -28 \\ \hline 39 \\ -36 \\ \hline 32 \\ -32 \\ \hline 0 \end{array}$$

- Think $4\overline{)32}$.
- Subtract 8×4.

Check: $\begin{array}{r} 798 \\ \times \quad 4 \\ \hline 3192 \end{array}$

The place-value chart can be used to show why this method works.

	HUNDREDS	TENS	ONES	
	7	9	8	
4) 3	1	9	2	
−2	8	0	0	7 hundreds × 4
	3	9	2	
	−3	6	0	9 tens × 4
		3	2	
		−3	2	8 ones × 4
			0	

There are other ways of expressing division.

54 divided by 9 equals 6.

54 ÷ 9 equals 6.

$\frac{54}{9}$ equals 6.

Copyright © Houghton Mifflin Company. All rights reserved.

Example 1 Divide 7)56 and check.

Solution
$$\frac{8}{7)56}$$

Check: $8 \times 7 = 56$

You Try It 1 Divide 9)63 and check.

Your solution

Example 2 Divide $2808 \div 8$ and check.

Solution
$$
\begin{array}{r}
351 \\
8)\overline{2808} \\
-24 \\
\hline
40 \\
-40 \\
\hline
08 \\
-8 \\
\hline
0
\end{array}
$$

Check: $351 \times 8 = 2808$

You Try It 2 Divide $4077 \div 9$ and check.

Your solution

Example 3 Divide 7)2856 and check.

Solution
$$
\begin{array}{r}
408 \\
7)\overline{2856} \\
-28 \\
\hline
05 \\
-0 \\
\hline
56 \\
-56 \\
\hline
0
\end{array}
$$

- Think 7)5. Place 0 in quotient.
- Subtract 0×7.
- Bring down the 6.

Check: $408 \times 7 = 2856$

You Try It 3 Divide 9)6345 and check.

Your solution

Solutions on pp. S2–S3

Objective B **To divide by a single digit with a remainder in the quotient**

Sometimes it is not possible to separate objects into a whole number of equal groups.

A baker has 14 muffins to pack into 3 boxes. Each box holds 4 muffins. From the diagram, we see that after the baker places 4 muffins in each box, there are 2 left over. The 2 is called the **remainder.**

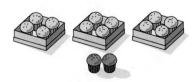

Copyright © Houghton Mifflin Company. All rights reserved.

The clerk's division problem could be written

$$
\begin{array}{r}
\textbf{Quotient} \\
\text{(Number in each box)} \\
4 \\
\text{Divisor} \longrightarrow 3\overline{)\,14} \longleftarrow \textbf{Dividend} \\
-12 \quad \text{(Total number of objects)} \\
\hline
2 \longleftarrow \textbf{Remainder} \\
\text{(Number left over)}
\end{array}
$$

Divisor (Number of boxes)

The answer to a division problem with a remainder is frequently written

$$3\overline{)14}^{\,4\ r2}$$

Note that $\boxed{\begin{array}{c}4\\ \text{Quotient}\end{array} \times \begin{array}{c}3\\ \text{Divisor}\end{array}} + \boxed{\begin{array}{c}2\\ \text{Remainder}\end{array}} = \boxed{\begin{array}{c}14\\ \text{Dividend}\end{array}}$.

Example 4   Divide $4\overline{)2522}$ and check.

Solution

$$
\begin{array}{r}
630 \ r2 \\
4\overline{)\,2522} \\
-24 \\
\hline
12 \\
-12 \\
\hline
02 \\
-\ 0 \\
\hline
2
\end{array}
$$

- Think $4\overline{)2}$. Place 0 in quotient.
- Subtract 0×4.

Check: $(630 \times 4) + 2 =$
$\qquad 2520 \ + 2 = 2522$

You Try It 4   Divide $6\overline{)5225}$ and check.

Your solution

Example 5   Divide $9\overline{)27,438}$ and check.

Solution

$$
\begin{array}{r}
3{,}048 \ r6 \\
9\overline{)\,27{,}438} \\
-27 \\
\hline
04 \\
-\ 0 \\
\hline
43 \\
-36 \\
\hline
78 \\
-72 \\
\hline
6
\end{array}
$$

- Think $9\overline{)4}$.
- Subtract 0×9.

Check: $(3048 \times 9) + 6 =$
$\qquad 27{,}432 \ + 6 = 27{,}438$

You Try It 5   Divide $7\overline{)21,409}$ and check.

Your solution

Solutions on p. S3

Copyright © Houghton Mifflin Company. All rights reserved.

Objective C **To divide by larger whole numbers**

When the divisor has more than one digit, estimate at each step by using the first digit of the divisor. If that product is too large, lower the guess by 1 and try again.

HOW TO Divide 34)1598 and check.

$$
\begin{array}{r}
5 \\
34)\overline{1598} \\
-170 \\
\end{array}
$$

- Think 3)15.
- Subtract 5 × 34.

$$
\begin{array}{r}
4 \\
34)\overline{1598} \\
-136 \\
\hline
238 \\
\end{array}
$$

- Subtract 4 × 34.

170 is too large. Lower the guess by 1 and try again.

$$
\begin{array}{r}
47 \\
34)\overline{1598} \\
-136 \\
\hline
238 \\
-238 \\
\hline
0 \\
\end{array}
$$

- Think 3)23.
- Subtract 7 × 34.

Check:
$$
\begin{array}{r}
47 \\
\times 34 \\
\hline
188 \\
141 \\
\hline
1598 \\
\end{array}
$$

Study Tip

One of the key instructional features of this text is the Example/You Try It pairs. Each Example is completely worked. You are to solve the You Try It problems. When you are ready, check your solution against the one in the Solutions section. The solution for You Try It 6 below is on page S3 (see the reference at the bottom right of the You Try It.) See *AIM for Success* at the front of the book.

The phrases below are used to indicate the operation of division. An example is shown at the right of each phrase.

the quotient of divided by	the quotient of 9 and 3 6 divided by 2	9 ÷ 3 6 ÷ 2

Example 6 Find 7077 divided by 34 and check.

Solution
$$
\begin{array}{r}
208 \text{ r5} \\
34)\overline{7077} \\
-68 \\
\hline
27 \\
-0 \\
\hline
277 \\
-272 \\
\hline
5 \\
\end{array}
$$

- Think 34)27.
- Place 0 in the quotient.
- Subtract 0 × 34.

Check: (208 × 34) + 5 =
7072   + 5 = 7077

You Try It 6 Divide 4578 ÷ 42 and check.

Your solution

Solution on p. S3

Copyright © Houghton Mifflin Company. All rights reserved.

Example 7   Find the quotient of 21,312 and 56 and check.

Solution

$$
\begin{array}{r}
380 \ \text{r32} \\
56\overline{)21{,}312} \\
-16\ 8 \\
\hline
4\ 51 \\
-4\ 48 \\
\hline
32 \\
-\ 0 \\
\hline
32
\end{array}
$$

• Think $5\overline{)21}$.
  4×56 is too large. Try 3.

Check: $(380 \times 56) + 32 =$
        $21{,}280 \ + 32 = 21{,}312$

You Try It 7   Divide $18{,}359 \div 39$ and check.

Your solution

Example 8   Divide $427\overline{)24{,}782}$ and check.

Solution

$$
\begin{array}{r}
58 \ \text{r16} \\
427\overline{)24{,}782} \\
-21\ 35 \\
\hline
3\ 432 \\
-3\ 416 \\
\hline
16
\end{array}
$$

Check: $(58 \times 427) + 16 =$
        $24{,}766 \ + 16 = 24{,}782$

You Try It 8   Divide $534\overline{)33{,}219}$ and check.

Your solution

Example 9   Divide $386\overline{)206{,}149}$ and check.

Solution

$$
\begin{array}{r}
534 \ \text{r25} \\
386\overline{)206{,}149} \\
-193\ 0 \\
\hline
13\ 14 \\
-11\ 58 \\
\hline
1\ 569 \\
-1\ 544 \\
\hline
25
\end{array}
$$

Check: $(534 \times 386) + 25 =$
        $206{,}124 \ + 25 = 206{,}149$

You Try It 9   Divide $515\overline{)216{,}848}$ and check.

Your solution

Copyright © Houghton Mifflin Company. All rights reserved.

Solutions on p. S3

Copyright © Houghton Mifflin Company. All rights reserved.

● E S T I M A T I O N ●

Estimating the Quotient of Two Whole Numbers

Calculate 36,936 ÷ 54. Then use estimation to determine whether the quotient is reasonable.

Divide to find the exact quotient.

$$36{,}936 \; \div \; 54 \; = \; 684$$

To estimate the quotient, round each number so there is one nonzero digit. Then divide. The estimated answer is 800, which is close to the exact quotient 684.

$$36{,}936 \div 54 \approx$$
$$40{,}000 \div 50 = 800$$

Objective D   **To solve application problems**

The **average** of several numbers is the sum of all the numbers divided by the number of those numbers.

$$\text{Average test score} = \frac{81 + 87 + 80 + 85 + 79 + 86}{6} = \frac{498}{6} = 83$$

HOW TO ● The table at the right shows what an upper-income family can expect to spend to raise a child to the age of 17 years. Find the average amount spent each year. Round to the nearest dollar.

Expenses to Raise a Child	
Housing	$89,580
Food	$35,670
Transportation	$32,760
Child care/education	$26,520
Clothing	$13,770
Health care	$13,380
Other	$30,090

Source: Department of Agriculture, *Expenditures on Children by Families*

Strategy
To find the average amount spent each year:

- Add all the numbers in the table to find the total amount spent during the 17 years.
- Divide the sum by 17.

Solution

```
  89,580            14,221
  35,670      17) 241,770
  32,760           −17
  26,520            71
  13,770           −68
  13,380            3 7
+ 30,090           −3 4
 ───────            37
 241,770          −34
                   30
Sum of all        −17
the costs          13
```

- When rounding to the nearest whole number, compare twice the remainder to the divisor. If twice the remainder is less than the divisor, drop the remainder. If twice the remainder is greater than or equal to the divisor, add 1 to the units digit of the quotient.

- Twice the remainder is 2 × 13 = 26. Because 26 > 17, add 1 to the units digit of the quotient.

The average amount spent each year to raise a child to the age of 17 is $14,222.

Example 10

Ngan Hui, a freight supervisor, shipped 192,600 bushels of wheat in 9 railroad cars. Find the amount of wheat shipped in each car.

Strategy

To find the amount of wheat shipped in each car, divide the number of bushels (192,600) by the number of cars (9).

Solution

$$
\begin{array}{r}
21{,}400 \\
9\overline{)\,192{,}600} \\
-18 \\
\hline
12 \\
-9 \\
\hline
36 \\
-36 \\
\hline
0
\end{array}
$$

Each car carried 21,400 bushels of wheat.

You Try It 10

Suppose a Michelin retail outlet can store 270 tires on 15 shelves. How many tires can be stored on each shelf?

Your strategy

Your solution

Example 11

The used car you are buying costs $11,216. A down payment of $2000 is required. The remaining balance is paid in 48 equal monthly payments. What is the monthly payment?

Strategy

To find the monthly payment:

- Find the remaining balance by subtracting the down payment (2000) from the total cost of the car (11,216).
- Divide the remaining balance by the number of equal monthly payments (48).

Solution

$$
\begin{array}{r}
11{,}216 \\
-\;2{,}000 \\
\hline
9{,}216
\end{array}
$$
Remaining balance

$$
\begin{array}{r}
192 \\
48\overline{)\,9216} \\
-48 \\
\hline
441 \\
-432 \\
\hline
96 \\
-96 \\
\hline
0
\end{array}
$$

The monthly payment is $192.

You Try It 11

A soft-drink manufacturer produces 12,600 cans of soft drink each hour. Cans are packed 24 to a case. How many cases of soft drink are produced in 8 hours?

Your strategy

Your solution

Solutions on pp. S3–S4

Copyright © Houghton Mifflin Company. All rights reserved.

1.5 Exercises

Objective A   **To divide by a single digit with no remainder in the quotient**

For Exercises 1 to 20, divide.

1. $4\overline{)8}$

2. $3\overline{)9}$

3. $6\overline{)36}$

4. $9\overline{)81}$

5. $7\overline{)49}$

6. $5\overline{)80}$

7. $6\overline{)96}$

8. $6\overline{)480}$

9. $4\overline{)840}$

10. $3\overline{)690}$

11. $7\overline{)308}$

12. $7\overline{)203}$

13. $9\overline{)6327}$

14. $4\overline{)2120}$

15. $8\overline{)7280}$

16. $9\overline{)8118}$

17. $3\overline{)64,680}$

18. $4\overline{)50,760}$

19. $6\overline{)21,480}$

20. $5\overline{)18,050}$

21. Find the quotient of 1446 and 3.

22. Find the quotient of 4123 and 7.

23. What is 7525 divided by 7?

24. What is 32,364 divided by 4?

For Exercises 25 to 28, use the relationship between multiplication and division to complete the multiplication problem.

25. ___ $\times 7 = 364$

26. $8 \times$ ___ $= 376$

27. $5 \times$ ___ $= 170$

28. ___ $\times 4 = 92$

Objective B   **To divide by a single digit with a remainder in the quotient**

For Exercises 29 to 51, divide.

29. $4\overline{)9}$

30. $2\overline{)7}$

31. $5\overline{)27}$

32. $9\overline{)88}$

33. $3\overline{)40}$

34. $6\overline{)97}$

35. $8\overline{)83}$

36. $5\overline{)54}$

37. $7\overline{)632}$

38. $4\overline{)363}$

Copyright © Houghton Mifflin Company. All rights reserved.

39. $4\overline{)921}$ **40.** $7\overline{)845}$ **41.** $8\overline{)1635}$ **42.** $5\overline{)1548}$ **43.** $7\overline{)9432}$

44. $7\overline{)8124}$ **45.** $3\overline{)5162}$ **46.** $5\overline{)3542}$ **47.** $8\overline{)3274}$

48. $4\overline{)15,301}$ **49.** $7\overline{)43,500}$ **50.** $8\overline{)72,354}$ **51.** $5\overline{)43,542}$

52. Find the quotient of 3107 and 8.

53. Find the quotient of 8642 and 8.

54. What is 45,738 divided by 4? Round to the nearest ten.

55. What is 37,896 divided by 9? Round to the nearest hundred.

56. What is 3572 divided by 7? Round to the nearest ten.

57. What is 78,345 divided by 4? Round to the nearest hundred.

Objective C   **To divide by larger whole numbers**

For Exercises 58 to 85, divide.

58. $27\overline{)96}$ **59.** $44\overline{)82}$ **60.** $42\overline{)87}$ **61.** $67\overline{)93}$

62. $41\overline{)897}$ **63.** $32\overline{)693}$ **64.** $23\overline{)784}$ **65.** $25\overline{)772}$

66. $74\overline{)600}$ **67.** $92\overline{)500}$ **68.** $70\overline{)329}$ **69.** $50\overline{)467}$

70. $36\overline{)7225}$ **71.** $44\overline{)8821}$ **72.** $19\overline{)3859}$ **73.** $32\overline{)9697}$

74. $88\overline{)3127}$ **75.** $92\overline{)6177}$ **76.** $33\overline{)8943}$ **77.** $27\overline{)4765}$

78. $22\overline{)98,654}$ **79.** $77\overline{)83,629}$ **80.** $64\overline{)38,912}$ **81.** $78\overline{)31,434}$

Copyright © Houghton Mifflin Company. All rights reserved.

82. $206\overline{)3097}$ **83.** $504\overline{)6504}$ **84.** $654\overline{)1217}$ **85.** $546\overline{)2344}$

86. Find the quotient of 5432 and 21.

87. Find the quotient of 8507 and 53.

88. What is 37,294 divided by 72?

89. What is 76,788 divided by 46?

90. Find 23,457 divided by 43. Round to the nearest hundred.

91. Find 341,781 divided by 43. Round to the nearest ten.

For Exercises 92 to 103, use a calculator to divide. Then use estimation to determine whether the quotient is reasonable.

92. $76\overline{)389,804}$ **93.** $53\overline{)117,925}$ **94.** $29\overline{)637,072}$ **95.** $67\overline{)738,072}$

96. $38\overline{)934,648}$ **97.** $34\overline{)906,304}$ **98.** $309\overline{)876,324}$ **99.** $642\overline{)323,568}$

100. $209\overline{)632,016}$ **101.** $614\overline{)332,174}$ **102.** $179\overline{)5,734,444}$ **103.** $374\overline{)7,712,254}$

Objective D   **To solve application problems**

Finance The graph at the right shows the annual expenditures, in a recent year, of the average household in the United States. Use this information for Exercises 104 to 106. Round answers to the nearest whole number.

104. What is the total amount spent annually by the average household in the United States?

105. What is the average monthly expense for housing?

106. What is the difference between the average monthly expense for food and the average monthly expense for health care?

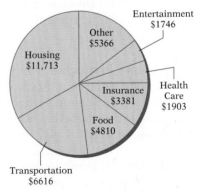

Average Annual Household Expenses

Source: Bureau of Labor Statistics Consumer Expenditure Survey

Copyright © Houghton Mifflin Company. All rights reserved.

Insurance   The table at the right shows the sources of insurance claims for losses of laptop computers in a recent year. Claims have been rounded to the nearest ten thousand dollars. Use this information for Exercises 107 and 108.

107. What was the average monthly claim for theft?

108. For all sources combined, find the average claims per month.

Source	Claims
Accidents	$560,000
Theft	$300,000
Power surge	$80,000
Lightning	$50,000
Transit	$20,000
Water/flood	$20,000
Other	$110,000

Source: Safeware, The Insurance Company

Work Hours   The table at the right shows, for different countries, the average number of hours per year that employees work. Use this information for Exercises 109 and 110. Use a 50-week year. Round answers to the nearest whole number.

109. What is the average number of hours worked per week by employees in Britain?

110. On average, how many more hours per week do employees in the United States work than employees in France?

Country	Annual Number of Hours Worked
Britain	1731
France	1656
Japan	1889
Norway	1399
United States	1966

Source: International Labor Organization

111.   **Coins**   The U.S. Mint estimates that about 114,000,000,000 of the 312,000,000,000 pennies it has minted over the last 30 years are in active circulation. That works out to how many pennies in circulation for each of the 300,000,000 people living in the United States?

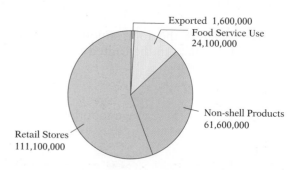

APPLYING THE CONCEPTS

112. Payroll Deductions   Your paycheck shows deductions of $225 for savings, $98 for taxes, and $27 for insurance. Find the total of the three deductions.

Dairy Products   The topic of the graph at the right is the eggs produced in the United States in a recent year. It shows where the eggs that were produced went or how they were used. Use this table for Exercises 113 and 114.

113. Use the graph to determine the total number of cases of eggs produced during the year.

114. How many more cases of eggs were sold by retail stores than were used for non-shell products?

Exported 1,600,000
Food Service Use
24,100,000

Non-shell Products
61,600,000

Retail Stores
111,100,000

Eggs Produced in the United States (in cases)
Source: American Egg Board

Copyright © Houghton Mifflin Company. All rights reserved.

115. **Fuel Consumption** The Energy Information Administration projects that in 2020, the average U.S. household will spend $1562 annually on gasoline. Use this estimate to determine how much the average U.S. household will spend on gasoline each month in 2020. Round to the nearest dollar.

Salaries The table at the right lists average starting salaries for students graduating with one of six different bachelor's degree. Use this table for Exercises 116 and 117.

Degree	Average Starting Salary
Accounting	$40,546
Biology	$29,554
Business	$37,122
Computer Sciences	$47,419
History	$32,108
Psychology	$27,454

Source: National Association of Colleges

116. Find the difference between the average starting salary of a computer sciences major and that of a biology major.

117. How much greater is the average starting salary of an accounting major than that of a psychology major?

The Military The graph at the right shows the basic monthly pay for Army commissioned officers with 20 years of service. Use this graph for Exercises 118 and 119.

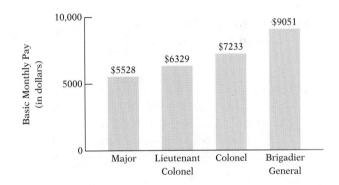

Basic Monthly Pay for Army Officers
Source: Department of Defense

118. What is a major's annual pay?

119. What is the difference between a colonel's annual pay and a lieutenant colonel's annual pay?

120. **Farming** A farmer harvested 48,290 pounds of avocados from one grove and 23,710 pounds of avocados from another grove. The avocados were packed in shipping boxes with 24 pounds in each box. How many boxes were needed to pack the avocados?

121. **Wages** A sales associate earns $374 for working a 40-hour week. Last week the associate worked an additional 9 hours at $13 an hour. Find the sales associate's total pay for last week's work.

122. **Finances** You purchase a used car with a down payment of $2500 and monthly payments of $195 for 48 months. Find the total amount paid for the car.

123. A palindromic number is a whole number that remains unchanged when its digits are written in reverse order. For instance, 292 is a palindromic number. Find the smallest three-digit palindromic number that is divisible by 4.

Copyright © Houghton Mifflin Company. All rights reserved.

1.6 Exponential Notation and the Order of Operations Agreement

Objective A **To simplify expressions that contain exponents**

Repeated multiplication of the same factor can be written in two ways:

$$3 \cdot 3 \cdot 3 \cdot 3 \cdot 3 \quad \text{or} \quad 3^5 \leftarrow \textbf{Exponent}$$

The **exponent** indicates how many times the factor occurs in the multiplication. The expression 3^5 is in **exponential notation.**

It is important to be able to read numbers written in exponential notation.

$6 = 6^1$ is read "six to the first **power**" or just "six." Usually the exponent 1 is not written.

$6 \cdot 6 = 6^2$ is read "six squared" or "six to the second power."

$6 \cdot 6 \cdot 6 = 6^3$ is read "six cubed" or "six to the third power."

$6 \cdot 6 \cdot 6 \cdot 6 = 6^4$ is read "six to the fourth power."

$6 \cdot 6 \cdot 6 \cdot 6 \cdot 6 = 6^5$ is read "six to the fifth power."

Each place value in the place-value chart can be expressed as a power of 10.

Ten =	10	=	10	$= 10^1$
Hundred =	100	=	$10 \cdot 10$	$= 10^2$
Thousand =	1000	=	$10 \cdot 10 \cdot 10$	$= 10^3$
Ten-thousand =	10,000	=	$10 \cdot 10 \cdot 10 \cdot 10$	$= 10^4$
Hundred-thousand =	100,000	=	$10 \cdot 10 \cdot 10 \cdot 10 \cdot 10$	$= 10^5$
Million =	1,000,000	=	$10 \cdot 10 \cdot 10 \cdot 10 \cdot 10 \cdot 10$	$= 10^6$

Integrating Technology

A calculator can be used to evaluate an exponential expression. The y^x key (or, on some calculators, a x^y or $\wedge$ key) is used to enter the exponent. For instance, for the example at the right, enter 4 y^x 3 = . The display reads 64.

To simplify a numerical expression containing exponents, write each factor as many times as indicated by the exponent and carry out the indicated multiplication.

$$4^3 = 4 \cdot 4 \cdot 4 = 64$$
$$2^2 \cdot 3^4 = (2 \cdot 2) \cdot (3 \cdot 3 \cdot 3 \cdot 3) = 4 \cdot 81 = 324$$

Example 1   Write $3 \cdot 3 \cdot 3 \cdot 5 \cdot 5$ in exponential notation.

Solution $3 \cdot 3 \cdot 3 \cdot 5 \cdot 5 = 3^3 \cdot 5^2$

You Try It 1   Write $2 \cdot 2 \cdot 2 \cdot 2 \cdot 3 \cdot 3 \cdot 3$ in exponential notation.

Your solution

Example 2   Write as a power of 10:
$10 \cdot 10 \cdot 10 \cdot 10$

Solution $10 \cdot 10 \cdot 10 \cdot 10 = 10^4$

You Try It 2   Write as a power of 10:
$10 \cdot 10 \cdot 10 \cdot 10 \cdot 10 \cdot 10 \cdot 10$

Your solution

Example 3   Simplify $3^2 \cdot 5^3$.

Solution $3^2 \cdot 5^3 = (3 \cdot 3) \cdot (5 \cdot 5 \cdot 5)$
$= 9 \cdot 125 = 1125$

You Try It 3   Simplify $2^3 \cdot 5^2$.

Your solution

Solutions on p. S4

Copyright © Houghton Mifflin Company. All rights reserved.

Objective B   **To use the Order of Operations Agreement to simplify expressions**

More than one operation may occur in a numerical expression. The answer may be different, depending on the order in which the operations are performed. For example, consider $3 + 4 \times 5$.

Multiply first, then add.

$$3 + \underbrace{4 \times 5}$$
$$\underbrace{3 + 20}$$
$$23$$

Add first, then multiply.

$$\underbrace{3 + 4} \times 5$$
$$\underbrace{7 \times 5}$$
$$35$$

An Order of Operations Agreement is used so that only one answer is possible.

The Order of Operations Agreement

Step 1.   Do all the operations inside parentheses.

Step 2.   Simplify any number expressions containing exponents.

Step 3.   Do multiplication and division as they occur from left to right.

Step 4.   Do addition and subtraction as they occur from left to right.

Integrating Technology

Many scientific calculators have an x^2 key. This key is used to square the displayed number. For example, after the user presses 2 x^2 = , the display reads 4.

HOW TO   Simplify $3 \times (2 + 1) - 2^2 + 4 \div 2$ by using the Order of Operations Agreement.

$$3 \times \underbrace{(2 + 1)} - 2^2 + 4 \div 2$$
$$\underbrace{3 \times 3} - 2^2 + 4 \div 2$$
$$3 \times 3 - \underbrace{4} + 4 \div 2$$
$$\underbrace{9} - 4 + \underbrace{4 \div 2}$$
$$\underbrace{9 - 4} + 2$$
$$\underbrace{5 + 2}$$
$$7$$

1. Perform operations in parentheses.

2. Simplify expressions with exponents.

3. Do multiplication and division as they occur from left to right.

4. Do addition and subtraction as they occur from left to right.

One or more of the foregoing steps may not be needed to simplify an expression. In that case, proceed to the next step in the Order of Operations Agreement.

HOW TO   Simplify $5 + 8 \div 2$.   There are no parentheses or exponents. Proceed to Step 3 of the agreement.

$$5 + \underbrace{8 \div 2}$$
$$\underbrace{5 + 4}$$
$$9$$

3. Do multiplication or division.

4. Do addition or subtraction.

Example 4   Simplify: $64 \div (8 - 4)^2 \cdot 9 - 5^2$

Solution

$$64 \div (8 - 4)^2 \cdot 9 - 5^2$$
$$= 64 \div 4^2 \cdot 9 - 5^2 \qquad \bullet \text{ Parentheses}$$
$$= 64 \div 16 \cdot 9 - 25 \qquad \bullet \text{ Exponents}$$
$$= 4 \cdot 9 - 25 \qquad\qquad \bullet \text{ Division and}$$
$$\qquad\qquad\qquad\qquad\qquad\quad \text{ multiplication}$$
$$= 36 - 25$$
$$= 11 \qquad\qquad\qquad\qquad \bullet \text{ Subtraction}$$

You Try It 4   Simplify: $5 \cdot (8 - 4)^2 \div 4 - 2$

Your solution

Solution on p. S4

Copyright © Houghton Mifflin Company. All rights reserved.

1.6 Exercises

Objective A   **To simplify expressions that contain exponents**

For Exercises 1 to 12, write the number in exponential notation.

1. $2 \cdot 2 \cdot 2$ **2.** $7 \cdot 7 \cdot 7 \cdot 7 \cdot 7$ **3.** $6 \cdot 6 \cdot 6 \cdot 7 \cdot 7 \cdot 7 \cdot 7$

4. $6 \cdot 6 \cdot 9 \cdot 9 \cdot 9 \cdot 9$ **5.** $2 \cdot 2 \cdot 2 \cdot 3 \cdot 3 \cdot 3$ **6.** $3 \cdot 3 \cdot 10 \cdot 10$

7. $5 \cdot 7 \cdot 7 \cdot 7 \cdot 7 \cdot 7$ **8.** $4 \cdot 4 \cdot 4 \cdot 5 \cdot 5 \cdot 5$ **9.** $3 \cdot 3 \cdot 3 \cdot 6 \cdot 6 \cdot 6 \cdot 6$

10. $2 \cdot 2 \cdot 5 \cdot 5 \cdot 5 \cdot 8$ **11.** $3 \cdot 3 \cdot 3 \cdot 5 \cdot 9 \cdot 9 \cdot 9$ **12.** $2 \cdot 2 \cdot 2 \cdot 4 \cdot 7 \cdot 7 \cdot 7$

For Exercises 13 to 37, simplify.

13. 2^3 **14.** 2^6 **15.** $2^4 \cdot 5^2$ **16.** $2^6 \cdot 3^2$ **17.** $3^2 \cdot 10^2$

18. $2^3 \cdot 10^4$ **19.** $6^2 \cdot 3^3$ **20.** $4^3 \cdot 5^2$ **21.** $5 \cdot 2^3 \cdot 3$ **22.** $6 \cdot 3^2 \cdot 4$

23. $2^2 \cdot 3^2 \cdot 10$ **24.** $3^2 \cdot 5^2 \cdot 10$ **25.** $0^2 \cdot 4^3$ **26.** $6^2 \cdot 0^3$ **27.** $3^2 \cdot 10^4$

28. $5^3 \cdot 10^3$ **29.** $2^2 \cdot 3^3 \cdot 5$ **30.** $5^2 \cdot 7^3 \cdot 2$ **31.** $2 \cdot 3^4 \cdot 5^2$ **32.** $6 \cdot 2^6 \cdot 7^2$

33. $5^2 \cdot 3^2 \cdot 7^2$ **34.** $4^2 \cdot 9^2 \cdot 6^2$ **35.** $3^4 \cdot 2^6 \cdot 5$ **36.** $4^3 \cdot 6^3 \cdot 7$ **37.** $4^2 \cdot 3^3 \cdot 10^4$

Objective B   **To use the Order of Operations Agreement to simplify expressions**

For Exercises 38 to 76, simplify by using the Order of Operations Agreement.

38. $4 - 2 + 3$ **39.** $6 - 3 + 2$ **40.** $6 \div 3 + 2$ **41.** $8 \div 4 + 8$

42. $6 \cdot 3 + 5$ **43.** $5 \cdot 9 + 2$ **44.** $3^2 - 4$ **45.** $5^2 - 17$

Copyright © Houghton Mifflin Company. All rights reserved.

46. $4 \cdot (5 - 3) + 2$ **47.** $3 + (4 + 2) \div 3$ **48.** $5 + (8 + 4) \div 6$ **49.** $8 - 2^2 + 4$

50. $16 \cdot (3 + 2) \div 10$ **51.** $12 \cdot (1 + 5) \div 12$ **52.** $10 - 2^3 + 4$ **53.** $5 \cdot 3^2 + 8$

54. $16 + 4 \cdot 3^2$ **55.** $12 + 4 \cdot 2^3$ **56.** $16 + (8 - 3) \cdot 2$ **57.** $7 + (9 - 5) \cdot 3$

58. $2^2 + 3 \cdot (6 - 2)^2$ **59.** $3^3 + 5 \cdot (8 - 6)^3$ **60.** $2^2 \cdot 3^2 + 2 \cdot 3$ **61.** $4 \cdot 6 + 3^2 \cdot 4^2$

62. $16 - 2 \cdot 4$ **63.** $12 + 3 \cdot 5$ **64.** $3 \cdot (6 - 2) + 4$

65. $5 \cdot (8 - 4) - 6$ **66.** $8 - (8 - 2) \div 3$ **67.** $12 - (12 - 4) \div 4$

68. $8 + 2 - 3 \cdot 2 \div 3$ **69.** $10 + 1 - 5 \cdot 2 \div 5$ **70.** $3 \cdot (4 + 2) \div 6$

71. $(7 - 3)^2 \div 2 - 4 + 8$ **72.** $20 - 4 \div 2 \cdot (3 - 1)^3$ **73.** $12 \div 3 \cdot 2^2 + (7 - 3)^2$

74. $(4 - 2) \cdot 6 \div 3 + (5 - 2)^2$ **75.** $18 - 2 \cdot 3 + (4 - 1)^3$ **76.** $100 \div (2 + 3)^2 - 8 \div 2$

APPLYING THE CONCEPTS

77. **Computers**   Memory in computers is measured in bytes. One kilobyte (1 K) is 2^{10} bytes. Write this number in standard form.

78. Explain the difference that the order of operations makes between **a.** $(14 - 2) \div 2 \cdot 3$ and **b.** $(14 - 2) \div (2 \cdot 3)$. Work the two problems. What is the difference between the larger answer and the smaller answer?

Copyright © Houghton Mifflin Company. All rights reserved.

1.7 Prime Numbers and Factoring

Objective A   To factor numbers

Whole-number **factors of a number** divide that number evenly (there is no remainder).

1, 2, 3, and 6 are whole-number factors of 6 because they divide 6 evenly.

$$\overset{6}{1)6} \qquad \overset{3}{2)6} \qquad \overset{2}{3)6} \qquad \overset{1}{6)6}$$

Note that both the divisor and the quotient are factors of the dividend.

To find the factors of a number, try dividing the number by 1, 2, 3, 4, 5, ... Those numbers that divide the number evenly are its factors. Continue this process until the factors start to repeat.

HOW TO   Find all the factors of 42.

$42 \div 1 = 42$	1 and 42 are factors.
$42 \div 2 = 21$	2 and 21 are factors.
$42 \div 3 = 14$	3 and 14 are factors.
$42 \div 4$	Will not divide evenly
$42 \div 5$	Will not divide evenly
$42 \div 6 = 7$	6 and 7 are factors. ⎤ Factors are repeating; all the
$42 \div 7 = 6$	7 and 6 are factors. ⎦ factors of 42 have been found.

1, 2, 3, 6, 7, 14, 21, and 42 are factors of 42.

The following rules are helpful in finding the factors of a number.

2 is a factor of a number if the last digit of the number is 0, 2, 4, 6, or 8.

436 ends in 6; therefore, 2 is a factor of 436. ($436 \div 2 = 218$)

3 is a factor of a number if the sum of the digits of the number is divisible by 3.

The sum of the digits of 489 is $4 + 8 + 9 = 21$. 21 is divisible by 3. Therefore, 3 is a factor of 489. ($489 \div 3 = 163$)

5 is a factor of a number if the last digit of the number is 0 or 5.

520 ends in 0; therefore, 5 is a factor of 520. ($520 \div 5 = 104$)

Example 1   Find all the factors of 30.

Solution

$30 \div 1 = 30$
$30 \div 2 = 15$
$30 \div 3 = 10$
$30 \div 4$ Will not divide evenly
$30 \div 5 = 6$
$30 \div 6 = 5$

1 2, 3, 5, 6, 10, 15, and 30 are factors of 30.

You Try It 1   Find all the factors of 40.

Your solution

Solution on p. S4

Copyright © Houghton Mifflin Company. All rights reserved.

Objective B   **To find the prime factorization of a number**

Point of Interest

Prime numbers are an important part of cryptology, the study of secret codes. To make it less likely that codes can be broken, cryptologists use prime numbers that have hundreds of digits.

A number is a **prime number** if its only whole-number factors are 1 and itself. 7 is prime because its only factors are 1 and 7. If a number is not prime, it is called a **composite number.** Because 6 has factors of 2 and 3, 6 is a composite number. The number 1 is not considered a prime number; therefore, it is not included in the following list of prime numbers less than 50.

$$2, 3, 5, 7, 11, 13, 17, 19, 23, 29, 31, 37, 41, 43, 47$$

The **prime factorization** of a number is the expression of the number as a product of its prime factors. We use a "T-diagram" to find the prime factors of 60. Begin with the smallest prime number as a trial divisor, and continue with prime numbers as trial divisors until the final quotient is 1.

$$
\begin{array}{c|c}
\multicolumn{2}{c}{60} \\
\hline
2 & 30 \\
2 & 15 \\
3 & 5 \\
5 & 1 \\
\end{array}
\qquad
\begin{aligned}
60 \div 2 &= 30 \\
30 \div 2 &= 15 \\
15 \div 3 &= 5 \\
5 \div 5 &= 1
\end{aligned}
$$

The prime factorization of 60 is $2 \cdot 2 \cdot 3 \cdot 5$.

Finding the prime factorization of larger numbers can be more difficult. Try each prime number as a trial divisor. Stop when the square of the trial divisor is greater than the number being factored.

HOW TO   Find the prime factorization of 106.

$$
\begin{array}{c|c}
\multicolumn{2}{c}{106} \\
\hline
2 & 53 \\
53 & 1 \\
\end{array}
$$

• 53 cannot be divided evenly by 2, 3, 5, 7, or 11. Prime numbers greater than 11 need not be tested because 11^2 is greater than 53.

The prime factorization of 106 is $2 \cdot 53$.

Example 2   Find the prime factorization of 315.

Solution

$$
\begin{array}{c|c}
\multicolumn{2}{c}{315} \\
\hline
3 & 105 \\
3 & 35 \\
5 & 7 \\
7 & 1 \\
\end{array}
$$

• $315 \div 3 = 105$
• $105 \div 3 = 35$
• $35 \div 5 = 7$
• $7 \div 7 = 1$

$315 = 3 \cdot 3 \cdot 5 \cdot 7$

You Try It 2   Find the prime factorization of 44.

Your solution

Example 3   Find the prime factorization of 201.

Solution

$$
\begin{array}{c|c}
\multicolumn{2}{c}{201} \\
\hline
3 & 67 \\
67 & 1 \\
\end{array}
$$

• Try only 2, 3, 5, 7, and 11 because $11^2 > 67$.

$201 = 3 \cdot 67$

You Try It 3   Find the prime factorization of 177.

Your solution

Solutions on p. S4

Copyright © Houghton Mifflin Company. All rights reserved.

1.7 Exercises

Copyright © Houghton Mifflin Company. All rights reserved.

Objective A To factor numbers

For Exercises 1 to 40, find all the factors of the number.

1. 4   **2.** 6   **3.** 10   **4.** 20

5. 7   **6.** 12   **7.** 9   **8.** 8

9. 13   **10.** 17   **11.** 18   **12.** 24

13. 56   **14.** 36   **15.** 45   **16.** 28

17. 29   **18.** 33   **19.** 22   **20.** 26

21. 52   **22.** 49   **23.** 82   **24.** 37

25. 57   **26.** 69   **27.** 48   **28.** 64

29. 95   **30.** 46   **31.** 54   **32.** 50

33. 66   **34.** 77   **35.** 80   **36.** 100

37. 96   **38.** 85   **39.** 90   **40.** 101

Objective B To find the prime factorization of a number

For Exercises 41 to 84, find the prime factorization.

41. 6   **42.** 14   **43.** 17   **44.** 83

45. 24   **46.** 12   **47.** 27   **48.** 9

49. 36   **50.** 40   **51.** 19   **52.** 37

53. 90   **54.** 65   **55.** 115   **56.** 80

57. 18   **58.** 26   **59.** 28   **60.** 49

61. 31   **62.** 42   **63.** 62   **64.** 81

65. 22   **66.** 39   **67.** 101   **68.** 89

69. 66   **70.** 86   **71.** 74   **72.** 95

73. 67   **74.** 78   **75.** 55   **76.** 46

77. 120   **78.** 144   **79.** 160   **80.** 175

81. 216   **82.** 400   **83.** 625   **84.** 225

APPLYING THE CONCEPTS

85. Twin primes are two prime numbers that differ by 2. For instance, 17 and 19 are twin primes. Find three sets of twin primes, not including 17 and 19.

86. In 1742, Christian Goldbach conjectured that every even number greater than 2 could be expressed as the sum of two prime numbers. Show that this conjecture is true for 8, 24, and 72. (*Note:* Mathematicians have not yet been able to determine whether Goldbach's conjecture is true or false.)

87. Explain why 2 is the only even prime number.

88. If a number is divisible by 6 and 10, is the number divisible by 30? If so, explain why. If not, give an example.

Copyright © Houghton Mifflin Company. All rights reserved.

Focus on Problem Solving

Questions to Ask

You encounter problem-solving situations every day. Some problems are easy to solve, and you may mentally solve these problems without considering the steps you are taking in order to draw a conclusion. Others may be more challenging and may require more thought and consideration.

Suppose a friend suggests that you both take a trip over spring break. You'd like to go. What questions go through your mind? You might ask yourself some of the following questions:

How much will the trip cost? What will be the cost for travel, hotel rooms, meals, and so on?

Are some costs going to be shared by both me and my friend?

Can I afford it?

How much money do I have in the bank?

How much more money than I have now do I need?

How much time is there to earn that much money?

How much can I earn in that amount of time?

How much money must I keep in the bank in order to pay the next tuition bill (or some other expense)?

These questions require different mathematical skills. Determining the cost of the trip requires **estimation;** for example, you must use your knowledge of air fares or the cost of gasoline to arrive at an estimate of these costs. If some of the costs are going to be shared, you need to **divide** those costs by 2 in order to determine your share of the expense. The question regarding how much more money you need requires **subtraction:** the amount needed minus the amount currently in the bank. To determine how much money you can earn in the given amount of time requires **multiplication**—for example, the amount you earn per week times the number of weeks to be worked. To determine if the amount you can earn in the given amount of time is sufficient, you need to use your knowledge of **order relations** to compare the amount you can earn with the amount needed.

Facing the problem-solving situation described above may not seem difficult to you. The reason may be that you have faced similar situations before and, therefore, know how to work through this one. You may feel better prepared to deal with a circumstance such as this one because you know what questions to ask. An important aspect of learning to solve problems is learning what questions to ask. As you work through application problems in this text, try to become more conscious of the mental process you are going through. You might begin the process by asking yourself the following questions whenever you are solving an application problem.

1. Have I read the problem enough times to be able to understand the situation being described?

2. Will restating the problem in different words help me to understand the problem situation better?

3. What facts are given? (You might make a list of the information contained in the problem.)

4. What information is being asked for?

Copyright © Houghton Mifflin Company. All rights reserved.

5. What relationship exists among the given facts? What relationship exists between the given facts and the solution?

6. What mathematical operations are needed in order to solve the problem?

Try to focus on the problem-solving situation, not on the computation or on getting the answer quickly. And remember, the more problems you solve, the better able you will be to solve other problems in the future, partly because you are learning what questions to ask.

Projects and Group Activities

Order of Operations

Does your calculator use the Order of Operations Agreement? To find out, try this problem:

$$2 + 4 \cdot 7$$

If your answer is 30, then the calculator uses the Order of Operations Agreement. If your answer is 42, it does not use that agreement.

Even if your calculator does not use the Order of Operations Agreement, you can still correctly evaluate numerical expressions. The parentheses keys, (and), are used for this purpose.

Remember that $2 + 4 \cdot 7$ means $2 + (4 \cdot 7)$ because the multiplication must be completed before the addition. To evaluate this expression, enter the following:

Enter:	2	+	(	4	×	7	)	=
Display:	2	2	(	4	4	7	28	30

When using your calculator to evaluate numerical expressions, insert parentheses around multiplications and around divisions. This has the effect of forcing the calculator to do the operations in the order you want.

For Exercises 1 to 10, evaluate.

1. $3 \cdot 8 - 5$

2. $6 + 8 \div 2$

3. $3 \cdot (8 - 2)^2$

4. $24 - (4 - 2)^2 \div 4$

5. $3 + (6 \div 2 + 4)^2 - 2$

6. $16 \div 2 + 4 \cdot (8 - 12 \div 4)^2 - 50$

7. $3 \cdot (15 - 2 \cdot 3) - 36 \div 3$

8. $4 \cdot 2^2 - (12 + 24 \div 6) + 5$

9. $16 \div 4 \cdot 3 + (3 \cdot 4 - 5) + 2$

10. $15 \cdot 3 \div 9 + (2 \cdot 6 - 3) + 4$

Copyright © Houghton Mifflin Company. All rights reserved.

Patterns in Mathematics

For the circle at the left, use a straight line to connect each dot on the circle with every other dot on the circle. How many different straight lines are there?

Follow the same procedure for each of the circles shown below. How many different straight lines are there in each?

Find a pattern to describe the number of dots on a circle and the corresponding number of different lines drawn. Use the pattern to determine the number of different lines that would be drawn in a circle with 7 dots and in a circle with 8 dots.

Now use the pattern to answer the following question. You are arranging a tennis tournament with 9 players. How many singles matches will be played among the 9 players if each player plays each of the other players only once?

Search the World Wide Web

Go to **www.census.gov** on the Internet.

1. Find a projection for the total U.S. population 10 years from now and a projection for the total population 20 years from now. Record the two numbers.
2. Use the data from Exercise 1 to determine the expected growth in the population over the next 10 years.
3. Use the answer from Exercise 2 to find the average increase in the U.S. population per year over the next 10 years. Round to the nearest million.
4. Use data in the population table you found to write two word problems. Then state whether addition, subtraction, multiplication, or division is required to solve each of the problems.

Chapter 1 Summary

Key Words

Examples

The *whole numbers* are 0, 1, 2, 3, 4, 5, 6, 7, 8, 9, 10, ... [1.1A, p. 3]

The *graph of a whole number* is shown by placing a heavy dot directly above that number on the number line. [1.1A, p. 3]

This is the graph of 4 on the number line.

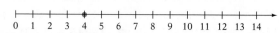

The symbol for *is less than* is <. The symbol for *is greater than* is >. These symbols are used to show the order relation between two numbers. [1.1A, p. 3]

3 < 7
9 > 2

Copyright © Houghton Mifflin Company. All rights reserved.

When a whole number is written using the digits 0, 1, 2, 3, 4, 5, 6, 7, 8, and 9, it is said to be in *standard form*. The position of each digit in the number determines the digit's *place value*. The place values are used to write the expanded form of a number. [1.1B, p. 4]

The number 598,317 is in standard form. The digit 8 is in the thousands place. The number 598,317 is written in expanded form as 500,000 + 90,000 + 8000 + 300 + 10 + 7.

Addition is the process of finding the total of two or more numbers. The numbers being added are called *addends*. The result is the *sum*. [1.2A, p. 9]

$$\begin{array}{r} \overset{1\ \ 11}{8,762} \\ +\ 1,359 \\ \hline 10,121 \end{array}$$

Subtraction is the process of finding the difference between two numbers. The *minuend* minus the *subtrahend* equals the *difference*. [1.3A, p. 17]

$$\begin{array}{r} \overset{4}{\cancel{5}}\ \overset{11}{\cancel{2}},\overset{11}{\cancel{1}}\ \overset{6}{\cancel{7}}\ \overset{13}{\cancel{3}} \\ -3\ 4\ ,9\ 6\ 8 \\ \hline 1\ 7,2\ 0\ 5 \end{array}$$

Multiplication is the repeated addition of the same number. The numbers that are multiplied are called *factors*. The result is the *product*. [1.4A, p. 25]

$$\begin{array}{r} \overset{4\ \ 5}{3\ 5\ 8} \\ \times\qquad 7 \\ \hline 2\ 5\ 0\ 6 \end{array}$$

Division is used to separate objects into equal groups. The *dividend* divided by the *divisor* equals the *quotient*. [1.5A, p. 33]

For any division problem,
(*quotient* · *divisor*) + *remainder* = *dividend*. [1.5B, p. 36]

$$\begin{array}{r} 93\ \text{r}3 \\ 7)\overline{654} \\ -63 \\ \hline 24 \\ -21 \\ \hline 3 \end{array}$$

Check: (7 · 93) + 3 = 651 + 3 = 654

The expression 4^3 is in *exponential notation*. The *exponent*, 3, indicates how many times 4 occurs as a factor in the multiplication. [1.6A, p. 46]

$5^4 = 5 \cdot 5 \cdot 5 \cdot 5 = 625$

Whole-number *factors of a number* divide that number evenly (there is no remainder). [1.7A, p. 50]

18 ÷ 1 = 18
18 ÷ 2 = 9
18 ÷ 3 = 6
18 ÷ 4   4 does not divide 18 evenly.
18 ÷ 5   5 does not divide 18 evenly.
18 ÷ 6 = 3 The factors are repeating.
The factors of 18 are 1, 2, 3, 6, 9, and 18.

A number greater than 1 is a *prime number* if its only whole-number factors are 1 and itself. If a number is not prime, it is a *composite number*. [1.7B, p. 51]

The prime numbers less than 20 are 2, 3, 5, 7, 11, 13, 17, and 19.
The composite numbers less than 20 are 4, 6, 8, 9, 10, 12, 14, 15, 16, and 18.

The *prime factorization* of a number is the expression of the number as a product of its prime factors. [1.7B, p. 51]

$$\begin{array}{r|l} \multicolumn{2}{c}{42} \\ \hline 2 & 21 \\ 3 & 7 \\ 7 & 1 \end{array}$$

The prime factorization of 42 is 2 · 3 · 7.

Copyright © Houghton Mifflin Company. All rights reserved.

Essential Rules and Procedures	**Examples**

To round a number to a given place value: If the digit to the right of the given place value is less than 5, replace that digit and all digits to the right by zeros. If the digit to the right of the given place value is greater than or equal to 5, increase the digit in the given place value by 1, and replace all other digits to the right by zeros. [1.1D, p. 6]

36,178 rounded to the nearest thousand is 36,000.
4592 rounded to the nearest thousand is 5000.

Properties of Addition [1.2A, p. 9]
Addition Property of Zero
Zero added to a number does not change the number.

$7 + 0 = 7$

Commutative Property of Addition
Two numbers can be added in either order; the sum will be the same.

$8 + 3 = 3 + 8$

Associative Property of Addition
Numbers to be added can be grouped in any order; the sum will be the same.

$(2 + 4) + 6 = 2 + (4 + 6)$

To estimate the answer to a calculation: Round each number to the same place value. Perform the calculation using the rounded numbers. [1.2A, p. 11]

$$\begin{array}{rr} 39,471 & 40,000 \\ 12,586 & +10,000 \\ \hline & 50,000 \end{array}$$

50,000 is an estimate of the sum of 39,471 and 12,586.

Properties of Multiplication [1.4A, p. 25]
Multiplication Property of Zero
The product of a number and zero is zero.

$3 \cdot 0 = 0$

Multiplication Property of One
The product of a number and one is the number.

$6 \cdot 1 = 6$

Commutative Property of Multiplication
Two numbers can be multiplied in either order; the product will be the same.

$2 \cdot 8 = 8 \cdot 2$

Associative Property of Multiplication
Grouping numbers to be multiplied in any order gives the same result.

$(2 \cdot 4) \cdot 6 = 2 \cdot (4 \cdot 6)$

Division Properties of Zero and One [1.5A, p. 33]
Any whole number, except zero, divided by itself is 1.
Any whole number divided by 1 is the whole number.
Zero divided by any other whole number is zero.
Division by zero is not allowed.

$3 \div 3 = 1$
$3 \div 1 = 3$
$0 \div 3 = 0$
$3 \div 0$ is not allowed.

Order of Operations Agreement [1.6B, p. 47]

Step 1   Do all the operations inside parentheses.

$5^2 - 3(2 + 4) = 5^2 - 3(6)$

Step 2   Simplify any number expressions containing exponents.

$= 25 - 3(6)$

Step 3   Do multiplications and divisions as they occur from left to right.

$= 25 - 18$

Step 4   Do addition and subtraction as they occur from left to right.

$= 7$

Copyright © Houghton Mifflin Company. All rights reserved.

Chapter 1 Review Exercises

1. Simplify: $3 \cdot 2^3 \cdot 5^2$

2. Write 10,327 in expanded form.

3. Find all the factors of 18.

4. Find the sum of 5894, 6301, and 298.

5. Subtract: $\begin{array}{r} 4926 \\ -\ 3177 \\ \hline \end{array}$

6. Divide: $7\overline{)14{,}945}$

7. Place the correct symbol, $<$ or $>$, between the two numbers: 101   87

8. Write $5 \cdot 5 \cdot 7 \cdot 7 \cdot 7 \cdot 7 \cdot 7$ in exponential notation.

$2 \cdot 7^5$

9. What is 2019 multiplied by 307?

10. What is 10,134 decreased by 4725?

11. Add: $\begin{array}{r} 298 \\ 461 \\ +\ 322 \\ \hline \end{array}$

12. Simplify: $2^3 - 3 \cdot 2$

13. Round 45,672 to the nearest hundred.

14. Write 276,057 in words.

15. Find the quotient of 109,763 and 84.

16. Write two million eleven thousand forty-four in standard form.

17. What is 3906 divided by 8?

18. Simplify: $3^2 + 2^2 \cdot (5 - 3)$

19. Simplify: $8 \cdot (6 - 2)^2 \div 4$

20. Find the prime factorization of 72.

Copyright © Houghton Mifflin Company. All rights reserved.

21. What is 3895 minus 1762?

22. Multiply: 843
            × 27

23. Wages Vincent Meyers, a sales assistant, earns $480 for working a 40-hour week. Last week Vincent worked an additional 12 hours at $24 an hour. Find Vincent's total pay for last week's work.

24. Fuel Efficiency Louis Reyes, a sales executive, drove a car 351 miles on 13 gallons of gas. Find the number of miles driven per gallon of gasoline.

25. Consumerism A car is purchased for $17,880, with a down payment of $3000. The balance is paid in 48 equal monthly payments. Find the monthly car payment.

26. Compensation An insurance account executive received commissions of $723, $544, $812, and $488 during a 4-week period. Find the total income from commissions for the 4 weeks.

27. Banking You had a balance of $516 in your checking account before making deposits of $88 and $213. Find the total amount deposited, and determine your new account balance.

28. Compensation You have a car payment of $246 per month. What is the total of the car payments over a 12-month period?

Athletics The table at the right shows the athletic participation by males and females at U.S. colleges in 1972 and 2001. Use this information for Exercises 29 to 32.

29. In which year, 1972 or 2001, were there more males involved in sports at U.S. colleges?

Year	Male Athletes	Female Athletes
1972	170,384	29,977
2001	208,866	150,916

Source: U.S. Department of Education commission report

30. What is the difference between the number of males involved in sports and the number of females involved in sports at U.S. colleges in 1972?

31. Find the increase in the number of females involved in sports in U.S. colleges from 1972 to 2001.

32. How many more U.S. college students were involved in athletics in 2001 than in 1972?

Copyright © Houghton Mifflin Company. All rights reserved.

Chapter 1 Test

1. Simplify: $3^3 \cdot 4^2$

2. Write 207,068 in words.

3. Subtract: $\begin{array}{r} 17,495 \\ -\ 8,162 \end{array}$

4. Find all the factors of 20.

5. Multiply: $\begin{array}{r} 9736 \\ \times\ 704 \end{array}$

6. Simplify: $4^2 \cdot (4-2) \div 8 + 5$

7. Write 906,378 in expanded form.

8. Round 74,965 to the nearest hundred.

9. Divide: $97\overline{)108,764}$

10. Write $3 \cdot 3 \cdot 3 \cdot 7 \cdot 7$ in exponential form.

11. Find the sum of 8756, 9094, and 37,065.

12. Find the prime factorization of 84.

13. Simplify: $16 \div 4 \cdot 2 - (7-5)^2$

14. Find the product of 8 and 90,763.

15. Write one million two hundred four thousand six in standard form.

16. Divide: $7\overline{)60,972}$

Copyright © Houghton Mifflin Company. All rights reserved.

17. Place the correct symbol, < or >, between the two numbers: 21   19

18. Find the quotient of 5624 and 8.

19. Add: 25,492
           +71,306

20. Find the difference between 29,736 and 9814.

Education   The table at the right shows the projected enrollment in public and private elementary and secondary schools in the fall of 2009 and the fall of 2012. Use this information for Exercises 21 and 22.

Year	Kindergarten through Grade 8	Grades 9 through 12
2009	37,726,000	15,812,000
2012	38,258,000	15,434,000

Source: The National Center for Education Statistics

21. Find the difference between the total enrollment in 2012 and that in 2009.

22. Find the average enrollment in each of grades 9 through 12 in 2012.

23. Farming   A farmer harvested 48,290 pounds of lemons from one grove and 23,710 pounds of lemons from another grove. The lemons were packed in boxes with 24 pounds of lemons in each box. How many boxes were needed to pack the lemons?

24. Investments   An investor receives $237 each month from a corporate bond fund. How much will the investor receive over a 12-month period?

25. Travel   A family drives 425 miles the first day, 187 miles the second day, and 243 miles the third day of their vacation. The odometer read 47,626 miles at the start of the vacation.
a. How many miles were driven during the 3 days?
b. What is the odometer reading at the end of the 3 days?

Copyright © Houghton Mifflin Company. All rights reserved.

Copyright © Houghton Mifflin Company. All rights reserved.

chapter

2 Fractions

Members of the Los Angeles Philharmonic follow conductor Esa-Pekka Salonen's movements to guarantee synchronized playing. By paying attention to both Salonen and the time signature for each musical piece, the musicians play in time with each other. The time signature appears as a fraction at the beginning of a piece of music and tells them how many beats to play per measure. The **project on page 115** demonstrates how to interpret the time signature.

Need help? For online student resources, such as section quizzes, visit this textbook's website at **math.college.hmco.com/students.**

OBJECTIVES

Section 2.1

A To find the least common multiple (LCM)
B To find the greatest common factor (GCF)

Section 2.2

A To write a fraction that represents part of a whole
B To write an improper fraction as a mixed number or a whole number, and a mixed number as an improper fraction

Section 2.3

A To find equivalent fractions by raising to higher terms
B To write a fraction in simplest form

Section 2.4

A To add fractions with the same denominator
B To add fractions with different denominators
C To add whole numbers, mixed numbers, and fractions
D To solve application problems

Section 2.5

A To subtract fractions with the same denominator
B To subtract fractions with different denominators
C To subtract whole numbers, mixed numbers, and fractions
D To solve application problems

Section 2.6

A To multiply fractions
B To multiply whole numbers, mixed numbers, and fractions
C To solve application problems

Section 2.7

A To divide fractions
B To divide whole numbers, mixed numbers, and fractions
C To solve application problems

Section 2.8

A To identify the order relation between two fractions
B To simplify expressions containing exponents
C To use the Order of Operations Agreement to simplify expressions

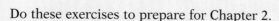
Do these exercises to prepare for Chapter 2.

For Exercises 1 to 6, add, subtract, multiply, or divide.

1. 4×5

2. $2 \cdot 2 \cdot 2 \cdot 3 \cdot 5$

3. 9×1

4. $6 + 4$

5. $10 - 3$

6. $63 \div 30$

7. Which of the following numbers divide evenly into 12?

1 2 3 4 5 6 7 8 9 10 11 12

8. Simplify: $8 \times 7 + 3$

9. Complete: $8 = ? + 1$

10. Place the correct symbol, $<$ or $>$, between the two numbers.

44   48

GO FIGURE • • •

You and a friend are swimming laps in a pool. You swim one lap every 4 minutes. Your friend swims one lap every 5 minutes. If you start at the same time from the same end of the pool, in how many minutes will both of you be at the starting point again? How many times will you have passed each other in the pool prior to that time?

Copyright © Houghton Mifflin Company. All rights reserved.

2.1

The Least Common Multiple and Greatest Common Factor

Objective A   **To find the least common multiple (LCM)**

S t u d y   T i p
Before you begin a new chapter, you should take some time to review previously learned skills. One way to do this is to complete the Prep Test. See page 64. This test focuses on the particular skills that will be required for the new chapter.

The **multiples of a number** are the products of that number and the numbers 1, 2, 3, 4, 5, ...

$3 \times 1 = 3$
$3 \times 2 = 6$
$3 \times 3 = 9$
$3 \times 4 = 12$ The multiples of 3 are 3, 6, 9, 12, 15, ...
$3 \times 5 = 15$

A number that is a multiple of two or more numbers is a **common multiple** of those numbers.

The multiples of 4 are 4, 8, 12, 16, 20, 24, 28, 32, 36, ...
The multiples of 6 are 6, 12, 18, 24, 30, 36, 42, ...
Some common multiples of 4 and 6 are 12, 24, and 36.

The **least common multiple (LCM)** is the smallest common multiple of two or more numbers.

The least common multiple of 4 and 6 is 12.

Listing the multiples of each number is one way to find the LCM. Another way to find the LCM uses the prime factorization of each number.

To find the LCM of 450 and 600, find the prime factorization of each number and write the factorization of each number in a table. Circle the greatest product in each column. The LCM is the product of the circled numbers.

	2	3	5
450 =	2	③ · ③	⑤ · ⑤
600 =	② · ② · ②	3	5 · 5

In the column headed by 5, the products are equal. Circle just one product.

The LCM is the product of the circled numbers.
The LCM = $2 \cdot 2 \cdot 2 \cdot 3 \cdot 3 \cdot 5 \cdot 5 = 1800$.

Example 1   Find the LCM of 24, 36, and 50.

Solution

	2	3	5
24 =	② · ② · ②	3	
36 =	2 · 2	③ · ③	
50 =	2		⑤ · ⑤

The LCM = $2 \cdot 2 \cdot 2 \cdot 3 \cdot 3 \cdot 5 \cdot 5$
          $= 1800$.

You Try It 1   Find the LCM of 12, 27, and 50.

Your solution

Solution on p. S4

Copyright © Houghton Mifflin Company. All rights reserved.

Objective B **To find the greatest common factor (GCF)**

Recall that a number that divides another number evenly is a factor of that number. The number 64 can be evenly divided by 1, 2, 4, 8, 16, 32, and 64, so the numbers 1, 2, 4, 8, 16, 32, and 64 are factors of 64.

A number that is a factor of two or more numbers is a **common factor** of those numbers.

The factors of 30 are 1, 2, 3, 5, 6, 10, 15, and 30.
The factors of 105 are 1, 3, 5, 7, 15, 21, 35, and 105.
The common factors of 30 and 105 are 1, 3, 5, and 15.

The **greatest common factor (GCF)** is the largest common factor of two or more numbers.

The greatest common factor of 30 and 105 is 15.

Listing the factors of each number is one way of finding the GCF. Another way to find the GCF uses the prime factorization of each number.

To find the GCF of 126 and 180, find the prime factorization of each number and write the factorization of each number in a table. Circle the least product in each column that does not have a blank. The GCF is the product of the circled numbers.

	2	3	5	7
126 =	②	③·3		7
180 =	2·2	3·3	5	

In the column headed by 3, the products are equal. Circle just one product.
Columns 5 and 7 have a blank, so 5 and 7 are not common factors of 126 and 180. Do not circle any number in these columns.

The GCF is the product of the circled numbers.
The GCF = 2 · 3 · 3 = 18.

Example 2 Find the GCF of 90, 168, and 420.

Solution

	2	3	5	7
90 =	②	3·3	5	
168 =	2·2·2	③		7
420 =	2·2	3	5	7

The GCF = 2 · 3 = 6.

You Try It 2 Find the GCF of 36, 60, and 72.

Your solution

Example 3 Find the GCF of 7, 12, and 20.

Solution

	2	3	5	7
7 =				7
12 =	2·2	3		
20 =	2·2		5	

Because no numbers are circled, the GCF = 1.

You Try It 3 Find the GCF of 11, 24, and 30.

Your solution

Solutions on p. S4

Copyright © Houghton Mifflin Company. All rights reserved.

2.1 Exercises

Objective A   **To find the least common multiple (LCM)**

For Exercises 1 to 34, find the LCM.

1. 5, 8   **2.** 3, 6   **3.** 3, 8   **4.** 2, 5   **5.** 5, 6

6. 5, 7   **7.** 4, 6   **8.** 6, 8   **9.** 8, 12   **10.** 12, 16

11. 5, 12   **12.** 3, 16   **13.** 8, 14   **14.** 6, 18   **15.** 3, 9

16. 4, 10   **17.** 8, 32   **18.** 7, 21   **19.** 9, 36   **20.** 14, 42

21. 44, 60   **22.** 120, 160   **23.** 102, 184   **24.** 123, 234   **25.** 4, 8, 12

26. 5, 10, 15   **27.** 3, 5, 10   **28.** 2, 5, 8   **29.** 3, 8, 12   **30.** 5, 12, 18

31. 9, 36, 64   **32.** 18, 54, 63   **33.** 16, 30, 84   **34.** 9, 12, 15

Objective B   **To find the greatest common factor (GCF)**

For Exercises 35 to 68, find the GCF.

35. 3, 5   **36.** 5, 7   **37.** 6, 9   **38.** 18, 24   **39.** 15, 25

40. 14, 49   **41.** 25, 100   **42.** 16, 80   **43.** 32, 51   **44.** 21, 44

45. 12, 80   **46.** 8, 36   **47.** 16, 140   **48.** 12, 76

Copyright © Houghton Mifflin Company. All rights reserved.

49. 24, 30

50. 48, 144

51. 44, 96

52. 18, 32

53. 3, 5, 11

54. 6, 8, 10

55. 7, 14, 49

56. 6, 15, 36

57. 10, 15, 20

58. 12, 18, 20

59. 24, 40, 72

60. 3, 17, 51

61. 17, 31, 81

62. 14, 42, 84

63. 25, 125, 625

64. 12, 68, 92

65. 28, 35, 70

66. 1, 49, 153

67. 32, 56, 72

68. 24, 36, 48

APPLYING THE CONCEPTS

69. Define the phrase *relatively prime numbers*. List three pairs of relatively prime numbers.

70. **Work Schedules**   Joe Salvo, a lifeguard, works 3 days and then has a day off. A friend works 5 days and then has a day off. How many days after Joe and his friend have a day off together will they have another day off together?

71. Find the LCM of each of the following pairs of numbers: 2 and 3, 5 and 7, and 11 and 19. Can you draw a conclusion about the LCM of two prime numbers? Suggest a way of finding the LCM of three distinct prime numbers.

72. Find the GCF of each of the following pairs of numbers: 3 and 5, 7 and 11, and 29 and 43. Can you draw a conclusion about the GCF of two prime numbers? What is the GCF of three prime distinct numbers?

73. Is the LCM of two numbers always divisible by the GCF of the two numbers? If so, explain why. If not, give an example.

74. Using the pattern for the first two triangles at the right, determine the center number of the last triangle.

Copyright © Houghton Mifflin Company. All rights reserved.

2.2 Introduction to Fractions

Copyright © Houghton Mifflin Company. All rights reserved.

Objective A   **To write a fraction that represents part of a whole**

TAKE NOTE

The **fraction bar** separates the numerator from the denominator. The **numerator** is the part of the fraction that appears above the fraction bar. The **denominator** is the part of the fraction that appears below the fraction bar.

Point of Interest

The fraction bar was first used in 1050 by al-Hassar. It is also called a vinculum.

A **fraction** can represent the number of equal parts of a whole. The shaded portion of the circle is represented by the fraction $\frac{4}{7}$. Four of the seven equal parts of the circle (that is, four-sevenths of it) are shaded.

Each part of a fraction has a name.

$$\text{Fraction bar} \rightarrow \frac{4}{7} \begin{array}{l}\leftarrow \textbf{Numerator} \\ \leftarrow \textbf{Denominator}\end{array}$$

A **proper fraction** is a fraction less than 1. The numerator of a proper fraction is smaller than the denominator. The shaded portion of the circle can be represented by the proper fraction $\frac{3}{4}$.

A **mixed number** is a number greater than 1 with a whole-number part and a fractional part. The shaded portion of the circles can be represented by the mixed number $2\frac{1}{4}$.

An **improper fraction** is a fraction greater than or equal to 1. The numerator of an improper fraction is greater than or equal to the denominator. The shaded portion of the circles can be represented by the improper fraction $\frac{9}{4}$. The shaded portion of the square can be represented by $\frac{4}{4}$.

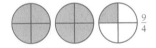

Example 1 Express the shaded portion of the circles as a mixed number.

Solution $3\frac{2}{5}$

You Try It 1 Express the shaded portion of the circles as a mixed number.

Your solution

Example 2 Express the shaded portion of the circles as an improper fraction.

Solution $\frac{17}{5}$

You Try It 2 Express the shaded portion of the circles as an improper fraction.

Your solution

Solutions on pp. S4–S5

Objective B
To write an improper fraction as a mixed number or a whole number, and a mixed number as an improper fraction

Note from the diagram that the mixed number $2\frac{3}{5}$ and the improper fraction $\frac{13}{5}$ both represent the shaded portion of the circles.

$$2\frac{3}{5} = \frac{13}{5}$$

An improper fraction can be written as a mixed number or a whole number.

$2\frac{3}{5}$

$\frac{13}{5}$

HOW TO   Write $\frac{13}{5}$ as a mixed number.

Divide the numerator by the denominator.

$$\begin{array}{r} 2 \\ 5)\overline{13} \\ -10 \\ \hline 3 \end{array}$$

To write the fractional part of the mixed number, write the remainder over the divisor.

$$\begin{array}{r} 2\frac{3}{5} \\ 5)\overline{13} \\ -10 \\ \hline 3 \end{array}$$

Write the answer.

$$\frac{13}{5} = 2\frac{3}{5}$$

Point of Interest

Archimedes (c. 287–212 B.C.) is the person who calculated that $\pi \approx 3\frac{1}{7}$. He actually showed that $3\frac{10}{71} < \pi < 3\frac{1}{7}$. The approximation $3\frac{10}{71}$ is more accurate but more difficult to use.

To write a mixed number as an improper fraction, multiply the denominator of the fractional part by the whole-number part. The sum of this product and the numerator of the fractional part is the numerator of the improper fraction. The denominator remains the same.

HOW TO   Write $7\frac{3}{8}$ as an improper fraction.

$$7\frac{3}{8} = \frac{(8 \times 7) + 3}{8} = \frac{56 + 3}{8} = \frac{59}{8} \qquad 7\frac{3}{8} = \frac{59}{8}$$

Example 3   Write $\frac{21}{4}$ as a mixed number.

Solution
$$\begin{array}{r} 5 \\ 4)\overline{21} \\ -20 \\ \hline 1 \end{array} \qquad \frac{21}{4} = 5\frac{1}{4}$$

You Try It 3   Write $\frac{22}{5}$ as a mixed number.

Your solution

Example 4   Write $\frac{18}{6}$ as a whole number.

Solution
$$\begin{array}{r} 3 \\ 6)\overline{18} \\ -18 \\ \hline 0 \end{array} \qquad \frac{18}{6} = 3$$

Note: The remainder is zero.

You Try It 4   Write $\frac{28}{7}$ as a whole number.

Your solution

Example 5   Write $21\frac{3}{4}$ as an improper fraction.

Solution
$$21\frac{3}{4} = \frac{84 + 3}{4} = \frac{87}{4}$$

You Try It 5   Write $14\frac{5}{8}$ as an improper fraction.

Your solution

Solutions on pp. S4–S5

Copyright © Houghton Mifflin Company. All rights reserved.

2.2 Exercises

Objective A   **To write a fraction that represents part of a whole**

For Exercises 1 to 4, identify the fraction as a proper fraction, an improper fraction, or a mixed number.

1. $\dfrac{12}{7}$

2. $5\dfrac{2}{11}$

3. $\dfrac{29}{40}$

4. $\dfrac{19}{13}$

For Exercises 5 to 8, express the shaded portion of the circle as a fraction.

5. 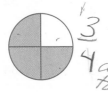 $\dfrac{3}{4}$ denom total

6. $\dfrac{4}{7}$

7.

8.

For Exercises 9 to 14, express the shaded portion of the circles as a mixed number.

9.

10.

11.

12.

13.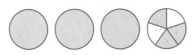

14.

For Exercises 15 to 20, express the shaded portion of the circles as an improper fraction.

15.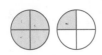

16.

17.

18.

19.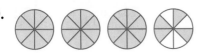

20.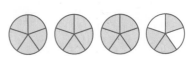

21. Shade $\dfrac{5}{6}$ of

22. Shade $\dfrac{3}{8}$ of

23. Shade $1\dfrac{2}{5}$ of

24. Shade $1\dfrac{3}{4}$ of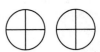

Copyright © Houghton Mifflin Company. All rights reserved.

25. Shade $\frac{6}{5}$ of

26. Shade $\frac{7}{3}$ of

Objective B **To write an improper fraction as a mixed number or a whole number, and a mixed number as an improper fraction**

For Exercises 27 to 50, write the improper fraction as a mixed number or a whole number.

27. $\frac{11}{4}$ **28.** $\frac{16}{3}$ **29.** $\frac{20}{4}$ **30.** $\frac{18}{9}$ **31.** $\frac{9}{8}$ **32.** $\frac{13}{4}$

33. $\frac{23}{10}$ **34.** $\frac{29}{2}$ **35.** $\frac{48}{16}$ **36.** $\frac{51}{3}$ **37.** $\frac{8}{7}$ **38.** $\frac{16}{9}$

39. $\frac{7}{3}$ **40.** $\frac{9}{5}$ **41.** $\frac{16}{1}$ **42.** $\frac{23}{1}$ **43.** $\frac{17}{8}$ **44.** $\frac{31}{16}$

45. $\frac{12}{5}$ **46.** $\frac{19}{3}$ **47.** $\frac{9}{9}$ **48.** $\frac{40}{8}$ **49.** $\frac{72}{8}$ **50.** $\frac{3}{3}$

For Exercises 51 to 74, write the mixed number as an improper fraction.

51. $2\frac{1}{3}$ **52.** $4\frac{2}{3}$ **53.** $6\frac{1}{2}$ **54.** $8\frac{2}{3}$ **55.** $6\frac{5}{6}$ **56.** $7\frac{3}{8}$

57. $9\frac{1}{4}$ **58.** $6\frac{1}{4}$ **59.** $10\frac{1}{2}$ **60.** $15\frac{1}{8}$ **61.** $8\frac{1}{9}$ **62.** $3\frac{5}{12}$

63. $5\frac{3}{11}$ **64.** $3\frac{7}{9}$ **65.** $2\frac{5}{8}$ **66.** $12\frac{2}{3}$ **67.** $1\frac{5}{8}$ **68.** $5\frac{3}{7}$

69. $11\frac{1}{9}$ **70.** $12\frac{3}{5}$ **71.** $3\frac{3}{8}$ **72.** $4\frac{5}{9}$ **73.** $6\frac{7}{13}$ **74.** $8\frac{5}{14}$

APPLYING THE CONCEPTS

75. Name three situations in which fractions are used. Provide an example of a fraction that is used in each situation.

Copyright © Houghton Mifflin Company. All rights reserved.

2.3 Writing Equivalent Fractions

Objective A **To find equivalent fractions by raising to higher terms**

Equal fractions with different denominators are called **equivalent fractions.**

$\frac{4}{6}$ is equivalent to $\frac{2}{3}$.

Remember that the Multiplication Property of One states that the product of a number and one is the number. This is true for fractions as well as whole numbers. This property can be used to write equivalent fractions.

$$\frac{2}{3} \times 1 = \frac{2}{3} \times \frac{1}{1} = \frac{2 \cdot 1}{3 \cdot 1} = \frac{2}{3}$$

$$\frac{2}{3} \times 1 = \frac{2}{3} \times \boxed{\frac{2}{2}} = \frac{2 \cdot 2}{3 \cdot 2} = \frac{4}{6} \qquad \frac{4}{6} \text{ is equivalent to } \frac{2}{3}.$$

$$\frac{2}{3} \times 1 = \frac{2}{3} \times \boxed{\frac{4}{4}} = \frac{2 \cdot 4}{3 \cdot 4} = \frac{8}{12} \qquad \frac{8}{12} \text{ is equivalent to } \frac{2}{3}.$$

$\frac{2}{3}$ was rewritten as the equivalent fractions $\frac{4}{6}$ and $\frac{8}{12}$.

HOW TO Write a fraction that is equivalent to $\frac{5}{8}$ and has a denominator of 32.

$32 \div 8 = 4$

$\frac{5}{8} = \frac{5 \cdot 4}{8 \cdot 4} = \frac{20}{32}$

- Divide the larger denominator by the smaller.
- Multiply the numerator and denominator of the given fraction by the quotient (4).

$\frac{20}{32}$ is equivalent to $\frac{5}{8}$.

Example 1 Write $\frac{2}{3}$ as an equivalent fraction that has a denominator of 42.

Solution $42 \div 3 = 14 \qquad \frac{2}{3} = \frac{2 \cdot 14}{3 \cdot 14} = \frac{28}{42}$

$\frac{28}{42}$ is equivalent to $\frac{2}{3}$.

You Try It 1 Write $\frac{3}{5}$ as an equivalent fraction that has a denominator of 45.

Your solution

Example 2 Write 4 as a fraction that has a denominator of 12.

Solution Write 4 as $\frac{4}{1}$.

$12 \div 1 = 12 \qquad 4 = \frac{4 \cdot 12}{1 \cdot 12} = \frac{48}{12}$

$\frac{48}{12}$ is equivalent to 4.

You Try It 2 Write 6 as a fraction that has a denominator of 18.

Your solution

Solutions on p. S5

Copyright © Houghton Mifflin Company. All rights reserved.

Objective B   **To write a fraction in simplest form**

Writing the **simplest form of a fraction** means writing it so that the numerator and denominator have no common factors other than 1.

The fractions $\frac{4}{6}$ and $\frac{2}{3}$ are equivalent fractions.

$\frac{4}{6}$ has been written in simplest form as $\frac{2}{3}$.

The Multiplication Property of One can be used to write fractions in simplest form. Write the numerator and denominator of the given fraction as a product of factors. Write factors common to both the numerator and denominator as an improper fraction equivalent to 1.

$$\frac{4}{6} = \frac{2 \cdot 2}{2 \cdot 3} = \left[\frac{2}{2} \cdot \frac{2}{3}\right] = \left[\frac{2}{2}\right] \cdot \frac{2}{3} = 1 \cdot \frac{2}{3} = \frac{2}{3}$$

The process of eliminating common factors is displayed with slashes through the common factors as shown at the right.

$$\frac{4}{6} = \frac{2 \cdot \overset{1}{\cancel{2}}}{\underset{1}{\cancel{2}} \cdot 3} = \frac{2}{3}$$

To write a fraction in simplest form, eliminate the common factors.

$$\frac{18}{30} = \frac{2 \cdot \overset{1}{\cancel{3}} \cdot 3}{\underset{1}{\cancel{2}} \cdot \underset{1}{\cancel{3}} \cdot 5} = \frac{3}{5}$$

An improper fraction can be changed to a mixed number.

$$\frac{22}{6} = \frac{\overset{1}{\cancel{2}} \cdot 11}{\underset{1}{\cancel{2}} \cdot 3} = \frac{11}{3} = 3\frac{2}{3}$$

Example 3 Write $\frac{15}{40}$ in simplest form.

Solution $\frac{15}{40} = \frac{3 \cdot \overset{1}{\cancel{5}}}{2 \cdot 2 \cdot 2 \cdot \underset{1}{\cancel{5}}} = \frac{3}{8}$

You Try It 3 Write $\frac{16}{24}$ in simplest form.

Your solution

Example 4 Write $\frac{6}{42}$ in simplest form.

Solution $\frac{6}{42} = \frac{\overset{1}{\cancel{2}} \cdot \overset{1}{\cancel{3}}}{\underset{1}{\cancel{2}} \cdot \underset{1}{\cancel{3}} \cdot 7} = \frac{1}{7}$

You Try It 4 Write $\frac{8}{56}$ in simplest form.

Your solution

Example 5 Write $\frac{8}{9}$ in simplest form.

Solution $\frac{8}{9} = \frac{2 \cdot 2 \cdot 2}{3 \cdot 3} = \frac{8}{9}$

$\frac{8}{9}$ is already in simplest form because there are no common factors in the numerator and denominator.

You Try It 5 Write $\frac{15}{32}$ in simplest form.

Your solution

Example 6 Write $\frac{30}{12}$ in simplest form.

Solution $\frac{30}{12} = \frac{\overset{1}{\cancel{2}} \cdot \overset{1}{\cancel{3}} \cdot 5}{\underset{1}{\cancel{2}} \cdot 2 \cdot \underset{1}{\cancel{3}}} = \frac{5}{2} = 2\frac{1}{2}$

You Try It 6 Write $\frac{48}{36}$ in simplest form.

Your solution

Solutions on p. S5

Copyright © Houghton Mifflin Company. All rights reserved.

2.3 Exercises

Objective A   **To find equivalent fractions by raising to higher terms**

For Exercises 1 to 40, write an equivalent fraction with the given denominator.

1. $\dfrac{1}{2} = \dfrac{}{10}$ **2.** $\dfrac{1}{4} = \dfrac{}{16}$ **3.** $\dfrac{3}{16} = \dfrac{}{48}$ **4.** $\dfrac{5}{9} = \dfrac{}{81}$ **5.** $\dfrac{3}{8} = \dfrac{}{32}$

6. $\dfrac{7}{11} = \dfrac{}{33}$ **7.** $\dfrac{3}{17} = \dfrac{}{51}$ **8.** $\dfrac{7}{10} = \dfrac{}{90}$ **9.** $\dfrac{3}{4} = \dfrac{}{16}$ **10.** $\dfrac{5}{8} = \dfrac{}{32}$

11. $3 = \dfrac{}{9}$ **12.** $5 = \dfrac{}{25}$ **13.** $\dfrac{1}{3} = \dfrac{}{60}$ **14.** $\dfrac{1}{16} = \dfrac{}{48}$ **15.** $\dfrac{11}{15} = \dfrac{}{60}$

16. $\dfrac{3}{50} = \dfrac{}{300}$ **17.** $\dfrac{2}{3} = \dfrac{}{18}$ **18.** $\dfrac{5}{9} = \dfrac{}{36}$ **19.** $\dfrac{5}{7} = \dfrac{}{49}$ **20.** $\dfrac{7}{8} = \dfrac{}{32}$

21. $\dfrac{5}{9} = \dfrac{}{18}$ **22.** $\dfrac{11}{12} = \dfrac{}{36}$ **23.** $7 = \dfrac{}{3}$ **24.** $9 = \dfrac{}{4}$ **25.** $\dfrac{7}{9} = \dfrac{}{45}$

26. $\dfrac{5}{6} = \dfrac{}{42}$ **27.** $\dfrac{15}{16} = \dfrac{}{64}$ **28.** $\dfrac{11}{18} = \dfrac{}{54}$ **29.** $\dfrac{3}{14} = \dfrac{}{98}$ **30.** $\dfrac{5}{6} = \dfrac{}{144}$

31. $\dfrac{5}{8} = \dfrac{}{48}$ **32.** $\dfrac{7}{12} = \dfrac{}{96}$ **33.** $\dfrac{5}{14} = \dfrac{}{42}$ **34.** $\dfrac{2}{3} = \dfrac{}{42}$ **35.** $\dfrac{17}{24} = \dfrac{}{144}$

36. $\dfrac{5}{13} = \dfrac{}{169}$ **37.** $\dfrac{3}{8} = \dfrac{}{408}$ **38.** $\dfrac{9}{16} = \dfrac{}{272}$ **39.** $\dfrac{17}{40} = \dfrac{}{800}$ **40.** $\dfrac{9}{25} = \dfrac{}{1000}$

Objective B   **To write a fraction in simplest form**

For Exercises 41 to 75, write the fraction in simplest form.

41. $\dfrac{4}{12}$ **42.** $\dfrac{8}{22}$ **43.** $\dfrac{22}{44}$ **44.** $\dfrac{2}{14}$ **45.** $\dfrac{2}{12}$

Copyright © Houghton Mifflin Company. All rights reserved.

46. $\dfrac{50}{75}$ **47.** $\dfrac{40}{36}$ **48.** $\dfrac{12}{8}$ **49.** $\dfrac{0}{30}$ **50.** $\dfrac{10}{10}$

51. $\dfrac{9}{22}$ **52.** $\dfrac{14}{35}$ **53.** $\dfrac{75}{25}$ **54.** $\dfrac{8}{60}$ **55.** $\dfrac{16}{84}$

56. $\dfrac{20}{44}$ **57.** $\dfrac{12}{35}$ **58.** $\dfrac{8}{36}$ **59.** $\dfrac{28}{44}$ **60.** $\dfrac{12}{16}$

61. $\dfrac{16}{12}$ **62.** $\dfrac{24}{18}$ **63.** $\dfrac{24}{40}$ **64.** $\dfrac{44}{60}$ **65.** $\dfrac{8}{88}$

66. $\dfrac{9}{90}$ **67.** $\dfrac{144}{36}$ **68.** $\dfrac{140}{297}$ **69.** $\dfrac{48}{144}$ **70.** $\dfrac{32}{120}$

71. $\dfrac{60}{100}$ **72.** $\dfrac{33}{110}$ **73.** $\dfrac{36}{16}$ **74.** $\dfrac{80}{45}$ **75.** $\dfrac{32}{160}$

APPLYING THE CONCEPTS

76. Make a list of five different fractions that are equivalent to $\dfrac{2}{3}$.

77. Make a list of five different fractions that are equivalent to 3.

78. Show that $\dfrac{15}{24} = \dfrac{5}{8}$ by using a diagram.

79. **a. Geography**   What fraction of the states in the United States of America have names that begin with the letter M?
   b. What fraction of the states have names that begin and end with a vowel?

Copyright © Houghton Mifflin Company. All rights reserved.

2.4 Addition of Fractions and Mixed Numbers

Copyright © Houghton Mifflin Company. All rights reserved.

Objective A   **To add fractions with the same denominator**

Fractions with the same denominator are added by adding the numerators and placing the sum over the common denominator. After adding, write the sum in simplest form.

HOW TO   Add: $\frac{2}{7} + \frac{4}{7}$

$$\begin{array}{r} \dfrac{2}{7} \\[2mm] +\ \dfrac{4}{7} \\[2mm] \hline \dfrac{6}{7} \end{array}$$

• Add the numerators and place the sum over the common denominator.

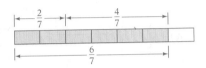

$$\frac{2}{7} + \frac{4}{7} = \frac{2+4}{7} = \frac{6}{7}$$

Example 1   Add: $\frac{5}{12} + \frac{11}{12}$

Solution

$$\begin{array}{r} \dfrac{5}{12} \\[2mm] +\ \dfrac{11}{12} \\[2mm] \hline \dfrac{16}{12} = \dfrac{4}{3} = 1\dfrac{1}{3} \end{array}$$

• The denominators are the same. Add the numerators. Place the sum over the common denominator.

You Try It 1   Add: $\frac{3}{8} + \frac{7}{8}$

Your solution

Solution on p. S5

Objective B   **To add fractions with different denominators**

Integrating Technology

Some scientific calculators have a fraction key, a^b/c. It is used to perform operations on fractions. To use this key to simplify the expression at the right, enter

1 a^b/c 2 + 1 a^b/c 3 =
$\underbrace{}_{\frac{1}{2}}$ $\underbrace{}_{\frac{1}{3}}$

To add fractions with different denominators, first rewrite the fractions as equivalent fractions with a common denominator. The common denominator is the LCM of the denominators of the fractions.

HOW TO   Find the total of $\frac{1}{2}$ and $\frac{1}{3}$.

The common denominator is the LCM of 2 and 3. The LCM = 6. The LCM of denominators is sometimes called the **least common denominator (LCD)**.

Write equivalent fractions using the LCM.

$$\begin{array}{r} \dfrac{1}{2} = \dfrac{3}{6} \\[2mm] +\ \dfrac{1}{3} = \dfrac{2}{6} \end{array}$$

Add the fractions.

$$\begin{array}{r} \dfrac{1}{2} = \dfrac{3}{6} \\[2mm] +\ \dfrac{1}{3} = \dfrac{2}{6} \\[2mm] \hline \dfrac{5}{6} \end{array}$$

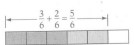

Example 2 Find $\frac{7}{12}$ more than $\frac{3}{8}$.

Solution

$$\frac{3}{8} = \frac{9}{24}$$

• The LCM of 8 and 12 is 24.

$$+\frac{7}{12} = \frac{14}{24}$$

$$\frac{23}{24}$$

You Try It 2 Find the sum of $\frac{5}{12}$ and $\frac{9}{16}$.

Your solution

Example 3 Add: $\frac{5}{8} + \frac{7}{9}$

Solution

$$\frac{5}{8} = \frac{45}{72}$$

$$+\frac{7}{9} = \frac{56}{72}$$

$$\frac{101}{72} = 1\frac{29}{72}$$

You Try It 3 Add: $\frac{7}{8} + \frac{11}{15}$

Your solution

Example 4 Add: $\frac{2}{3} + \frac{3}{5} + \frac{5}{6}$

Solution

$$\frac{2}{3} = \frac{20}{30}$$

• The LCM of 3, 5, and 6 is 30.

$$\frac{3}{5} = \frac{18}{30}$$

$$+\frac{5}{6} = \frac{25}{30}$$

$$\frac{63}{30} = 2\frac{3}{30} = 2\frac{1}{10}$$

You Try It 4 Add: $\frac{3}{4} + \frac{4}{5} + \frac{5}{8}$

Your solution

Solutions on p. S5

Objective C **To add whole numbers, mixed numbers, and fractions**

TAKE NOTE

The procedure at the right illustrates why $2 + \frac{2}{3} = 2\frac{2}{3}$. You do not need to show these steps when adding a whole number and a fraction. Here are two more examples:

$$7 + \frac{1}{5} = 7\frac{1}{5}$$

$$6 + \frac{3}{4} = 6\frac{3}{4}$$

The sum of a whole number and a fraction is a mixed number.

HOW TO Add: $2 + \frac{2}{3}$

$$\boxed{2} + \frac{2}{3} = \boxed{\frac{6}{3}} + \frac{2}{3} = \frac{8}{3} = 2\frac{2}{3}$$

To add a whole number and a mixed number, write the fraction and then add the whole numbers.

HOW TO Add: $7\frac{2}{5} + 4$

Write the fraction.

$$7\frac{2}{5}$$
$$+4$$
$$\overline{\frac{2}{5}}$$

Add the whole numbers.

$$7\frac{2}{5}$$
$$+4$$
$$\overline{11\frac{2}{5}}$$

Copyright © Houghton Mifflin Company. All rights reserved.

To add two mixed numbers, add the fractional parts and then add the whole numbers. Remember to reduce the sum to simplest form.

Integrating Technology

Use the fraction key on a calculator to enter mixed numbers. For the example at the right, enter

5 a^{b}/c 4 a^{b}/c 9 +

$5\frac{4}{9}$

6 a^{b}/c 14 a^{b}/c 15 =

$6\frac{14}{15}$

HOW TO What is $6\frac{14}{15}$ added to $5\frac{4}{9}$?

The LCM of 9 and 15 is 45.

Add the fractional parts.

$$5\frac{4}{9} = 5\frac{20}{45}$$
$$+ 6\frac{14}{15} = 6\frac{42}{45}$$
$$\overline{\qquad\qquad \frac{62}{45}}$$

Add the whole numbers.

$$5\frac{4}{9} = 5\frac{20}{45}$$
$$+ 6\frac{14}{15} = 6\frac{42}{45}$$
$$\overline{\qquad 11\frac{62}{45} = 11 + 1\frac{17}{45} = 12\frac{17}{45}}$$

Example 5 Add: $5 + \frac{3}{8}$

Solution $5 + \frac{3}{8} = 5\frac{3}{8}$

You Try It 5 What is 7 added to $\frac{6}{11}$?

Your solution

Example 6 Find 17 increased by $3\frac{3}{8}$.

Solution
$$17$$
$$+ \; 3\frac{3}{8}$$
$$\overline{20\frac{3}{8}}$$

You Try It 6 Find the sum of 29 and $17\frac{5}{12}$.

Your solution

Example 7 Add: $5\frac{2}{3} + 11\frac{5}{6} + 12\frac{7}{9}$

Solution
$$5\frac{2}{3} = \; 5\frac{12}{18} \qquad \bullet \; \text{LCM} = 18$$
$$11\frac{5}{6} = 11\frac{15}{18}$$
$$+ \; 12\frac{7}{9} = 12\frac{14}{18}$$
$$\overline{28\frac{41}{18} = 30\frac{5}{18}}$$

You Try It 7 Add: $7\frac{4}{5} + 6\frac{7}{10} + 13\frac{11}{15}$

Your solution

Example 8 Add: $11\frac{5}{8} + 7\frac{5}{9} + 8\frac{7}{15}$

Solution
$$11\frac{5}{8} = 11\frac{225}{360} \qquad \bullet \; \text{LCM} = 360$$
$$7\frac{5}{9} = \; 7\frac{200}{360}$$
$$+ \; 8\frac{7}{15} = \; 8\frac{168}{360}$$
$$\overline{26\frac{593}{360} = 27\frac{233}{360}}$$

You Try It 8 Add: $9\frac{3}{8} + 17\frac{7}{12} + 10\frac{14}{15}$

Your solution

Copyright © Houghton Mifflin Company. All rights reserved.

Solutions on p. S5

Objective D   **To solve application problems**

Example 9

A rain gauge collected $2\frac{1}{3}$ inches of rain in October, $5\frac{1}{2}$ inches in November, and $3\frac{3}{8}$ inches in December. Find the total rainfall for the 3 months.

Strategy
To find the total rainfall for the 3 months, add the three amounts of rainfall $\left(2\frac{1}{3}, 5\frac{1}{2}, \text{and } 3\frac{3}{8}\right)$.

Solution

$$2\frac{1}{3} = 2\frac{8}{24}$$
$$5\frac{1}{2} = 5\frac{12}{24}$$
$$+\ 3\frac{3}{8} = 3\frac{9}{24}$$
$$\overline{\qquad\qquad}$$
$$10\frac{29}{24} = 11\frac{5}{24}$$

The total rainfall for the 3 months was $11\frac{5}{24}$ inches.

You Try It 9

On Monday, you spent $4\frac{1}{2}$ hours in class, $3\frac{3}{4}$ hours studying, and $1\frac{1}{3}$ hours driving. Find the number of hours spent on these three activities.

Your strategy

Your solution

Example 10

Barbara Walsh worked 4 hours, $2\frac{1}{3}$ hours, and $5\frac{2}{3}$ hours this week at a part-time job. Barbara is paid $9 an hour. How much did she earn this week?

Strategy
To find how much Barbara earned:
• Find the total number of hours worked.
• Multiply the total number of hours worked by the hourly wage (9).

Solution

$$
\begin{array}{ll}
4 & \qquad\quad 12 \\
2\frac{1}{3} & \qquad\ \underline{\times\ 9} \\
+\ 5\frac{2}{3} & \qquad\quad 108 \\
\overline{\qquad\quad} & \\
11\frac{3}{3} = 12 \text{ hours worked}
\end{array}
$$

Barbara earned $108 this week.

You Try It 10

Jeff Sapone, a carpenter, worked $1\frac{2}{3}$ hours of overtime on Monday, $3\frac{1}{3}$ hours of overtime on Tuesday, and 2 hours of overtime on Wednesday. At an overtime hourly rate of $36, find Jeff's overtime pay for these 3 days.

Your strategy

Your solution

Solutions on pp. S5–S6

Copyright © Houghton Mifflin Company. All rights reserved.

2.4 Exercises

Objective A   To add fractions with the same denominator

For Exercises 1 to 20, add.

1. $\dfrac{2}{7} + \dfrac{1}{7}$ **2.** $\dfrac{3}{11} + \dfrac{5}{11}$ **3.** $\dfrac{1}{2} + \dfrac{1}{2}$ **4.** $\dfrac{1}{3} + \dfrac{2}{3}$

5. $\dfrac{8}{11} + \dfrac{7}{11}$ **6.** $\dfrac{9}{13} + \dfrac{7}{13}$ **7.** $\dfrac{8}{5} + \dfrac{9}{5}$ **8.** $\dfrac{5}{3} + \dfrac{7}{3}$

9. $\dfrac{3}{5} + \dfrac{8}{5} + \dfrac{3}{5}$ **10.** $\dfrac{3}{8} + \dfrac{5}{8} + \dfrac{7}{8}$ **11.** $\dfrac{3}{4} + \dfrac{1}{4} + \dfrac{5}{4}$ **12.** $\dfrac{2}{7} + \dfrac{4}{7} + \dfrac{5}{7}$

13. $\dfrac{3}{8} + \dfrac{7}{8} + \dfrac{1}{8}$ **14.** $\dfrac{5}{12} + \dfrac{7}{12} + \dfrac{1}{12}$ **15.** $\dfrac{4}{15} + \dfrac{7}{15} + \dfrac{11}{15}$ **16.** $\dfrac{3}{4} + \dfrac{3}{4} + \dfrac{1}{4}$

17. $\dfrac{3}{16} + \dfrac{5}{16} + \dfrac{7}{16}$ **18.** $\dfrac{5}{18} + \dfrac{11}{18} + \dfrac{17}{18}$ **19.** $\dfrac{3}{11} + \dfrac{5}{11} + \dfrac{7}{11}$ **20.** $\dfrac{5}{7} + \dfrac{4}{7} + \dfrac{5}{7}$

21. Find the sum of $\dfrac{4}{9}$ and $\dfrac{5}{9}$.   **22.** Find the sum of $\dfrac{5}{12}$, $\dfrac{1}{12}$, and $\dfrac{11}{12}$.

23. Find the total of $\dfrac{5}{8}$, $\dfrac{3}{8}$, and $\dfrac{7}{8}$.   **24.** Find the total of $\dfrac{4}{13}$, $\dfrac{7}{13}$, and $\dfrac{11}{13}$.

Objective B   To add fractions with different denominators

For Exercises 25 to 48, add.

25. $\dfrac{1}{2} + \dfrac{2}{3}$ **26.** $\dfrac{2}{3} + \dfrac{1}{4}$ **27.** $\dfrac{3}{14} + \dfrac{5}{7}$ **28.** $\dfrac{3}{5} + \dfrac{7}{10}$

29. $\dfrac{8}{15} + \dfrac{7}{20}$ **30.** $\dfrac{1}{6} + \dfrac{7}{9}$ **31.** $\dfrac{3}{8} + \dfrac{9}{14}$ **32.** $\dfrac{5}{12} + \dfrac{5}{16}$

33. $\dfrac{3}{20} + \dfrac{7}{30}$ **34.** $\dfrac{5}{12} + \dfrac{7}{30}$ **35.** $\dfrac{2}{3} + \dfrac{6}{19}$ **36.** $\dfrac{1}{2} + \dfrac{3}{29}$

Copyright © Houghton Mifflin Company. All rights reserved.

37. $\dfrac{1}{3} + \dfrac{5}{6} + \dfrac{7}{9}$ **38.** $\dfrac{2}{3} + \dfrac{5}{6} + \dfrac{7}{12}$ **39.** $\dfrac{5}{6} + \dfrac{1}{12} + \dfrac{5}{16}$ **40.** $\dfrac{2}{9} + \dfrac{7}{15} + \dfrac{4}{21}$

41. $\dfrac{2}{3} + \dfrac{1}{5} + \dfrac{7}{12}$ **42.** $\dfrac{3}{4} + \dfrac{4}{5} + \dfrac{7}{12}$ **43.** $\dfrac{1}{4} + \dfrac{4}{5} + \dfrac{5}{9}$ **44.** $\dfrac{2}{3} + \dfrac{3}{5} + \dfrac{7}{8}$

45. $\dfrac{5}{16} + \dfrac{11}{18} + \dfrac{17}{24}$ **46.** $\dfrac{3}{10} + \dfrac{14}{15} + \dfrac{9}{25}$ **47.** $\dfrac{2}{3} + \dfrac{5}{8} + \dfrac{7}{9}$ **48.** $\dfrac{1}{3} + \dfrac{2}{9} + \dfrac{7}{8}$

49. What is $\dfrac{3}{8}$ added to $\dfrac{3}{5}$?   **50.** What is $\dfrac{5}{9}$ added to $\dfrac{7}{12}$?

51. Find the sum of $\dfrac{3}{8}$, $\dfrac{5}{6}$, and $\dfrac{7}{12}$.   **52.** Find the sum of $\dfrac{11}{12}$, $\dfrac{13}{24}$, and $\dfrac{4}{15}$.

53. Find the total of $\dfrac{1}{2}$, $\dfrac{5}{8}$, and $\dfrac{7}{9}$.   **54.** Find the total of $\dfrac{5}{14}$, $\dfrac{3}{7}$, and $\dfrac{5}{21}$.

> **Objective C**   **To add whole numbers, mixed numbers, and fractions**

For Exercises 55 to 82, add.

55. $\begin{array}{r} 1\dfrac{1}{2} \\ + 2\dfrac{1}{6} \\ \hline \end{array}$ **56.** $\begin{array}{r} 2\dfrac{2}{5} \\ + 3\dfrac{3}{10} \\ \hline \end{array}$ **57.** $\begin{array}{r} 4\dfrac{1}{2} \\ + 5\dfrac{7}{12} \\ \hline \end{array}$ **58.** $\begin{array}{r} 3\dfrac{3}{8} \\ + 2\dfrac{5}{16} \\ \hline \end{array}$

59. $\begin{array}{r} 4 \\ + 5\dfrac{2}{7} \\ \hline \end{array}$ **60.** $\begin{array}{r} 6\dfrac{8}{9} \\ + 12 \\ \hline \end{array}$ **61.** $\begin{array}{r} 3\dfrac{5}{8} \\ + 2\dfrac{11}{20} \\ \hline \end{array}$ **62.** $\begin{array}{r} 4\dfrac{5}{12} \\ + 6\dfrac{11}{18} \\ \hline \end{array}$

63. $7\dfrac{5}{12} + 2\dfrac{9}{16}$ **64.** $9\dfrac{1}{2} + 3\dfrac{3}{11}$ **65.** $6 + 2\dfrac{3}{13}$ **66.** $8\dfrac{21}{40} + 6$

67. $8\dfrac{29}{30} + 7\dfrac{11}{40}$ **68.** $17\dfrac{5}{16} + 3\dfrac{11}{24}$ **69.** $17\dfrac{3}{8} + 7\dfrac{7}{20}$ **70.** $14\dfrac{7}{12} + 29\dfrac{13}{21}$

Copyright © Houghton Mifflin Company. All rights reserved.

71. $5\frac{7}{8} + 27\frac{5}{12}$

72. $7\frac{5}{6} + 3\frac{5}{9}$

73. $7\frac{5}{9} + 2\frac{7}{12}$

74. $3\frac{1}{2} + 2\frac{3}{4} + 1\frac{5}{6}$

75. $2\frac{1}{2} + 3\frac{2}{3} + 4\frac{1}{4}$

76. $3\frac{1}{3} + 7\frac{1}{5} + 2\frac{1}{7}$

77. $3\frac{1}{2} + 3\frac{1}{5} + 8\frac{1}{9}$

78. $6\frac{5}{9} + 6\frac{5}{12} + 2\frac{5}{18}$

79. $2\frac{3}{8} + 4\frac{7}{12} + 3\frac{5}{16}$

80. $2\frac{1}{8} + 4\frac{2}{9} + 5\frac{17}{18}$

81. $6\frac{5}{6} + 17\frac{2}{9} + 18\frac{5}{27}$

82. $4\frac{7}{20} + \frac{17}{80} + 25\frac{23}{60}$

83. Find the sum of $2\frac{4}{9}$ and $5\frac{7}{12}$.

84. Find $5\frac{5}{6}$ more than $3\frac{3}{8}$.

85. What is $4\frac{3}{4}$ added to $9\frac{1}{3}$?

86. What is $4\frac{8}{9}$ added to $9\frac{1}{6}$?

87. Find the total of 2, $4\frac{5}{8}$, and $2\frac{2}{9}$.

88. Find the total of $1\frac{5}{8}$, 3, and $7\frac{7}{24}$.

Objective D   **To solve application problems**

89. Mechanics   Find the length of the shaft.

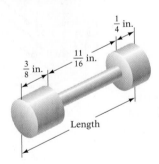

Length

90. Mechanics   Find the length of the shaft.

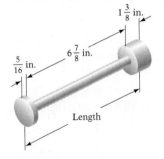

Length

91. Carpentry   A table 30 inches high has a top that is $1\frac{1}{8}$ inches thick. Find the total thickness of the table top after a $\frac{3}{16}$-inch veneer is applied.

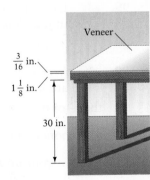

Veneer

$\frac{3}{16}$ in.

$1\frac{1}{8}$ in.

30 in.

92. Wages   Fred Thomson worked $2\frac{2}{3}$ hours of overtime on Monday, $1\frac{1}{4}$ hours on Wednesday, $1\frac{1}{3}$ hours on Friday, and $6\frac{3}{4}$ hours on Saturday.
   a.   Find the total number of overtime hours worked during the week.
   b.   At an overtime hourly wage of $22 per hour, how much overtime pay does Fred receive?

Copyright © Houghton Mifflin Company. All rights reserved.

93. Wages You are working a part-time job that pays $11 an hour. You worked 5, $3\frac{3}{4}$, $2\frac{1}{3}$, $1\frac{1}{4}$, and $7\frac{2}{3}$ hours during the last five days.

a. Find the total number of hours you worked during the last five days.
b. Find your total wages for the five days.

94. Sports The course of a yachting race is in the shape of a triangle with sides that measure $4\frac{3}{10}$ miles, $3\frac{7}{10}$ miles, and $2\frac{1}{2}$ miles. Find the total length of the course.

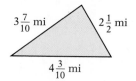
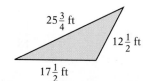

95. Landscaping A flower garden in the yard of a colonial home is in the shape of a triangle as shown at the right. The wooden beams lining the edge of the garden need to be replaced. Find the total length of wood beams that must be purchased in order to replace the old beams.

96. **Food Preferences** When the Heller Research Group conducted a survey to determine favorite doughnut flavors, $\frac{2}{5}$ of the respondents named glazed doughnuts, $\frac{8}{25}$ named filled doughnuts, and $\frac{3}{20}$ named frosted doughnuts. What fraction of the respondents named glazed, filled, *or* frosted as their favorite type of doughnut?

97. **Mobility** During a recent year, over 42 million Americans changed homes. The graph at the right shows what fractions of the people moved within the same county, moved to a different county in the same state, and moved to a different state. (Source: Census Bureau, *Geographical Mobility*) What fractional part of those who changed homes moved outside the county they had been living in?

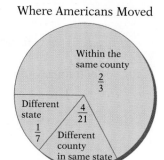

Where Americans Moved

APPLYING THE CONCEPTS

98. What is a unit fraction? Find the sum of the three largest unit fractions. Is there a smallest unit fraction? If so, write it down. If not, explain why.

99. Use a model to illustrate and explain the addition of fractions with unlike denominators.

100. A survey was conducted to determine people's favorite color from among blue, green, red, purple, and other. The surveyor claims that $\frac{1}{3}$ of the people responded blue, $\frac{1}{6}$ responded green, $\frac{1}{8}$ responded red, $\frac{1}{12}$ responded purple, and $\frac{2}{5}$ responded some other color. Is this possible? Explain your answer.

Copyright © Houghton Mifflin Company. All rights reserved.

2.5 Subtraction of Fractions and Mixed Numbers

Objective A   **To subtract fractions with the same denominator**

Fractions with the same denominator are subtracted by subtracting the numerators and placing the difference over the common denominator. After subtracting, write the fraction in simplest form.

HOW TO   Subtract: $\dfrac{5}{7} - \dfrac{3}{7}$

$$\begin{array}{r} \dfrac{5}{7} \\[2mm] -\dfrac{3}{7} \\[2mm] \hline \dfrac{2}{7} \end{array}$$

- Subtract the numerators and place the difference over the common denominator.

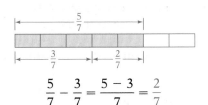

$$\frac{5}{7} - \frac{3}{7} = \frac{5-3}{7} = \frac{2}{7}$$

Example 1   Find $\dfrac{17}{30}$ less $\dfrac{11}{30}$.

Solution

$$\begin{array}{r} \dfrac{17}{30} \\[2mm] -\dfrac{11}{30} \\[2mm] \hline \dfrac{6}{30} = \dfrac{1}{5} \end{array}$$

- The denominators are the same. Subtract the numerators. Place the difference over the common denominator.

You Try It 1   Subtract: $\dfrac{16}{27} - \dfrac{7}{27}$

Your solution

Solution on p. S6

Objective B   **To subtract fractions with different denominators**

To subtract fractions with different denominators, first rewrite the fractions as equivalent fractions with a common denominator. As with adding fractions, the common denominator is the LCM of the denominators of the fractions.

HOW TO   Subtract: $\dfrac{5}{6} - \dfrac{1}{4}$

The common denominator is the LCM of 6 and 4. The LCM = 12.

Write equivalent fractions using the LCM.

$$\dfrac{5}{6} = \dfrac{10}{12}$$
$$-\dfrac{1}{4} = \dfrac{3}{12}$$

Subtract the fractions.

$$\begin{array}{r} \dfrac{5}{6} = \dfrac{10}{12} \\[2mm] -\dfrac{1}{4} = \dfrac{3}{12} \\[2mm] \hline \dfrac{7}{12} \end{array}$$

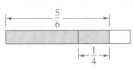

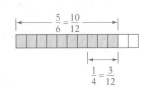

Copyright © Houghton Mifflin Company. All rights reserved.

Example 2 Subtract: $\dfrac{11}{16} - \dfrac{5}{12}$

Solution

$$\dfrac{11}{16} = \dfrac{33}{48}$$

$$-\dfrac{5}{12} = \dfrac{20}{48}$$

$$\dfrac{13}{48}$$

• LCM = 48

You Try It 2 Subtract: $\dfrac{13}{18} - \dfrac{7}{24}$

Your solution

Solution on p. S6

Objective C   **To subtract whole numbers, mixed numbers, and fractions**

To subtract mixed numbers without borrowing, subtract the fractional parts and then subtract the whole numbers.

HOW TO Subtract: $5\dfrac{5}{6} - 2\dfrac{3}{4}$

Subtract the fractional parts.

$$5\dfrac{5}{6} = 5\dfrac{10}{12}$$

$$-2\dfrac{3}{4} = 2\dfrac{9}{12}$$

$$\dfrac{1}{12}$$

• The LCM of 6 and 4 is 12.

Subtract the whole numbers.

$$5\dfrac{5}{6} = 5\dfrac{10}{12}$$

$$-2\dfrac{3}{4} = 2\dfrac{9}{12}$$

$$3\dfrac{1}{12}$$

Subtraction of mixed numbers sometimes involves borrowing.

HOW TO Subtract: $5 - 2\dfrac{5}{8}$

Borrow 1 from 5.

$$5 = \overset{4}{\not5}\,1$$

$$-2\dfrac{5}{8} = 2\dfrac{5}{8}$$

Write 1 as a fraction so that the fractions have the same denominators.

$$5 = 4\dfrac{8}{8}$$

$$-2\dfrac{5}{8} = 2\dfrac{5}{8}$$

Subtract the mixed numbers.

$$5 = 4\dfrac{8}{8}$$

$$-2\dfrac{5}{8} = 2\dfrac{5}{8}$$

$$2\dfrac{3}{8}$$

HOW TO Subtract: $7\dfrac{1}{6} - 2\dfrac{5}{8}$

Write equivalent fractions using the LCM.

$$7\dfrac{1}{6} = 7\dfrac{4}{24}$$

$$-2\dfrac{5}{8} = 2\dfrac{15}{24}$$

Borrow 1 from 7. Add the 1 to $\dfrac{4}{24}$. Write $1\dfrac{4}{24}$ as $\dfrac{28}{24}$.

$$7\dfrac{1}{6} = \overset{6}{\not7}\,1\dfrac{4}{24} = 6\dfrac{28}{24}$$

$$-2\dfrac{5}{8} = \quad 2\dfrac{15}{24} = 2\dfrac{15}{24}$$

Subtract the mixed numbers.

$$7\dfrac{1}{6} = 6\dfrac{28}{24}$$

$$-2\dfrac{5}{8} = 2\dfrac{15}{24}$$

$$4\dfrac{13}{24}$$

Copyright © Houghton Mifflin Company. All rights reserved.

Example 3   Subtract: $15\frac{7}{8} - 12\frac{2}{3}$

Solution

$$15\frac{7}{8} = 15\frac{21}{24} \quad \bullet \text{ LCM} = 24$$

$$-\ 12\frac{2}{3} = 12\frac{16}{24}$$

$$\rule{4cm}{0.4pt}$$

$$3\frac{5}{24}$$

You Try It 3   Subtract: $17\frac{5}{9} - 11\frac{5}{12}$

Your solution

Example 4   Subtract: $9 - 4\frac{3}{11}$

Solution

$$9 \quad\ = 8\frac{11}{11} \quad \bullet \text{ LCM} = 11$$

$$-\ 4\frac{3}{11} = 4\frac{3}{11}$$

$$\rule{4cm}{0.4pt}$$

$$4\frac{8}{11}$$

You Try It 4   Subtract: $8 - 2\frac{4}{13}$

Your solution

Example 5   Find $11\frac{5}{12}$ decreased by $2\frac{11}{16}$.

Solution

$$11\frac{5}{12} = 11\frac{20}{48} = 10\frac{68}{48} \quad \bullet \text{ LCM} = 48$$

$$-\ 2\frac{11}{16} = \ 2\frac{33}{48} = \ 2\frac{33}{48}$$

$$\rule{6cm}{0.4pt}$$

$$8\frac{35}{48}$$

You Try It 5   What is $21\frac{7}{9}$ minus $7\frac{11}{12}$?

Your solution

Solutions on p. S6

Objective D   **To solve application problems**

HOW TO   The outside diameter of a bushing is $3\frac{3}{8}$ inches and the wall thickness is $\frac{1}{4}$ inch. Find the inside diameter of the bushing.

Outside Diameter

Inside Diameter

$$\frac{1}{4} + \frac{1}{4} = \frac{2}{4} = \frac{1}{2}$$

- Add $\frac{1}{4}$ to $\frac{1}{4}$ to find the total thickness of the two walls.

$$3\frac{3}{8} = 3\frac{3}{8} = 2\frac{11}{8}$$

$$-\ \frac{1}{2} = \ \frac{4}{8} = \ \frac{4}{8}$$

$$\rule{4cm}{0.4pt}$$

$$2\frac{7}{8}$$

- Subtract the total thickness of the two walls to find the inside diameter.

The inside diameter of the bushing is $2\frac{7}{8}$ inches.

Copyright © Houghton Mifflin Company. All rights reserved.

Example 6

A $2\frac{2}{3}$-inch piece is cut from a $6\frac{5}{8}$-inch board. How much of the board is left?

Strategy

To find the length remaining, subtract the length of the piece cut from the total length of the board.

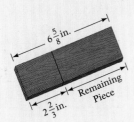

$6\frac{5}{8}$ in.

$2\frac{2}{3}$ in.

Remaining Piece

Solution

$$6\frac{5}{8} = 6\frac{15}{24} = 5\frac{39}{24}$$
$$-2\frac{2}{3} = 2\frac{16}{24} = 2\frac{16}{24}$$
$$\overline{\phantom{-2\frac{2}{3} = 2\frac{16}{24} =\;} 3\frac{23}{24}}$$

$3\frac{23}{24}$ inches of the board are left.

You Try It 6

A flight from New York to Los Angeles takes $5\frac{1}{2}$ hours. After the plane has been in the air for $2\frac{3}{4}$ hours, how much flight time remains?

Your strategy

Your solution

Example 7

Two painters are staining a house. In 1 day one painter stained $\frac{1}{3}$ of the house, and the other stained $\frac{1}{4}$ of the house. How much of the job remains to be done?

Strategy

To find how much of the job remains:
• Find the total amount of the house already stained $\left(\frac{1}{3} + \frac{1}{4}\right)$.
• Subtract the amount already stained from 1, which represents the complete job.

Solution

$$\frac{1}{3} = \frac{4}{12} \qquad\qquad 1 = \frac{12}{12}$$
$$+\frac{1}{4} = \frac{3}{12} \qquad\qquad -\frac{7}{12} = \frac{7}{12}$$
$$\overline{\frac{7}{12}} \qquad\qquad \overline{\frac{5}{12}}$$

$\frac{5}{12}$ of the house remains to be stained.

You Try It 7

A patient is put on a diet to lose 24 pounds in 3 months. The patient lost $7\frac{1}{2}$ pounds the first month and $5\frac{3}{4}$ pounds the second month. How much weight must be lost the third month to achieve the goal?

Your strategy

Your solution

Solutions on p. S6

Copyright © Houghton Mifflin Company. All rights reserved.

2.5 Exercises

Objective A   **To subtract fractions with the same denominator**

For Exercises 1 to 10, subtract.

1. $\dfrac{9}{17}$
    $-\dfrac{7}{17}$

2. $\dfrac{11}{15}$
    $-\dfrac{3}{15}$

3. $\dfrac{11}{12}$
    $-\dfrac{7}{12}$

4. $\dfrac{13}{15}$
    $-\dfrac{4}{15}$

5. $\dfrac{9}{20}$
    $-\dfrac{7}{20}$

6. $\dfrac{48}{55}$
    $-\dfrac{13}{55}$

7. $\dfrac{42}{65}$
    $-\dfrac{17}{65}$

8. $\dfrac{11}{24}$
    $-\dfrac{5}{24}$

9. $\dfrac{23}{30}$
    $-\dfrac{13}{30}$

10. $\dfrac{17}{42}$
     $-\dfrac{5}{42}$

11. What is $\dfrac{5}{14}$ less than $\dfrac{13}{14}$?

12. What is $\dfrac{7}{19}$ less than $\dfrac{17}{19}$?

13. Find the difference between $\dfrac{7}{8}$ and $\dfrac{5}{8}$.

14. Find the difference between $\dfrac{7}{12}$ and $\dfrac{5}{12}$.

15. What is $\dfrac{18}{23}$ minus $\dfrac{9}{23}$?

16. What is $\dfrac{7}{9}$ minus $\dfrac{3}{9}$?

17. Find $\dfrac{17}{24}$ decreased by $\dfrac{11}{24}$.

18. Find $\dfrac{19}{30}$ decreased by $\dfrac{11}{30}$.

Objective B   **To subtract fractions with different denominators**

For Exercises 19 to 33, subtract.

19. $\dfrac{2}{3}$
     $-\dfrac{1}{6}$

20. $\dfrac{7}{8}$
     $-\dfrac{5}{16}$

21. $\dfrac{5}{8}$
     $-\dfrac{2}{7}$

22. $\dfrac{5}{6}$
     $-\dfrac{3}{7}$

23. $\dfrac{5}{7}$
     $-\dfrac{3}{14}$

24. $\dfrac{5}{9}$
     $-\dfrac{7}{15}$

25. $\dfrac{8}{15}$
     $-\dfrac{7}{20}$

26. $\dfrac{7}{9}$
     $-\dfrac{1}{6}$

27. $\dfrac{9}{14}$
     $-\dfrac{3}{8}$

28. $\dfrac{5}{12}$
     $-\dfrac{5}{16}$

Copyright © Houghton Mifflin Company. All rights reserved.

29. $\frac{46}{51}$
$-\frac{3}{17}$

30. $\frac{9}{16}$
$-\frac{17}{32}$

31. $\frac{21}{35}$
$-\frac{5}{14}$

32. $\frac{19}{40}$
$-\frac{3}{16}$

33. $\frac{29}{60}$
$-\frac{3}{40}$

34. What is $\frac{3}{5}$ less than $\frac{11}{12}$?

35. What is $\frac{5}{9}$ less than $\frac{11}{15}$?

36. Find the difference between $\frac{11}{24}$ and $\frac{7}{18}$.

37. Find the difference between $\frac{9}{14}$ and $\frac{5}{42}$.

38. Find $\frac{11}{12}$ decreased by $\frac{11}{15}$.

39. Find $\frac{17}{20}$ decreased by $\frac{7}{15}$.

40. What is $\frac{13}{20}$ minus $\frac{1}{6}$?

41. What is $\frac{5}{6}$ minus $\frac{7}{9}$?

Objective C   **To subtract whole numbers, mixed numbers, and fractions**

For Exercises 42 to 61, subtract.

42. $5\frac{7}{12}$
$-2\frac{5}{12}$

43. $16\frac{11}{15}$
$-11\frac{8}{15}$

44. $72\frac{21}{23}$
$-16\frac{17}{23}$

45. $19\frac{16}{17}$
$-9\frac{7}{17}$

46. $6\frac{1}{3}$
-2

47. $5\frac{7}{8}$
-1

48. 10
$-6\frac{1}{3}$

49. $3\frac{\cancel{2}1}{\cancel{2}}$
$-2\frac{5}{21}\frac{\cancel{12}1}{\cancel{21}}$
$\frac{6}{21}$

50. $6\frac{2}{5}$
$-4\frac{4}{5}$

51. $16\frac{3}{8}$
$-10\frac{7}{8}$

52. $25\frac{4}{9}$
$-16\frac{7}{9}$

53. $8\frac{3}{7}$
$-2\frac{6}{7}$

54. $16\frac{2}{5}$
$-8\frac{4}{9}$

55. $23\frac{7}{8}\frac{21}{24}$
$-16\frac{2}{3}\frac{16}{24}$
$\boxed{7\frac{1}{8}}\ \frac{\cancel{6}}{24}\boxed{\frac{1}{8}}$

56. 6
$-4\frac{3}{5}$

57. $\overset{64}{65}\frac{8}{35}\Big|\frac{16}{70}$
$-16\frac{11}{14}\ \frac{55}{70}$

58. $82\frac{4}{33}$
$-16\frac{5}{22}$

59. $101\frac{2}{9}$
-16

60. $77\frac{5}{18}$
-61

61. $\overset{16}{17}\overset{13}{1}/13$
$-7\frac{8}{13}$

Copyright © Houghton Mifflin Company. All rights reserved.

62. What is $5\frac{3}{8}$ less than $8\frac{1}{9}$?

63. What is $7\frac{3}{5}$ less than $23\frac{3}{20}$?

64. Find the difference between $9\frac{2}{7}$ and $3\frac{1}{4}$.

65. Find the difference between $12\frac{3}{8}$ and $7\frac{5}{12}$.

66. What is $10\frac{5}{9}$ minus $5\frac{11}{15}$?

67. Find $6\frac{1}{3}$ decreased by $3\frac{3}{5}$.

Objective D   **To solve application problems**

68. Mechanics   Find the missing dimension.

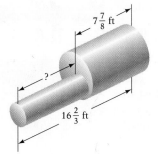

69. Mechanics   Find the missing dimension.

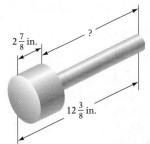

70. 🥧 **Sports**   In the Kentucky Derby the horses run $1\frac{1}{4}$ miles. In the Belmont Stakes they run $1\frac{1}{2}$ miles, and in the Preakness Stakes they run $1\frac{3}{16}$ miles. How much farther do the horses run in the Kentucky Derby than in the Preakness Stakes? How much farther do they run in the Belmont Stakes than in the Preakness Stakes?

71. 🥧 **Sports**   In the running high jump in the 1948 Summer Olympic Games, Alice Coachman's distance was $66\frac{1}{8}$ inches. In the same event in the 1972 Summer Olympics, Urika Meyfarth jumped $75\frac{1}{2}$ inches, and in the 1996 Olympic Games, Stefka Kostadinova jumped $80\frac{3}{4}$ inches. Find the difference between Meyfarth's distance and Coachman's distance. Find the difference between Kostadinova's distance and Meyfarth's distance.

72. Hiking   Two hikers plan a 3-day, $27\frac{1}{2}$-mile backpack trip carrying a total of 80 pounds. The hikers plan to travel $7\frac{3}{8}$ miles the first day and $10\frac{1}{3}$ miles the second day.
a. How many miles do the hikers plan to travel the first two days?
b. How many miles will be left to travel on the third day?

Copyright © Houghton Mifflin Company. All rights reserved.

73. Fundraising A 12-mile walkathon has three checkpoints. The first is $3\frac{3}{8}$ miles from the starting point. The second checkpoint is $4\frac{1}{3}$ miles from the first.

   a. How many miles is it from the starting point to the second checkpoint?

   b. How many miles is it from the second checkpoint to the finish line?

74. Health A patient with high blood pressure who weighs 225 pounds is put on a diet to lose 25 pounds in 3 months. The patient loses $8\frac{3}{4}$ pounds the first month and $11\frac{5}{8}$ pounds the second month. How much weight must be lost the third month for the goal to be achieved?

75. Sports A wrestler is entered in the 172-pound weight class in the conference finals coming up in 3 weeks. The wrestler needs to lose $12\frac{3}{4}$ pounds. The wrestler loses $5\frac{1}{4}$ pounds the first week and $4\frac{1}{4}$ pounds the second week.

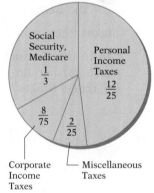

   a. Without doing the calculations, determine whether the wrestler can reach his weight class by losing less in the third week than was lost in the second week.

   b. How many pounds must be lost in the third week for the desired weight to be reached?

76. **Government** The figure at the right shows the sources of federal income. The category "Social Security, Medicare" includes unemployment and other retirement taxes. The category "Miscellaneous Taxes" includes excise, customs, estate, and gift taxes.

   a. What is the difference between the fraction of federal income derived from personal income taxes and the fraction derived from corporate income taxes?

   b. What is the difference between the fraction of federal income derived from the Social Security category and the fraction derived from miscellaneous taxes?

Sources of Federal Income

Social Security, Medicare $\frac{1}{3}$

Personal Income Taxes $\frac{12}{25}$

$\frac{8}{75}$

$\frac{2}{25}$

Corporate Income Taxes

Miscellaneous Taxes

APPLYING THE CONCEPTS

77. Fill in the square to produce a true statement: $5\frac{1}{3} - \boxed{} = 2\frac{1}{2}$.

78. Fill in the square to produce a true statement: $\boxed{} - 4\frac{1}{2} = 1\frac{5}{8}$.

79. Fill in the blank squares at the right so that the sum of the numbers is the same along any row, column, or diagonal. The resulting square is called a magic square.

		$\frac{3}{4}$
1	$\frac{5}{8}$	
$\frac{1}{2}$		$\frac{7}{8}$

80. Finances If $\frac{4}{15}$ of an electrician's income is spent for housing, what fraction of the electrician's income is not spent for housing?

Copyright © Houghton Mifflin Company. All rights reserved.

2.6 Multiplication of Fractions and Mixed Numbers

Objective A   **To multiply fractions**

The product of two fractions is the product of the numerators over the product of the denominators.

Study Tip

Before the class meeting in which your professor begins a new section, you should read each objective statement for that section. Next, browse through the material in that objective. The purpose of browsing through the material is so that your brain will be prepared to accept and organize the new information when it is presented to you. See *AIM for Success* at the front of the book.

HOW TO   Multiply: $\frac{2}{3} \times \frac{4}{5}$

$$\frac{2}{3} \times \frac{4}{5} = \frac{2 \cdot 4}{3 \cdot 5} = \frac{8}{15}$$

- Multiply the numerators.
- Multiply the denominators.

The product $\frac{2}{3} \times \frac{4}{5}$ can be read "$\frac{2}{3}$ times $\frac{4}{5}$" or "$\frac{2}{3}$ of $\frac{4}{5}$."

Reading the times sign as "of" is useful in application problems.

$\frac{4}{5}$ of the bar is shaded.

Shade $\frac{2}{3}$ of the $\frac{4}{5}$ already shaded.

$\frac{8}{15}$ of the bar is then shaded light yellow.

$$\frac{2}{3} \text{ of } \frac{4}{5} = \frac{2}{3} \times \frac{4}{5} = \frac{8}{15}$$

After multiplying two fractions, write the product in simplest form.

HOW TO   Multiply: $\frac{3}{4} \times \frac{14}{15}$

$$\frac{3}{4} \times \frac{14}{15} = \frac{3 \cdot 14}{4 \cdot 15}$$

- Multiply the numerators.
- Multiply the denominators.

$$= \frac{3 \cdot 2 \cdot 7}{2 \cdot 2 \cdot 3 \cdot 5}$$

- Write the prime factorization of each number.

$$= \frac{\overset{1}{\cancel{3}} \cdot \overset{1}{\cancel{2}} \cdot 7}{2 \cdot 2 \cdot \underset{1}{\cancel{3}} \cdot 5} = \frac{7}{10}$$

- Eliminate the common factors. Then multiply the remaining factors in the numerator and denominator.

This example could also be worked by using the GCF.

$$\frac{3}{4} \times \frac{14}{15} = \frac{42}{60}$$

- Multiply the numerators.
- Multiply the denominators.

$$= \frac{6 \cdot 7}{6 \cdot 10}$$

- The GCF of 42 and 60 is 6. Factor 6 from 42 and 60.

$$= \frac{\overset{1}{\cancel{6}} \cdot 7}{\underset{1}{\cancel{6}} \cdot 10} = \frac{7}{10}$$

- Eliminate the GCF.

Copyright © Houghton Mifflin Company. All rights reserved.

Example 1

Multiply $\frac{4}{15}$ and $\frac{5}{28}$.

Solution

$$\frac{4}{15} \times \frac{5}{28} = \frac{4 \cdot 5}{15 \cdot 28} = \frac{\overset{1}{\cancel{2}} \cdot \overset{1}{\cancel{2}} \cdot \overset{1}{\cancel{5}}}{3 \cdot \underset{1}{\cancel{5}} \cdot \underset{1}{\cancel{2}} \cdot \underset{1}{\cancel{2}} \cdot 7} = \frac{1}{21}$$

You Try It 1

Multiply $\frac{4}{21}$ and $\frac{7}{44}$.

Your solution

Example 2

Find the product of $\frac{9}{20}$ and $\frac{33}{35}$.

Solution

$$\frac{9}{20} \times \frac{33}{35} = \frac{9 \cdot 33}{20 \cdot 35} = \frac{3 \cdot 3 \cdot 3 \cdot 11}{2 \cdot 2 \cdot 5 \cdot 5 \cdot 7} = \frac{297}{700}$$

You Try It 2

Find the product of $\frac{2}{21}$ and $\frac{10}{33}$.

Your solution

Example 3

What is $\frac{14}{9}$ times $\frac{12}{7}$?

Solution

$$\frac{14}{9} \times \frac{12}{7} = \frac{14 \cdot 12}{9 \cdot 7} = \frac{2 \cdot \overset{1}{\cancel{7}} \cdot 2 \cdot 2 \cdot \overset{1}{\cancel{3}}}{3 \cdot \underset{1}{\cancel{3}} \cdot \underset{1}{\cancel{7}}} = \frac{8}{3} = 2\frac{2}{3}$$

You Try It 3

What is $\frac{16}{5}$ times $\frac{15}{24}$?

Your solution

Solutions on pp. S6–S7

Objective B   **To multiply whole numbers, mixed numbers, and fractions**

To multiply a whole number by a fraction or mixed number, first write the whole number as a fraction with a denominator of 1.

> **HOW TO**   Multiply: $4 \times \frac{3}{7}$
>
> $$4 \times \frac{3}{7} = \frac{4}{1} \times \frac{3}{7} = \frac{4 \cdot 3}{1 \cdot 7} = \frac{2 \cdot 2 \cdot 3}{7} = \frac{12}{7} = 1\frac{5}{7}$$

• Write 4 with a denominator of 1; then multiply the fractions.

When one or more of the factors in a product is a mixed number, write the mixed number as an improper fraction before multiplying.

> **HOW TO**   Multiply: $2\frac{1}{3} \times \frac{3}{14}$
>
> $$2\frac{1}{3} \times \frac{3}{14} = \frac{7}{3} \times \frac{3}{14} = \frac{7 \cdot 3}{3 \cdot 14} = \frac{\overset{1}{\cancel{7}} \cdot \overset{1}{\cancel{3}}}{\underset{1}{\cancel{3}} \cdot 2 \cdot \underset{1}{\cancel{7}}} = \frac{1}{2}$$

• Write $2\frac{1}{3}$ as an improper fraction; then multiply the fractions.

Copyright © Houghton Mifflin Company. All rights reserved.

Example 4

Multiply: $4\frac{5}{6} \times \frac{12}{13}$

Solution

$$4\frac{5}{6} \times \frac{12}{13} = \frac{29}{6} \times \frac{12}{13} = \frac{29 \cdot 12}{6 \cdot 13}$$

$$= \frac{29 \cdot \overset{1}{\cancel{2}} \cdot 2 \cdot \overset{1}{\cancel{3}}}{\underset{1}{\cancel{2}} \cdot \underset{1}{\cancel{3}} \cdot 13} = \frac{58}{13} = 4\frac{6}{13}$$

You Try It 4

Multiply: $5\frac{2}{5} \times \frac{5}{9}$

Your solution

Example 5

Find $5\frac{2}{3}$ times $4\frac{1}{2}$.

Solution

$$5\frac{2}{3} \times 4\frac{1}{2} = \frac{17}{3} \times \frac{9}{2} = \frac{17 \cdot 9}{3 \cdot 2}$$

$$= \frac{17 \cdot \overset{1}{\cancel{3}} \cdot 3}{\underset{1}{\cancel{3}} \cdot 2} = \frac{51}{2} = 25\frac{1}{2}$$

You Try It 5

Multiply: $3\frac{2}{5} \times 6\frac{1}{4}$

Your solution

Example 6

Multiply: $4\frac{2}{5} \times 7$

Solution

$$4\frac{2}{5} \times 7 = \frac{22}{5} \times \frac{7}{1} = \frac{22 \cdot 7}{5 \cdot 1}$$

$$= \frac{2 \cdot 11 \cdot 7}{5} = \frac{154}{5} = 30\frac{4}{5}$$

You Try It 6

Multiply: $3\frac{2}{7} \times 6$

Your solution

Solutions on p. S7

Objective C   **To solve application problems**

Length (ft)	Weight (lb/ft)
$6\frac{1}{2}$	$\frac{3}{8}$
$8\frac{5}{8}$	$1\frac{1}{4}$
$10\frac{3}{4}$	$2\frac{1}{2}$
$12\frac{7}{12}$	$4\frac{1}{3}$

The table at the left lists the length of steel rods and the weight per foot. The weight per foot is measured in pounds for each foot of rod and is abbreviated as lb/ft.

HOW TO   Find the weight of the steel bar that is $10\frac{3}{4}$ feet long.

Strategy
To find the weight of the steel bar, multiply its length by the weight per foot.

Solution $10\frac{3}{4} \times 2\frac{1}{2} = \frac{43}{4} \times \frac{5}{2} = \frac{43 \cdot 5}{4 \cdot 2} = \frac{215}{8} = 26\frac{7}{8}$

The weight of the $10\frac{3}{4}$-foot rod is $26\frac{7}{8}$ pounds.

Copyright © Houghton Mifflin Company. All rights reserved.

Example 7

An electrician earns $206 for each day worked. What are the electrician's earnings for working $4\frac{1}{2}$ days?

Strategy

To find the electrician's total earnings, multiply the daily earnings (206) by the number of days worked $\left(4\frac{1}{2}\right)$.

Solution

$$206 \times 4\frac{1}{2} = \frac{206}{1} \times \frac{9}{2}$$
$$= \frac{206 \cdot 9}{1 \cdot 2}$$
$$= 927$$

The electrician's earnings are $927.

You Try It 7

Over the last 10 years, a house increased in value by $2\frac{1}{2}$ times. The price of the house 10 years ago was $170,000. What is the value of the house today?

Your strategy

Your solution

Example 8

The value of a small office building and the land on which it is built is $290,000. The value of the land is $\frac{1}{4}$ the total value. What is the dollar value of the building?

Strategy

To find the value of the building:

• Find the value of the land $\left(\frac{1}{4} \times 290,000\right)$.

• Subtract the value of the land from the total value (290,000).

Solution

$$\frac{1}{4} \times 290,000 = \frac{290,000}{4}$$
$$= 72,500 \qquad \bullet \text{ Value of the land}$$
$$290,000 - 72,500 = 217,500$$

The value of the building is $217,500.

You Try It 8

A paint company bought a drying chamber and an air compressor for spray painting. The total cost of the two items was $160,000. The drying chamber's cost was $\frac{4}{5}$ of the total cost. What was the cost of the air compressor?

Your strategy

Your solution

Solutions on p. S7

Copyright © Houghton Mifflin Company. All rights reserved.

2.6 Exercises

Objective A   **To multiply fractions**

For Exercises 1 to 36, multiply.

1. $\dfrac{2}{3} \times \dfrac{7}{8}$

2. $\dfrac{1}{2} \times \dfrac{2}{3}$

3. $\dfrac{5}{16} \times \dfrac{7}{15}$

4. $\dfrac{3}{8} \times \dfrac{6}{7}$

5. $\dfrac{1}{6} \times \dfrac{1}{8}$

6. $\dfrac{2}{5} \times \dfrac{5}{6}$

7. $\dfrac{11}{12} \times \dfrac{6}{7}$

8. $\dfrac{11}{12} \times \dfrac{3}{5}$

9. $\dfrac{1}{6} \times \dfrac{6}{7}$

10. $\dfrac{3}{5} \times \dfrac{10}{11}$

11. $\dfrac{1}{5} \times \dfrac{5}{8}$

12. $\dfrac{6}{7} \times \dfrac{14}{15}$

13. $\dfrac{8}{9} \times \dfrac{27}{4}$

14. $\dfrac{3}{5} \times \dfrac{3}{10}$

15. $\dfrac{5}{6} \times \dfrac{1}{2}$

16. $\dfrac{3}{8} \times \dfrac{5}{12}$

17. $\dfrac{16}{9} \times \dfrac{27}{8}$

18. $\dfrac{5}{8} \times \dfrac{16}{15}$

19. $\dfrac{3}{2} \times \dfrac{4}{9}$

20. $\dfrac{5}{3} \times \dfrac{3}{7}$

21. $\dfrac{7}{8} \times \dfrac{3}{14}$

22. $\dfrac{2}{9} \times \dfrac{1}{5}$

23. $\dfrac{1}{10} \times \dfrac{3}{8}$

24. $\dfrac{5}{12} \times \dfrac{6}{7}$

25. $\dfrac{15}{8} \times \dfrac{16}{3}$

26. $\dfrac{5}{6} \times \dfrac{4}{15}$

27. $\dfrac{1}{2} \times \dfrac{2}{15}$

28. $\dfrac{3}{8} \times \dfrac{5}{16}$

29. $\dfrac{5}{7} \times \dfrac{14}{15}$

30. $\dfrac{3}{8} \times \dfrac{15}{41}$

31. $\dfrac{5}{12} \times \dfrac{42}{65}$

32. $\dfrac{16}{33} \times \dfrac{55}{72}$

33. $\dfrac{12}{5} \times \dfrac{5}{3}$

34. $\dfrac{17}{9} \times \dfrac{81}{17}$

35. $\dfrac{16}{85} \times \dfrac{125}{84}$

36. $\dfrac{19}{64} \times \dfrac{48}{95}$

Copyright © Houghton Mifflin Company. All rights reserved.

37. Multiply $\frac{7}{12}$ and $\frac{15}{42}$.

38. Multiply $\frac{32}{9}$ and $\frac{3}{8}$.

39. Find the product of $\frac{5}{9}$ and $\frac{3}{20}$.

40. Find the product of $\frac{7}{3}$ and $\frac{15}{14}$.

41. What is $\frac{1}{2}$ times $\frac{8}{15}$?

42. What is $\frac{3}{8}$ times $\frac{12}{17}$?

Objective B   **To multiply whole numbers, mixed numbers, and fractions**

For Exercises 43 to 82, multiply.

43. $4 \times \frac{3}{8}$

44. $14 \times \frac{5}{7}$

45. $\frac{2}{3} \times 6$

46. $\frac{5}{12} \times 40$

47. $\frac{1}{3} \times 1\frac{1}{3}$

48. $\frac{2}{5} \times 2\frac{1}{2}$

49. $1\frac{7}{8} \times \frac{4}{15}$

50. $2\frac{1}{5} \times \frac{5}{22}$

51. $55 \times \frac{3}{10}$

52. $\frac{5}{14} \times 49$

53. $4 \times 2\frac{1}{2}$

54. $9 \times 3\frac{1}{3}$

55. $2\frac{1}{7} \times 3$

56. $5\frac{1}{4} \times 8$

57. $3\frac{2}{3} \times 5$

58. $4\frac{2}{9} \times 3$

59. $\frac{1}{2} \times 3\frac{3}{7}$

60. $\frac{3}{8} \times 4\frac{4}{5}$

61. $6\frac{1}{8} \times \frac{4}{7}$

62. $5\frac{1}{3} \times \frac{5}{16}$

63. $5\frac{1}{8} \times 5$

64. $6\frac{1}{9} \times 2$

65. $\frac{3}{8} \times 4\frac{1}{2}$

66. $\frac{5}{7} \times 2\frac{1}{3}$

67. $6 \times 2\frac{2}{3}$

68. $6\frac{1}{8} \times 0$

69. $1\frac{1}{3} \times 2\frac{1}{4}$

70. $2\frac{5}{8} \times \frac{3}{23}$

71. $2\frac{5}{8} \times 3\frac{2}{5}$

72. $5\frac{3}{16} \times 5\frac{1}{3}$

73. $3\frac{1}{7} \times 2\frac{1}{8}$

74. $16\frac{5}{8} \times 1\frac{1}{16}$

Copyright © Houghton Mifflin Company. All rights reserved.

75. $2\frac{2}{5} \times 3\frac{1}{12}$ **76.** $2\frac{2}{3} \times \frac{3}{20}$ **77.** $5\frac{1}{5} \times 3\frac{1}{13}$ **78.** $3\frac{3}{4} \times 2\frac{3}{20}$

79. $10\frac{1}{4} \times 3\frac{1}{5}$ **80.** $12\frac{3}{5} \times 1\frac{3}{7}$ **81.** $5\frac{3}{7} \times 5\frac{1}{4}$ **82.** $6\frac{1}{2} \times 1\frac{3}{13}$

83. Multiply $2\frac{1}{2}$ and $3\frac{3}{5}$.

84. Multiply $4\frac{3}{8}$ and $3\frac{3}{5}$.

85. Find the product of $2\frac{1}{8}$ and $\frac{5}{17}$.

86. Find the product of $12\frac{2}{5}$ and $3\frac{7}{31}$.

87. What is $1\frac{3}{8}$ times $2\frac{1}{5}$?

88. What is $3\frac{1}{8}$ times $2\frac{4}{7}$?

Objective C   **To solve application problems**

89. Consumerism   Salmon costs \$4 per pound. Find the cost of $2\frac{3}{4}$ pounds of salmon.

90. Exercise   Maria Rivera can walk $3\frac{1}{2}$ miles in 1 hour. At this rate, how far can Maria walk in $\frac{1}{3}$ hour?

91. Carpentry   A board that costs \$6 is $9\frac{1}{4}$ feet long. One-third of the board is cut off.

   a. Without doing the calculation, is the piece being cut off at least 4 feet long?
   b. What is the length of the piece cut off?

92. Geometry   The perimeter of a square is equal to four times the length of a side of the square. Find the perimeter of a square whose side measures $16\frac{3}{4}$ inches.

93. Geometry   To find the area of a square, multiply the length of one side of the square times itself. What is the area of a square whose side measures $5\frac{1}{4}$ feet? The area of the square will be in square feet.

94. Geometry   The area of a rectangle is equal to the product of the length of the rectangle times its width. Find the area of a rectangle that has a length of $4\frac{2}{5}$ miles and a width of $3\frac{3}{10}$ miles. The area will be in square miles.

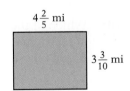

95. Finances   A family budgets $\frac{2}{5}$ of its monthly income of \$4200 per month for housing and utilities.
   a. What amount is budgeted for housing and utilities?
   b. What amount remains for purposes other than housing and utilities?

Copyright © Houghton Mifflin Company. All rights reserved.

96. Education $\frac{5}{6}$ of a chemistry class of 36 students has passing grades. $\frac{1}{5}$ of the students with passing grades received an A.
a. How many students passed the chemistry course?
b. How many students received A grades?

97. Sewing The Booster Club is making 22 capes for the members of the high school marching band. Each cape is $1\frac{3}{8}$ yards of material at a cost of $12 per yard. Find the total cost of the material.

Measurement The table at the right shows the length of steel rods and their weight per foot. Use this table for Exercises 98 to 100.

98. Find the weight of the $6\frac{1}{2}$-foot steel rod.

99. Find the weight of the $12\frac{7}{12}$-foot steel rod.

100. Find the total weight of the $8\frac{5}{8}$-foot and the $10\frac{3}{4}$-foot steel rods.

Length (ft)	Weight (lb/ft)
$6\frac{1}{2}$	$\frac{3}{8}$
$8\frac{5}{8}$	$1\frac{1}{4}$
$10\frac{3}{4}$	$2\frac{1}{2}$
$12\frac{7}{12}$	$4\frac{1}{3}$

101. Investments The manager of a mutual fund has $\frac{1}{2}$ of the portfolio invested in bonds. Of the amount invested in bonds, $\frac{3}{8}$ is invested in corporate bonds. What fraction of the total portfolio is invested in corporate bonds?

APPLYING THE CONCEPTS

102. The product of 1 and a number is $\frac{1}{2}$. Find the number.

103. **Time** Our calendar is based on the solar year, which is $365\frac{1}{4}$ days. Use this fact to explain leap years.

104. Is the product of two positive fractions always greater than either one of the two numbers? If so, explain why. If not, give an example.

105. Which of the labeled points on the number line at the right could be the graph of the product of B and C?

$\begin{array}{ccccccccc} \vdash & \bullet & \bullet & \bullet & + & \bullet & + & \bullet & \dashv \\ 0 & A & B & C & 1 & D & 2 & E & 3 \end{array}$

106. Fill in the circles on the square at the right with the fractions $\frac{1}{6}$, $\frac{5}{18}$, $\frac{4}{9}$, $\frac{5}{9}$, $\frac{2}{3}$, $\frac{3}{4}$, $1\frac{1}{9}$, $1\frac{1}{2}$, and $2\frac{1}{4}$ so that the product of any row is equal to $\frac{5}{18}$. (*Note:* There is more than one possible answer.)

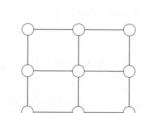

Copyright © Houghton Mifflin Company. All rights reserved.

2.7 Division of Fractions and Mixed Numbers

Objective A **To divide fractions**

The **reciprocal of a fraction** is the fraction with the numerator and denominator interchanged.

The reciprocal of $\frac{2}{3}$ is $\frac{3}{2}$.

The process of interchanging the numerator and denominator is called **inverting a fraction.**

To find the reciprocal of a whole number, first write the whole number as a fraction with a denominator of 1; then find the reciprocal of that fraction.

The reciprocal of 5 is $\frac{1}{5}$. $\left(\text{Think } 5 = \frac{5}{1}.\right)$

Reciprocals are used to rewrite division problems as related multiplication problems. Look at the following two problems:

$$8 \div 2 = 4 \qquad\qquad 8 \times \frac{1}{2} = 4$$

8 divided by 2 is **4.**   8 times the reciprocal of 2 is **4.**

"Divided by" means the same as "times the reciprocal of." Thus "÷ 2" can be replaced with "$\times \frac{1}{2}$," and the answer will be the same. Fractions are divided by making this replacement.

> **HOW TO** Divide: $\frac{2}{3} \div \frac{3}{4}$
>
> $$\frac{2}{3} \div \frac{3}{4} = \frac{2}{3} \times \frac{4}{3} = \frac{2 \cdot 4}{3 \cdot 3} = \frac{2 \cdot 2 \cdot 2}{3 \cdot 3} = \frac{8}{9}$$ • Multiply the first fraction by the reciprocal of the second fraction.

Example 1 Divide: $\frac{5}{8} \div \frac{4}{9}$

Solution
$$\frac{5}{8} \div \frac{4}{9} = \frac{5}{8} \times \frac{9}{4} = \frac{5 \cdot 9}{8 \cdot 4}$$
$$= \frac{5 \cdot 3 \cdot 3}{2 \cdot 2 \cdot 2 \cdot 2 \cdot 2} = \frac{45}{32} = 1\frac{13}{32}$$

You Try It 1 Divide: $\frac{3}{7} \div \frac{2}{3}$

Your solution

Example 2 Divide: $\frac{3}{5} \div \frac{12}{25}$

Solution
$$\frac{3}{5} \div \frac{12}{25} = \frac{3}{5} \times \frac{25}{12} = \frac{3 \cdot 25}{5 \cdot 12}$$
$$= \frac{\overset{1}{\cancel{3}} \cdot \overset{1}{\cancel{5}} \cdot 5}{\underset{1}{\cancel{5}} \cdot 2 \cdot 2 \cdot \underset{1}{\cancel{3}}} = \frac{5}{4} = 1\frac{1}{4}$$

You Try It 2 Divide: $\frac{3}{4} \div \frac{9}{10}$

Your solution

Solutions on p. S7

Copyright © Houghton Mifflin Company. All rights reserved.

Objective B   **To divide whole numbers, mixed numbers, and fractions**

To divide a fraction and a whole number, first write the whole number as a fraction with a denominator of 1.

> **HOW TO** Divide: $\dfrac{3}{7} \div 5$
>
> $$\dfrac{3}{7} \div \boxed{5} = \dfrac{3}{7} \div \boxed{\dfrac{5}{1}} = \dfrac{3}{7} \times \dfrac{1}{5} = \dfrac{3 \cdot 1}{7 \cdot 5} = \dfrac{3}{35}$$
>
> • Write 5 with a denominator of 1. Then divide the fractions.

When a number in a quotient is a mixed number, write the mixed number as an improper fraction before dividing.

> **HOW TO** Divide: $1\dfrac{13}{15} \div 4\dfrac{4}{5}$
>
> Write the mixed numbers as improper fractions. Then divide the fractions.
>
> $$1\dfrac{13}{15} \div 4\dfrac{4}{5} = \dfrac{28}{15} \div \dfrac{24}{5} = \dfrac{28}{15} \times \dfrac{5}{24} = \dfrac{28 \cdot 5}{15 \cdot 24} = \dfrac{\overset{1}{2} \cdot \overset{1}{2} \cdot 7 \cdot \overset{1}{\cancel{5}}}{3 \cdot \underset{1}{\cancel{5}} \cdot \underset{1}{2} \cdot \underset{1}{2} \cdot 2 \cdot 3} = \dfrac{7}{18}$$

Example 3 Divide $\dfrac{4}{9}$ by 5.

Solution

$$\dfrac{4}{9} \div 5 = \dfrac{4}{9} \div \dfrac{5}{1} = \dfrac{4}{9} \times \dfrac{1}{5}$$

• $5 = \dfrac{5}{1}$. The reciprocal of $\dfrac{5}{1}$ is $\dfrac{1}{5}$.

$$= \dfrac{4 \cdot 1}{9 \cdot 5} = \dfrac{2 \cdot 2}{3 \cdot 3 \cdot 5} = \dfrac{4}{45}$$

You Try It 3 Divide $\dfrac{5}{7}$ by 6.

Your solution

Example 4 Find the quotient of $\dfrac{3}{8}$ and $2\dfrac{1}{10}$.

Solution

$$\dfrac{3}{8} \div 2\dfrac{1}{10} = \dfrac{3}{8} \div \dfrac{21}{10} = \dfrac{3}{8} \times \dfrac{10}{21}$$

$$= \dfrac{3 \cdot 10}{8 \cdot 21} = \dfrac{\overset{1}{\cancel{3}} \cdot \overset{1}{\cancel{2}} \cdot 5}{\underset{1}{\cancel{2}} \cdot 2 \cdot 2 \cdot \underset{1}{\cancel{3}} \cdot 7} = \dfrac{5}{28}$$

You Try It 4 Find the quotient of $12\dfrac{3}{5}$ and 7.

Your solution

Example 5 Divide: $2\dfrac{3}{4} \div 1\dfrac{5}{7}$

Solution

$$2\dfrac{3}{4} \div 1\dfrac{5}{7} = \dfrac{11}{4} \div \dfrac{12}{7} = \dfrac{11}{4} \times \dfrac{7}{12} = \dfrac{11 \cdot 7}{4 \cdot 12}$$

$$= \dfrac{11 \cdot 7}{2 \cdot 2 \cdot 2 \cdot 2 \cdot 3} = \dfrac{77}{48} = 1\dfrac{29}{48}$$

You Try It 5 Divide: $3\dfrac{2}{3} \div 2\dfrac{2}{5}$

Your solution

Solutions on p. S7

Copyright © Houghton Mifflin Company. All rights reserved.

Example 6   Divide: $1\frac{13}{15} \div 4\frac{1}{5}$

Solution

$$1\frac{13}{15} \div 4\frac{1}{5} = \frac{28}{15} \div \frac{21}{5} = \frac{28}{15} \times \frac{5}{21} = \frac{28 \cdot 5}{15 \cdot 21}$$

$$= \frac{2 \cdot 2 \cdot \overset{1}{\cancel{7}} \cdot \overset{1}{\cancel{5}}}{3 \cdot \underset{1}{\cancel{5}} \cdot 3 \cdot \underset{1}{\cancel{7}}} = \frac{4}{9}$$

You Try It 6   Divide: $2\frac{5}{6} \div 8\frac{1}{2}$

Your solution

Example 7   Divide: $4\frac{3}{8} \div 7$

Solution

$$4\frac{3}{8} \div 7 = \frac{35}{8} \div \frac{7}{1} = \frac{35}{8} \times \frac{1}{7}$$

$$= \frac{35 \cdot 1}{8 \cdot 7} = \frac{5 \cdot \overset{1}{\cancel{7}}}{2 \cdot 2 \cdot 2 \cdot \underset{1}{\cancel{7}}} = \frac{5}{8}$$

You Try It 7   Divide: $6\frac{2}{5} \div 4$

Your solution

Solutions on p. S7

Objective C   **To solve application problems**

Example 8

A car used $15\frac{1}{2}$ gallons of gasoline on a 310-mile trip. How many miles can this car travel on 1 gallon of gasoline?

Strategy
To find the number of miles, divide the number of miles traveled by the number of gallons of gasoline used.

Solution

$$310 \div 15\frac{1}{2} = \frac{310}{1} \div \frac{31}{2}$$

$$= \frac{310}{1} \times \frac{2}{31} = \frac{310 \cdot 2}{1 \cdot 31}$$

$$= \frac{2 \cdot 5 \cdot \overset{1}{\cancel{31}} \cdot 2}{1 \cdot \underset{1}{\cancel{31}}} = \frac{20}{1} = 20$$

The car travels 20 miles on 1 gallon of gasoline.

You Try It 8

A factory worker can assemble a product in $7\frac{1}{2}$ minutes. How many products can the worker assemble in 1 hour?

Your strategy

Your solution

Solution on p. S7

Copyright © Houghton Mifflin Company. All rights reserved.

Example 9

A 12-foot board is cut into pieces $2\frac{1}{4}$ feet long for use as bookshelves. What is the length of the remaining piece after as many shelves as possible have been cut?

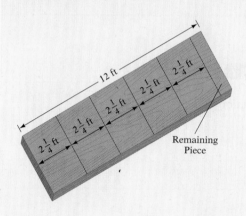

Strategy

To find the length of the remaining piece:

- Divide the total length of the board (12) by the length of each shelf $\left(2\frac{1}{4}\right)$. This will give you the number of shelves cut, with a certain fraction of a shelf left over.
- Multiply the fractional part of the result in step 1 by the length of one shelf to determine the length of the remaining piece.

Solution

$$12 \div 2\frac{1}{4} = \frac{12}{1} \div \frac{9}{4} = \frac{12}{1} \times \frac{4}{9}$$

$$= \frac{12 \cdot 4}{1 \cdot 9} = \frac{16}{3} = 5\frac{1}{3}$$

There are 5 pieces $2\frac{1}{4}$ feet long.

There is 1 piece that is $\frac{1}{3}$ of $2\frac{1}{4}$ feet long.

$$\frac{1}{3} \times 2\frac{1}{4} = \frac{1}{3} \times \frac{9}{4} = \frac{1 \cdot 9}{3 \cdot 4} = \frac{3}{4}$$

The length of the piece remaining is $\frac{3}{4}$ foot.

You Try It 9

A 16-foot board is cut into pieces $3\frac{1}{3}$ feet long for shelves for a bookcase. What is the length of the remaining piece after as many shelves as possible have been cut?

Your strategy

Your solution

Solution on p. S8

Copyright © Houghton Mifflin Company. All rights reserved.

2.7 Exercises

Objective A   **To divide fractions**

For Exercises 1 to 32, divide.

1. $\dfrac{1}{3} \div \dfrac{2}{5}$

2. $\dfrac{3}{7} \div \dfrac{3}{2}$

3. $\dfrac{3}{7} \div \dfrac{3}{7}$

4. $0 \div \dfrac{1}{2}$

5. $0 \div \dfrac{3}{4}$

6. $\dfrac{16}{33} \div \dfrac{4}{11}$

7. $\dfrac{5}{24} \div \dfrac{15}{36}$

8. $\dfrac{11}{15} \div \dfrac{1}{12}$

9. $\dfrac{15}{16} \div \dfrac{16}{39}$

10. $\dfrac{2}{15} \div \dfrac{3}{5}$

11. $\dfrac{8}{9} \div \dfrac{4}{5}$

12. $\dfrac{11}{15} \div \dfrac{5}{22}$

13. $\dfrac{1}{9} \div \dfrac{2}{3}$

14. $\dfrac{10}{21} \div \dfrac{5}{7}$

15. $\dfrac{2}{5} \div \dfrac{4}{7}$

16. $\dfrac{3}{8} \div \dfrac{5}{12}$

17. $\dfrac{1}{2} \div \dfrac{1}{4}$

18. $\dfrac{1}{3} \div \dfrac{1}{9}$

19. $\dfrac{1}{5} \div \dfrac{1}{10}$

20. $\dfrac{4}{15} \div \dfrac{2}{5}$

21. $\dfrac{7}{15} \div \dfrac{14}{5}$

22. $\dfrac{5}{8} \div \dfrac{15}{2}$

23. $\dfrac{14}{3} \div \dfrac{7}{9}$

24. $\dfrac{7}{4} \div \dfrac{9}{2}$

25. $\dfrac{5}{9} \div \dfrac{25}{3}$

26. $\dfrac{5}{16} \div \dfrac{3}{8}$

27. $\dfrac{2}{3} \div \dfrac{1}{3}$

28. $\dfrac{4}{9} \div \dfrac{1}{9}$

29. $\dfrac{5}{7} \div \dfrac{2}{7}$

30. $\dfrac{5}{6} \div \dfrac{1}{9}$

31. $\dfrac{2}{3} \div \dfrac{2}{9}$

32. $\dfrac{5}{12} \div \dfrac{5}{6}$

33. Divide $\dfrac{7}{8}$ by $\dfrac{3}{4}$.

34. Divide $\dfrac{7}{12}$ by $\dfrac{3}{4}$.

35. Find the quotient of $\dfrac{5}{7}$ and $\dfrac{3}{14}$.

36. Find the quotient of $\dfrac{6}{11}$ and $\dfrac{9}{32}$.

Copyright © Houghton Mifflin Company. All rights reserved.

> **Objective B**   **To divide whole numbers, mixed numbers, and fractions**

For Exercises 37 to 81, divide.

37. $4 \div \dfrac{2}{3}$

38. $\dfrac{2}{3} \div 4$

39. $\dfrac{3}{2} \div 3$

40. $3 \div \dfrac{3}{2}$

41. $\dfrac{5}{6} \div 25$

42. $22 \div \dfrac{3}{11}$

43. $6 \div 3\dfrac{1}{3}$

44. $5\dfrac{1}{2} \div 11$

45. $6\dfrac{1}{2} \div \dfrac{1}{2}$

46. $\dfrac{3}{8} \div 2\dfrac{1}{4}$

47. $\dfrac{5}{12} \div 4\dfrac{4}{5}$

48. $1\dfrac{1}{2} \div 1\dfrac{3}{8}$

49. $8\dfrac{1}{4} \div 2\dfrac{3}{4}$

50. $3\dfrac{5}{9} \div 32$

51. $4\dfrac{1}{5} \div 21$

52. $6\dfrac{8}{9} \div \dfrac{31}{36}$

53. $\dfrac{11}{12} \div 2\dfrac{1}{3}$

54. $\dfrac{7}{8} \div 3\dfrac{1}{4}$

55. $\dfrac{5}{16} \div 5\dfrac{3}{8}$

56. $\dfrac{9}{14} \div 3\dfrac{1}{7}$

57. $35 \div \dfrac{7}{24}$

58. $\dfrac{3}{8} \div 2\dfrac{3}{4}$

59. $\dfrac{11}{18} \div 2\dfrac{2}{9}$

60. $\dfrac{21}{40} \div 3\dfrac{3}{10}$

61. $2\dfrac{1}{16} \div 2\dfrac{1}{2}$

62. $7\dfrac{3}{5} \div 1\dfrac{7}{12}$

63. $1\dfrac{2}{3} \div \dfrac{3}{8}$

64. $16 \div \dfrac{2}{3}$

65. $1\dfrac{5}{8} \div 4$

66. $13\dfrac{3}{8} \div \dfrac{1}{4}$

67. $16 \div 1\dfrac{1}{2}$

68. $9 \div \dfrac{7}{8}$

69. $16\dfrac{5}{8} \div 1\dfrac{2}{3}$

70. $24\dfrac{4}{5} \div 2\dfrac{3}{5}$

71. $1\dfrac{1}{3} \div 5\dfrac{8}{9}$

72. $13\dfrac{2}{3} \div 0$

Copyright © Houghton Mifflin Company. All rights reserved.

73. $82\frac{3}{5} \div 19\frac{1}{10}$

74. $45\frac{3}{5} \div 15$

75. $102 \div 1\frac{1}{2}$

76. $0 \div 3\frac{1}{2}$

77. $8\frac{2}{7} \div 1$

78. $6\frac{9}{16} \div 1\frac{3}{32}$

79. $8\frac{8}{9} \div 2\frac{13}{18}$

80. $10\frac{1}{5} \div 1\frac{7}{10}$

81. $7\frac{3}{8} \div 1\frac{27}{32}$

82. Divide $7\frac{7}{9}$ by $5\frac{5}{6}$.

83. Divide $2\frac{3}{4}$ by $1\frac{23}{32}$.

84. Find the quotient of $8\frac{1}{4}$ and $1\frac{5}{11}$.

85. Find the quotient of $\frac{14}{17}$ and $3\frac{1}{9}$.

Objective C **To solve application problems**

86. **Consumerism** Individual cereal boxes contain $\frac{3}{4}$ ounce of cereal. How many boxes can be filled with 600 ounces of cereal?

87. **Consumerism** A box of Post's Great Grains cereal costing $4 contains 16 ounces of cereal. How many $1\frac{1}{3}$-ounce portions can be served from this box?

88. **Gemology** A $\frac{5}{8}$-karat diamond was purchased for $1200. What would a similar diamond weighing 1 karat cost?

89. **Real Estate** The Inverness Investor Group bought $8\frac{1}{3}$ acres of land for $200,000. What was the cost of each acre?

90. **Fuel Efficiency** A car used $12\frac{1}{2}$ gallons of gasoline on a 275-mile trip. How many miles can their car travel on 1 gallon of gasoline?

91. **Mechanics** A nut moves $\frac{5}{32}$ inch for each turn. Find the number of turns it will take for the nut to move $1\frac{7}{8}$ inches.

92. **Real Estate** The Hammond Company purchased $9\frac{3}{4}$ acres for a housing project. One and one-half acres were set aside for a park.
  a. How many acres are available for housing?
  b. How many $\frac{1}{4}$-acre parcels of land can be sold after the land for the park is set aside?

Copyright © Houghton Mifflin Company. All rights reserved.

93.   The Food Industry   A chef purchased a roast that weighed $10\frac{3}{4}$ pounds. After the fat was trimmed and the bone removed, the roast weighed $9\frac{1}{3}$ pounds.

    a. What was the total weight of the fat and bone?

    b. How many $\frac{1}{3}$-pound servings can be cut from the trimmed roast?

94.   Carpentry   A 15-foot board is cut into pieces $3\frac{1}{2}$ feet long for a bookcase. What is the length of the piece remaining after as many shelves as possible have been cut?

95.   Architecture   A scale of $\frac{1}{2}$ inch to 1 foot is used to draw the plans for a house. The scale measurements for three walls are given in the table at the right. Complete the table to determine the actual wall lengths for the three walls a, b, and c.

Wall	Scale	Actual Wall Length
a	$6\frac{1}{4}$ in.	?
b	9 in.	?
c	$7\frac{7}{8}$ in.	?

APPLYING THE CONCEPTS

Loans   The figure at the right shows how the money borrowed on home equity loans is spent. Use this graph for Exercises 96 and 97.

96.   What fractional part of the money borrowed on home equity loans is spent on debt consolidation and home improvement?

97.   What fractional part of the money borrowed on home equity loans is spent on home improvement, cars, and tuition?

Real Estate · Auto Purchase · Tuition

$\frac{1}{25}$ · $\frac{1}{20}$ · $\frac{1}{20}$

Debt Consolidation $\frac{19}{50}$

Home Improvement $\frac{6}{25}$

Other $\frac{6}{25}$

How Money Borrowed on Home Equity Loans Is Spent

Source: Consumer Bankers Association

98.   Finances   A bank recommends that the maximum monthly payment for a home be $\frac{1}{3}$ of your total monthly income. Your monthly income is $4500. What would the bank recommend as your maximum monthly house payment?

99.   Entertainment   On Friday evening, 1200 people attended a concert at the music center. The center was $\frac{2}{3}$ full. What is the capacity of the music center?

100.   Sports   During the second half of the 1900s, greenskeepers mowed the grass on golf putting surfaces progressively lower. The table at the right shows the average grass height by decade. What was the difference between the average height of the grass in the 1980s and its average height in the 1950s?

Average Height of Grass on Golf Putting Surfaces	
Decade	Height (in inches)
1950s	$\frac{1}{4}$
1960s	$\frac{7}{32}$
1970s	$\frac{3}{16}$
1980s	$\frac{5}{32}$
1990s	$\frac{1}{8}$

Source: Golf Course Superintendents Association of America

Copyright © Houghton Mifflin Company. All rights reserved.

101. Puzzles   You completed $\frac{1}{3}$ of a jigsaw puzzle yesterday and $\frac{1}{2}$ of the puzzle today. What fraction of the puzzle is left to complete?

102. Board Games   A wooden travel game board has hinges that allow the board to be folded in half. If the dimensions of the open board are 14 inches by 14 inches by $\frac{7}{8}$ inch, what are the dimensions of the board when it is closed?

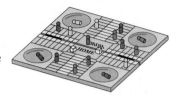

103. Wages   You have a part-time job that pays $9 an hour. You worked 5 hours, $3\frac{3}{4}$ hours, $1\frac{1}{4}$ hours, and $2\frac{1}{3}$ hours during the four days you worked last week. Find your total earnings for last week's work.

Nutrition   According to the Center for Science in the Public Interest, the average teenage boy drinks $3\frac{1}{3}$ cans of soda per day. The average teenage girl drinks $2\frac{1}{3}$ cans of soda per day. Use this information for Exercises 104 to 106.

104. The average teenage boy drinks how many cans of soda per week?

105. If a can of soda contains 150 calories, how many calories does the average teenage boy consume each week in soda?

106. How many more cans of soda per week does the average teenage boy drink than the average teenage girl?

107. Maps   On a map, two cities are $4\frac{5}{8}$ inches apart. If $\frac{3}{8}$ inch on the map represents 60 miles, what is the number of miles between the two cities?

108. Is the quotient always less than the dividend in a division problem? Explain.

109. Fill in the box to make a true statement.

   a. $\frac{3}{4} \cdot \boxed{} = \frac{1}{2}$ **b.** $\frac{2}{3} \cdot \boxed{} = 1\frac{3}{4}$

110. Publishing   A page of type in a certain textbook is $7\frac{1}{2}$ inches wide. If the page is divided into three equal columns, with $\frac{3}{8}$ inch between columns, how wide is each column?

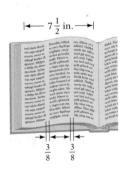

111. A whole number is both multiplied and divided by the same proper fraction. What is greater, the product or the quotient?

Copyright © Houghton Mifflin Company. All rights reserved.

2.8 Order, Exponents, and the Order of Operations Agreement

Copyright © Houghton Mifflin Company. All rights reserved.

Objective A To identify the order relation between two fractions

Point of Interest

Leonardo of Pisa, who was also called Fibonacci (c. 1175 –1250), is credited with bringing the Hindu-Arabic number system to the Western world and promoting its use in place of the cumbersome Roman numeral system.

He was also influential in promoting the idea of the fraction bar. His notation, however, was very different from what we have today.

For instance, he wrote $\dfrac{3}{4}\dfrac{5}{7}$

to mean $\dfrac{5}{7} + \dfrac{3}{7 \cdot 4}$, which

equals $\dfrac{23}{28}$.

Recall that whole numbers can be graphed as points on the number line. Fractions can also be graphed as points on the number line.

The graph of $\dfrac{3}{4}$ on the number line

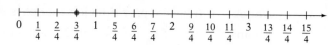

The number line can be used to determine the order relation between two fractions. A fraction that appears to the left of a given fraction is less than the given fraction. A fraction that appears to the right of a given fraction is greater than the given fraction.

$\dfrac{1}{8} < \dfrac{3}{8}$ $\dfrac{6}{8} > \dfrac{3}{8}$

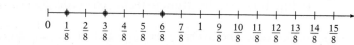

To find the order relation between two fractions with the same denominator, compare the numerators. The fraction that has the smaller numerator is the smaller fraction. When the denominators are different, begin by writing equivalent fractions with a common denominator; then compare numerators.

HOW TO Find the order relation between $\dfrac{11}{18}$ and $\dfrac{5}{8}$.

The LCM of 18 and 8 is 72.

$\dfrac{11}{18} = \dfrac{44}{72} \leftarrow$ Smaller numerator $\dfrac{11}{18} < \dfrac{5}{8}$ or $\dfrac{5}{8} > \dfrac{11}{18}$

$\dfrac{5}{8} = \dfrac{45}{72} \leftarrow$ Larger numerator

Example 1 Place the correct symbol, $<$ or $>$, between the two numbers.

$\dfrac{5}{12} \quad \dfrac{7}{18}$

Solution $\dfrac{5}{12} = \dfrac{15}{36}$ $\dfrac{7}{18} = \dfrac{14}{36}$

$\dfrac{5}{12} > \dfrac{7}{18}$

You Try It 1 Place the correct symbol, $<$ or $>$, between the two numbers.

$\dfrac{9}{14} \quad \dfrac{13}{21}$

Your solution

Solution on p. S8

Objective B To simplify expressions containing exponents

Repeated multiplication of the same fraction can be written in two ways:

$\dfrac{1}{2} \cdot \dfrac{1}{2} \cdot \dfrac{1}{2} \cdot \dfrac{1}{2}$ or $\left(\dfrac{1}{2}\right)^{4} \leftarrow$ Exponent

The exponent indicates how many times the fraction occurs as a factor in the multiplication. The expression $\left(\dfrac{1}{2}\right)^{4}$ is in exponential notation.

Example 2 Simplify: $\left(\frac{5}{6}\right)^3 \cdot \left(\frac{3}{5}\right)^2$

Solution

$$\left(\frac{5}{6}\right)^3 \cdot \left(\frac{3}{5}\right)^2 = \left(\frac{5}{6} \cdot \frac{5}{6} \cdot \frac{5}{6}\right) \cdot \left(\frac{3}{5} \cdot \frac{3}{5}\right)$$

$$= \frac{\overset{1}{\cancel{5}} \cdot \overset{1}{\cancel{5}} \cdot 5 \cdot \overset{1}{\cancel{3}} \cdot \overset{1}{\cancel{3}}}{2 \cdot \underset{1}{\cancel{3}} \cdot 2 \cdot \underset{1}{\cancel{3}} \cdot 2 \cdot 3 \cdot \underset{1}{\cancel{5}} \cdot \underset{1}{\cancel{5}}} = \frac{5}{24}$$

You Try It 2 Simplify: $\left(\frac{7}{11}\right)^2 \cdot \left(\frac{2}{7}\right)$

Your solution

Solution on p. S8

Objective C   **To use the Order of Operations Agreement to simplify expressions**

The Order of Operations Agreement is used for fractions as well as whole numbers.

> **The Order of Operations Agreement**
>
> **Step 1.** Do all the operations inside parentheses.
> **Step 2.** Simplify any number expressions containing exponents.
> **Step 3.** Do multiplications and divisions as they occur from left to right.
> **Step 4.** Do additions and subtractions as they occur from left to right.

HOW TO Simplify $\frac{14}{15} - \left(\frac{1}{2}\right)^2 \times \left(\frac{2}{3} + \frac{4}{5}\right)$.

$$\frac{14}{15} - \left(\frac{1}{2}\right)^2 \times \left(\frac{2}{3} + \frac{4}{5}\right)$$

1. Perform operations in parentheses.

$$\frac{14}{15} - \left(\frac{1}{2}\right)^2 \times \underbrace{\frac{22}{15}}$$

2. Simplify expressions with exponents.

$$\frac{14}{15} - \underbrace{\frac{1}{4} \times \frac{22}{15}}$$

3. Do multiplication and division as they occur from left to right.

$$\frac{14}{15} - \underbrace{\frac{11}{30}}$$

4. Do addition and subtraction as they occur from left to right.

$$\frac{17}{30}$$

One or more of the above steps may not be needed to simplify an expression. In that case, proceed to the next step in the Order of Operations agreement.

Example 3 Simplify: $\left(\frac{3}{4}\right)^2 \div \left(\frac{3}{8} - \frac{1}{12}\right)$

Solution

$$\left(\frac{3}{4}\right)^2 \div \left(\frac{3}{8} - \frac{1}{12}\right)$$

$$= \left(\frac{3}{4}\right)^2 \div \left(\frac{7}{24}\right) = \frac{9}{16} \div \frac{7}{24}$$

$$= \frac{9}{16} \cdot \frac{24}{7} = \frac{27}{14} = 1\frac{13}{14}$$

You Try It 3 Simplify: $\left(\frac{1}{13}\right)^2 \cdot \left(\frac{1}{4} + \frac{1}{6}\right) \div \frac{5}{13}$

Your solution

Solution on p. S8

Copyright © Houghton Mifflin Company. All rights reserved.

2.8 Exercises

Objective A **To identify the order relation between two fractions**

For Exercises 1 to 12, place the correct symbol, $<$ or $>$, between the two numbers.

1. $\dfrac{11}{40} \quad \dfrac{19}{40}$

2. $\dfrac{92}{103} \quad \dfrac{19}{103}$

3. $\dfrac{2}{3} \quad \dfrac{5}{7}$

4. $\dfrac{2}{5} \quad \dfrac{3}{8}$

5. $\dfrac{5}{8} \quad \dfrac{7}{12}$

6. $\dfrac{11}{16} \quad \dfrac{17}{24}$

7. $\dfrac{7}{9} \quad \dfrac{11}{12}$

8. $\dfrac{5}{12} \quad \dfrac{7}{15}$

9. $\dfrac{13}{14} \quad \dfrac{19}{21}$

10. $\dfrac{13}{18} \quad \dfrac{7}{12}$

11. $\dfrac{7}{24} \quad \dfrac{11}{30}$

12. $\dfrac{13}{36} \quad \dfrac{19}{48}$

Objective B **To simplify expressions containing exponents**

For Exercises 13 to 28, simplify.

13. $\left(\dfrac{3}{8}\right)^2$

14. $\left(\dfrac{5}{12}\right)^2$

15. $\left(\dfrac{2}{9}\right)^3$

16. $\left(\dfrac{1}{2}\right) \cdot \left(\dfrac{2}{3}\right)^2$

17. $\left(\dfrac{2}{3}\right) \cdot \left(\dfrac{1}{2}\right)^4$

18. $\left(\dfrac{1}{3}\right)^2 \cdot \left(\dfrac{3}{5}\right)^3$

19. $\left(\dfrac{2}{5}\right)^3 \cdot \left(\dfrac{5}{7}\right)^2$

20. $\left(\dfrac{5}{9}\right)^3 \cdot \left(\dfrac{18}{25}\right)^2$

21. $\left(\dfrac{1}{3}\right)^4 \cdot \left(\dfrac{9}{11}\right)^2$

22. $\left(\dfrac{1}{2}\right)^6 \cdot \left(\dfrac{32}{35}\right)^2$

23. $\left(\dfrac{2}{3}\right)^4 \cdot \left(\dfrac{81}{100}\right)^2$

24. $\left(\dfrac{1}{6}\right) \cdot \left(\dfrac{6}{7}\right)^2 \cdot \left(\dfrac{2}{3}\right)$

25. $\left(\dfrac{2}{7}\right) \cdot \left(\dfrac{7}{8}\right)^2 \cdot \left(\dfrac{8}{9}\right)$

26. $3 \cdot \left(\dfrac{3}{5}\right)^3 \cdot \left(\dfrac{1}{3}\right)^2$

27. $4 \cdot \left(\dfrac{3}{4}\right)^3 \cdot \left(\dfrac{4}{7}\right)^2$

28. $11 \cdot \left(\dfrac{3}{8}\right)^3 \cdot \left(\dfrac{8}{11}\right)^2$

Objective C **To use the Order of Operations Agreement to simplify expressions**

For Exercises 29 to 47, simplify.

29. $\dfrac{1}{2} - \dfrac{1}{3} + \dfrac{2}{3}$

30. $\dfrac{2}{5} + \dfrac{3}{10} - \dfrac{2}{3}$

31. $\dfrac{1}{3} \div \dfrac{1}{2} + \dfrac{3}{4}$

32. $\dfrac{4}{5} + \dfrac{3}{7} \cdot \dfrac{14}{15}$

Copyright © Houghton Mifflin Company. All rights reserved.

33. $\left(\dfrac{3}{4}\right)^2 - \dfrac{5}{12}$

34. $\left(\dfrac{3}{5}\right)^3 - \dfrac{3}{25}$

35. $\dfrac{5}{6} \cdot \left(\dfrac{2}{3} - \dfrac{1}{6}\right) + \dfrac{7}{18}$

36. $\dfrac{3}{4} \cdot \left(\dfrac{11}{12} - \dfrac{7}{8}\right) + \dfrac{5}{16}$

37. $\dfrac{7}{12} - \left(\dfrac{2}{3}\right)^2 + \dfrac{5}{8}$

38. $\dfrac{11}{16} - \left(\dfrac{3}{4}\right)^2 + \dfrac{7}{12}$

39. $\dfrac{3}{4} \cdot \left(\dfrac{4}{9}\right)^2 + \dfrac{1}{2}$

40. $\dfrac{9}{10} \cdot \left(\dfrac{2}{3}\right)^3 + \dfrac{2}{3}$

41. $\left(\dfrac{1}{2} + \dfrac{3}{4}\right) \div \dfrac{5}{8}$

42. $\left(\dfrac{2}{3} + \dfrac{5}{6}\right) \div \dfrac{5}{9}$

43. $\dfrac{3}{8} \div \left(\dfrac{5}{12} + \dfrac{3}{8}\right)$

44. $\dfrac{7}{12} \div \left(\dfrac{2}{3} + \dfrac{5}{9}\right)$

45. $\left(\dfrac{3}{8}\right)^2 \div \left(\dfrac{3}{7} + \dfrac{3}{14}\right)$

46. $\left(\dfrac{5}{6}\right)^2 \div \left(\dfrac{5}{12} + \dfrac{2}{3}\right)$

47. $\dfrac{2}{5} \div \dfrac{3}{8} \cdot \dfrac{4}{5}$

APPLYING THE CONCEPTS

48. **The Food Industry**   The table at the right shows the results of a survey that asked fast-food patrons their criteria for choosing where to go for fast food. For example, 3 out of every 25 people surveyed said that the speed of the service was most important.
  a. According to the survey, do more people choose a fast-food restaurant on the basis of its location or the quality of the food?
  b. Which criterion was cited by the most people?

Fast-Food Patrons' Top Criteria for Fast-Food Restaurants	
Food quality	$\dfrac{1}{4}$
Location	$\dfrac{13}{50}$
Menu	$\dfrac{4}{25}$
Price	$\dfrac{2}{25}$
Speed	$\dfrac{3}{25}$
Other	$\dfrac{3}{100}$

Source: Maritz Marketing Research, Inc.

49.   A farmer died and left 17 horses to be divided among 3 children. The first child was to receive $\dfrac{1}{2}$ of the horses, the second child $\dfrac{1}{3}$ of the horses, and the third child $\dfrac{1}{9}$ of the horses. The executor for the family's estate realized that 17 horses could not be divided by halves, thirds, or ninths and so added a neighbor's horse to the farmer's. With 18 horses, the executor gave 9 horses to the first child, 6 horses to the second child, and 2 horses to the third child. This accounted for the 17 horses, so the executor returned the borrowed horse to the neighbor. Explain why this worked.

50. $\dfrac{2}{3} < \dfrac{3}{4}$. Is $\dfrac{2+3}{3+4}$ less than $\dfrac{2}{3}$, greater than $\dfrac{2}{3}$, or between $\dfrac{2}{3}$ and $\dfrac{3}{4}$?

Copyright © Houghton Mifflin Company. All rights reserved.

Focus on Problem Solving

Common Knowledge
An application problem may not provide all the information that is needed to solve the problem. Sometimes, however, the necessary information is common knowledge.

> **HOW TO** You are traveling by bus from Boston to New York. The trip is 4 hours long. If the bus leaves Boston at 10 A.M., what time should you arrive in New York?
>
> What other information do you need to solve this problem?
>
> You need to know that, using a 12-hour clock, the hours run
>
> 10 A.M.
> 11 A.M.
> 12 P.M.
> 1 P.M.
> 2 P.M.
>
> Four hours after 10 A.M. is 2 P.M.

You should arrive in New York at 2 P.M.

> **HOW TO** You purchase a 37¢ stamp at the Post Office and hand the clerk a one-dollar bill. How much change do you receive?
>
> What information do you need to solve this problem?
>
> You need to know that there are 100¢ in one dollar.
>
> Your change is 100¢ − 37¢.
>
> 100 − 37 = 63

You receive 63¢ in change.

What information do you need to know to solve each of the following problems?

1. You sell a dozen tickets to a fundraiser. Each ticket costs $10. How much money do you collect?

2. The weekly lab period for your science course is 1 hour and 20 minutes long. Find the length of the science lab period in minutes.

3. An employee's monthly salary is $3750. Find the employee's annual salary.

4. A survey revealed that eighth graders spend an average of 3 hours each day watching television. Find the total time an eighth grader spends watching TV each week.

5. You want to buy a carpet for a room that is 15 feet wide and 18 feet long. Find the amount of carpet that you need.

Copyright © Houghton Mifflin Company. All rights reserved.

Projects and Group Activities

Music   In musical notation, notes are printed on a **staff,** which is a set of five horizontal lines and the spaces between them. The notes of a musical composition are grouped into **measures,** or **bars.** Vertical lines separate measures on a staff. The shape of a note indicates how long it should be held. The whole note has the longest time value of any note. Each time value is divided by 2 in order to find the next smallest time value.

The **time signature** is a fraction that appears at the beginning of a piece of music. The numerator of the fraction indicates the number of beats in a measure. The denominator indicates what kind of note receives 1 beat. For example, music written in $\frac{2}{4}$ time has 2 beats to a measure, and a quarter note receives 1 beat. One measure in $\frac{2}{4}$ time may have 1 half note, 2 quarter notes, 4 eighth notes, or

any other combination of notes totaling 2 beats. Other common time signatures are $\frac{4}{4}$, $\frac{3}{4}$, and $\frac{6}{8}$.

1. Explain the meaning of the 6 and the 8 in the time signature $\frac{6}{8}$.

2. Give some possible combinations of notes in one measure of a piece written in $\frac{4}{4}$ time.

3. What does a dot at the right of a note indicate? What is the effect of a dot at the right of a half note? At the right of a quarter note? At the right of an eighth note?

4. Symbols called rests are used to indicate periods of silence in a piece of music. What symbols are used to indicate the different time values of rests?

5. Find some examples of musical compositions written in different time signatures. Use a few measures from each to show that the sum of the time values of the notes and rests in each measure equals the numerator of the time signature.

Construction   Suppose you are involved in building your own home. Design a stairway from the first floor of the house to the second floor. Some of the questions you will need to answer follow.

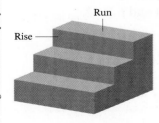

What is the distance from the floor of the first story to the floor of the second story?

Typically, what is the number of steps in a stairway?

What is a reasonable length for the run of each step?

What is the width of the wood being used to build the staircase?

In designing the stairway, remember that each riser should be the same height, that each run should be the same length, and that the width of the wood used for the steps will have to be incorporated into the calculation.

Copyright © Houghton Mifflin Company. All rights reserved.

Fractions of Diagrams The diagram that follows has been broken up into nine areas separated by heavy lines. Eight of the areas have been labeled *A* through *H*. The ninth area is shaded. Determine which lettered areas would have to be shaded so that half of the entire diagram is shaded and half is not shaded. Write down the strategy that you or your group use to arrive at the solution. Compare your strategy with that of other individual students or groups.

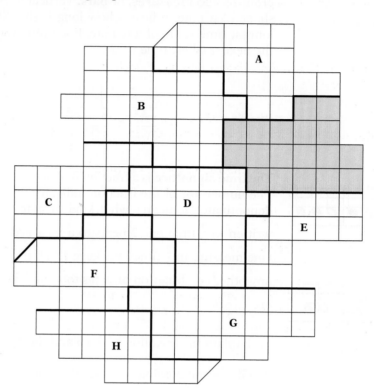

Study Tip

Three important features of this text that can be used to prepare for a test are the
• Chapter Summary
• Chapter Review Exercises
• Chapter Test
See *AIM for Success* at the front of the book.

Chapter 2 Summary

Key Words

A number that is a multiple of two or more numbers is a *common multiple* of those numbers. The *least common multiple* (*LCM*) is the smallest common multiple of two or more numbers. [2.1A, p. 65]

A number that is a factor of two or more numbers is a *common factor* of those numbers. The *greatest common factor* (*GCF*) is the largest common factor of two or more numbers. [2.1B, p. 66]

A *fraction* can represent the number of equal parts of a whole. In a fraction, the *fraction bar* separates the *numerator* and the *denominator*. [2.2A, p. 69]

Examples

12, 24, 36, 48, . . . are common multiples of 4 and 6.
The LCM of 4 and 6 is 12.

The common factors of 12 and 16 are 1, 2, and 4.
The GCF of 12 and 16 is 4.

In the fraction $\frac{3}{4}$, the numerator is 3 and the denominator is 4.

Copyright © Houghton Mifflin Company. All rights reserved.

In a *proper fraction*, the numerator is smaller than the denominator; a proper fraction is a number less than 1. In an *improper fraction*, the numerator is greater than or equal to the denominator; an improper fraction is a number greater than or equal to 1. A *mixed number* is a number greater than 1 with a whole-number part and a fractional part. [2.2A, p. 69]

$\frac{2}{5}$ is proper fraction.

$\frac{7}{6}$ is an improper fraction.

$4\frac{1}{10}$ is a mixed number; 4 is the whole-number part and $\frac{1}{10}$ is the fractional part.

Equal fractions with different denominators are called *equivalent fractions*. [2.3A, p. 73]

$\frac{3}{4}$ and $\frac{6}{8}$ are equivalent fractions.

A fraction is in *simplest form* when the numerator and denominator have no common factors other than 1. [2.3B, p. 74]

The fraction $\frac{11}{12}$ is in simplest form.

The *reciprocal* of a fraction is the fraction with the numerator and denominator interchanged. [2.7A, p. 101]

The reciprocal of $\frac{3}{8}$ is $\frac{8}{3}$.

The reciprocal of 5 is $\frac{1}{5}$.

Essential Rules and Procedures

Examples

To find the LCM of two or more numbers, find the prime factorization of each number and write the factorization of each number in a table. Circle the greatest product in each column. The LCM is the product of the circled numbers. [2.1A, p. 65]

The LCM of 12 and 18 is
$2 \cdot 2 \cdot 3 \cdot 3 = 36$.

To find the GCF of two or more numbers, find the prime factorization of each number and write the factorization of each number in a table. Circle the least product in each column that does not have a blank. The GCF is the product of the circled numbers. [2.1B, p. 66]

The GCF of 12 and 18 is $2 \cdot 3 = 6$.

To write an improper fraction as a mixed number or a whole number, divide the numerator by the denominator. [2.2B, p. 70]

$\frac{29}{6} = 29 \div 6 = 4\frac{5}{6}$

To write a mixed number as an improper fraction, multiply the denominator of the fractional part of the mixed number by the whole-number part. Add this product and the numerator of the fractional part. The sum is the numerator of the improper fraction. The denominator remains the same. [2.2B, p. 70]

$3\frac{2}{5} = \frac{5 \times 3 + 2}{5} = \frac{17}{5}$

To find equivalent fractions by raising to higher terms, multiply the numerator and denominator of the fraction by the same number. [2.3A, p. 73]

$\frac{3}{4} = \frac{3 \cdot 5}{4 \cdot 5} = \frac{15}{20}$

$\frac{3}{4}$ and $\frac{15}{20}$ are equivalent fractions.

To write a fraction in simplest form, factor the numerator and denominator of the fraction; then eliminate the common factors. [2.3B, p. 74]

$\frac{30}{45} = \frac{2 \cdot \overset{1}{\cancel{3}} \cdot \overset{1}{\cancel{5}}}{\underset{1}{\cancel{3}} \cdot 3 \cdot \underset{1}{\cancel{5}}} = \frac{2}{3}$

Copyright © Houghton Mifflin Company. All rights reserved.

To add fractions with the same denominator, add the numerators and place the sum over the common denominator. [2.4A, p. 77]

$$\frac{5}{12} + \frac{11}{12} = \frac{16}{12} = 1\frac{4}{12} = 1\frac{1}{3}$$

To add fractions with different denominators, first rewrite the fractions as equivalent fractions with a common denominator. (The common denominator is the LCM of the denominators of the fractions.) Then add the fractions. [2.4B, p. 77]

$$\frac{1}{4} + \frac{2}{5} = \frac{5}{20} + \frac{8}{20} = \frac{13}{20}$$

To subtract fractions with the same denominator, subtract the numerators and place the difference over the common denominator. [2.5A, p. 85]

$$\frac{9}{16} - \frac{5}{16} = \frac{4}{16} = \frac{1}{4}$$

To subtract fractions with different denominators, first rewrite the fractions as equivalent fractions with a common denominator. (The common denominator is the LCM of the denominators of the fractions.) Then subtract the fractions. [2.5B, p. 85]

$$\frac{2}{3} - \frac{7}{16} = \frac{32}{48} - \frac{21}{48} = \frac{11}{48}$$

To multiply two fractions, multiply the numerators; this is the numerator of the product. Multiply the denominators; this is the denominator of the product. [2.6A, p. 93]

$$\frac{3}{4} \cdot \frac{2}{9} = \frac{3 \cdot 2}{4 \cdot 9} = \frac{\overset{1}{\cancel{3}} \cdot \overset{1}{\cancel{2}}}{\underset{1}{\cancel{2}} \cdot 2 \cdot \underset{1}{\cancel{3}} \cdot 3} = \frac{1}{6}$$

To divide two fractions, multiply the first fraction by the reciprocal of the second fraction. [2.7A, p. 101]

$$\frac{8}{15} \div \frac{4}{5} = \frac{8}{15} \cdot \frac{5}{4} = \frac{8 \cdot 5}{15 \cdot 4}$$
$$= \frac{\overset{1}{\cancel{2}} \cdot \overset{1}{\cancel{2}} \cdot 2 \cdot \overset{1}{\cancel{5}}}{3 \cdot \underset{1}{\cancel{5}} \cdot \underset{1}{\cancel{2}} \cdot \underset{1}{\cancel{2}}} = \frac{2}{3}$$

The find the order relation between two fractions with the same denominator, compare the numerators. The fraction that has the smaller numerator is the smaller fraction. [2.8A, p. 110]

$$\frac{17}{25} \leftarrow \text{Smaller numerator}$$
$$\frac{19}{25} \leftarrow \text{Larger numerator}$$
$$\frac{17}{25} < \frac{19}{25}$$

To find the order relation between two fractions with different denominators, first rewrite the fractions with a common denominator. The fraction that has the smaller numerator is the smaller fraction. [2.8A, p. 110]

$$\frac{3}{5} = \frac{24}{40} \qquad \frac{5}{8} = \frac{25}{40}$$
$$\frac{24}{40} < \frac{25}{40}$$
$$\frac{3}{5} < \frac{5}{8}$$

Order of Operations Agreement [2.8C, p. 111]

Step 1 Do all the operations inside parentheses.

Step 2 Simplify any numerical expressions containing exponents.

Step 3 Do multiplication and division as they occur from left to right.

Step 4 Do addition and subtraction as they occur from left to right.

$$\left(\frac{1}{3}\right)^2 + \left(\frac{5}{6} - \frac{7}{12}\right) \cdot (4)$$
$$= \left(\frac{1}{3}\right)^2 + \left(\frac{1}{4}\right) \cdot (4)$$
$$= \frac{1}{9} + \left(\frac{1}{4}\right) \cdot (4)$$
$$= \frac{1}{9} + 1 = 1\frac{1}{9}$$

Copyright © Houghton Mifflin Company. All rights reserved.

Chapter 2 Review Exercises

1. Write $\frac{30}{45}$ in simplest form.

2. Simplify: $\left(\frac{3}{4}\right)^3 \cdot \frac{20}{27}$

3. Express the shaded portion of the circles as an improper fraction.

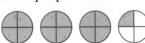

4. Find the total of $\frac{2}{3}$, $\frac{5}{6}$, and $\frac{2}{9}$.

5. Place the correct symbol, $<$ or $>$, between the two numbers.

$\frac{11}{18}$ $\quad$ $\frac{17}{24}$

6. Subtract: $\begin{aligned} 18\frac{1}{6} \\ -\ 3\frac{5}{7} \\ \hline \end{aligned}$

7. Simplify: $\frac{2}{7}\left(\frac{5}{8} - \frac{1}{3}\right) \div \frac{3}{5}$

8. Multiply: $2\frac{1}{3} \times 3\frac{7}{8}$

9. Divide: $1\frac{1}{3} \div \frac{2}{3}$

10. Find $\frac{17}{24}$ decreased by $\frac{3}{16}$.

11. Divide: $8\frac{2}{3} \div 2\frac{3}{5}$

12. Find the GCF of 20 and 48.

13. Write an equivalent fraction with the given denominator.

$\frac{2}{3} = \frac{}{36}$

14. What is $\frac{15}{28}$ divided by $\frac{5}{7}$?

15. Write an equivalent fraction with the given denominator.

$\frac{8}{11} = \frac{}{44}$

16. Multiply: $2\frac{1}{4} \times 7\frac{1}{3}$

17. Find the LCM of 18 and 12.

18. Write $\frac{16}{44}$ in simplest form.

Copyright © Houghton Mifflin Company. All rights reserved.

19. Add: $\frac{3}{8} + \frac{5}{8} + \frac{1}{8}$

20. Subtract: 16
        $-\ 5\frac{7}{8}$
        _____

21. Add: $4\frac{4}{9} + 2\frac{1}{6} + 11\frac{17}{27}$

22. Find the GCF of 15 and 25.

23. Write $\frac{17}{5}$ as a mixed number.

24. Simplify: $\left(\frac{4}{5} - \frac{2}{3}\right)^2 \div \frac{4}{15}$

25. Add: $\frac{3}{8} + 1\frac{2}{3} + 3\frac{5}{6}$

26. Find the LCM of 18 and 27.

27. Subtract: $\frac{11}{18} - \frac{5}{18}$

28. Write $2\frac{5}{7}$ as an improper fraction.

29. Divide: $\frac{5}{6} \div \frac{5}{12}$

30. Multiply: $\frac{5}{12} \times \frac{4}{25}$

31. What is $\frac{11}{50}$ multiplied by $\frac{25}{44}$?

32. Express the shaded portion of the circles as a mixed number.

33. **Meteorology**   During 3 months of the rainy season, $5\frac{7}{8}$, $6\frac{2}{3}$, and $8\frac{3}{4}$ inches of rain fell. Find the total rainfall for the 3 months.

34. **Real Estate**   A home building contractor bought $4\frac{2}{3}$ acres for \$168,000. What was the cost of each acre?

35. **Sports**   A 15-mile race has three checkpoints. The first checkpoint is $4\frac{1}{2}$ miles from the starting point. The second checkpoint is $5\frac{3}{4}$ miles from the first checkpoint. How many miles is the second checkpoint from the finish line?

36. **Fuel Efficiency**   A compact car gets 36 miles on each gallon of gasoline. How many miles can the car travel on $6\frac{3}{4}$ gallons of gasoline?

Copyright © Houghton Mifflin Company. All rights reserved.

Chapter 2 Test

1. Multiply: $\dfrac{9}{11} \times \dfrac{44}{81}$

2. Find the GCF of 24 and 80.

3. Divide: $\dfrac{5}{9} \div \dfrac{7}{18}$

4. Simplify: $\left(\dfrac{3}{4}\right)^2 \div \left(\dfrac{2}{3} + \dfrac{5}{6}\right) - \dfrac{1}{12}$

5. Write $9\dfrac{4}{5}$ as an improper fraction.

6. What is $5\dfrac{2}{3}$ multiplied by $1\dfrac{7}{17}$?

7. Write $\dfrac{40}{64}$ in simplest form.

8. Place the correct symbol, $<$ or $>$, between the two numbers.

    $\dfrac{3}{8} \quad \dfrac{5}{12}$

9. Simplify: $\left(\dfrac{1}{4}\right)^3 \div \left(\dfrac{1}{8}\right)^2 - \dfrac{1}{6}$

10. Find the LCM of 24 and 40.

11. Subtract: $\dfrac{17}{24} - \dfrac{11}{24}$

12. Write $\dfrac{18}{5}$ as a mixed number.

13. Find the quotient of $6\dfrac{2}{3}$ and $3\dfrac{1}{6}$.

14. Write an equivalent fraction with the given denominator.

    $\dfrac{5}{8} = \dfrac{}{72}$

Copyright © Houghton Mifflin Company. All rights reserved.

15. Add: $\dfrac{5}{6}$

$\dfrac{7}{9}$

$+\dfrac{1}{15}$

16. Subtract: $23\dfrac{1}{8}$

$-\ 9\dfrac{9}{44}$

17. What is $\dfrac{9}{16}$ minus $\dfrac{5}{12}$?

18. Simplify: $\left(\dfrac{2}{3}\right)^4 \cdot \dfrac{27}{32}$

19. Add: $\dfrac{7}{12} + \dfrac{11}{12} + \dfrac{5}{12}$

20. What is $12\dfrac{5}{12}$ more than $9\dfrac{17}{20}$?

21. Express the shaded portion of the circles as an improper fraction.

22. **Compensation** An electrician earns \$240 for each day worked. What is the total of the electrician's earnings for working $3\dfrac{1}{2}$ days?

23. **Real Estate** Grant Miura bought $7\dfrac{1}{4}$ acres of land for a housing project. One and three-fourths acres were set aside for a park, and the remaining land was developed into $\dfrac{1}{2}$-acre lots. How many lots were available for sale?

24. **Real Estate** A land developer purchases $25\dfrac{1}{2}$ acres of land and plans to set aside 3 acres for an entranceway to a housing development to be built on the property. Each house will be built on a $\dfrac{3}{4}$-acre plot of land. How many houses does the developer plan to build on the property?

25. **Meterology** In 3 successive months, the rainfall measured $11\dfrac{1}{2}$ inches, $7\dfrac{5}{8}$ inches, and $2\dfrac{1}{3}$ inches. Find the total rainfall for the 3 months.

Copyright © Houghton Mifflin Company. All rights reserved.

Cumulative Review Exercises

1. Round 290,496 to the nearest thousand.

2. Subtract: 390,047
            − 98,769

3. Find the product of 926 and 79.

4. Divide: $57\overline{)30{,}792}$

5. Simplify: $4 \cdot (6 - 3) \div 6 - 1$

6. Find the prime factorization of 44.

7. Find the LCM of 30 and 42.

8. Find the GCF of 60 and 80.

9. Write $7\frac{2}{3}$ as an improper fraction.

10. Write $\frac{25}{4}$ as a mixed number.

11. Write an equivalent fraction with the given denominator.
$$\frac{5}{16} = \frac{}{48}$$

12. Write $\frac{24}{60}$ in simplest form.

13. What is $\frac{9}{16}$ more than $\frac{7}{12}$?

14. Add: $3\frac{7}{8}$
         $7\frac{5}{12}$
       $+\ 2\frac{15}{16}$

15. Find $\frac{3}{8}$ less than $\frac{11}{12}$.

16. Subtract: $5\frac{1}{6}$
           $-\ 3\frac{7}{18}$

Copyright © Houghton Mifflin Company. All rights reserved.

17. Multiply: $\dfrac{3}{8} \times \dfrac{14}{15}$

18. Multiply: $3\dfrac{1}{8} \times 2\dfrac{2}{5}$

19. Divide: $\dfrac{7}{16} \div \dfrac{5}{12}$

20. Find the quotient of $6\dfrac{1}{8}$ and $2\dfrac{1}{3}$.

21. Simplify: $\left(\dfrac{1}{2}\right)^3 \cdot \dfrac{8}{9}$

22. Simplify: $\left(\dfrac{1}{2} + \dfrac{1}{3}\right) \div \left(\dfrac{2}{5}\right)^2$

23. **Banking** Molly O'Brien had $1359 in a checking account. During the week, Molly wrote checks for $128, $54, and $315. Find the amount in the checking account at the end of the week.

24. **Entertainment** The tickets for a movie were $10 for an adult and $4 for a student. Find the total income from the sale of 87 adult tickets and 135 student tickets.

25. **Measurement** Find the total weight of three packages that weigh $1\dfrac{1}{2}$ pounds, $7\dfrac{7}{8}$ pounds, and $2\dfrac{2}{3}$ pounds.

26. **Carpentry** A board $2\dfrac{5}{8}$ feet long is cut from a board $7\dfrac{1}{3}$ feet long. What is the length of the remaining piece?

27. **Fuel Efficiency** A car travels 27 miles on each gallon of gasoline. How many miles can the car travel on $8\dfrac{1}{3}$ gallons of gasoline?

28. **Real Estate** Jimmy Santos purchased $10\dfrac{1}{3}$ acres of land to build a housing development. Jimmy donated 2 acres for a park. How many $\dfrac{1}{3}$-acre parcels can be sold from the remaining land?

Copyright © Houghton Mifflin Company. All rights reserved.

Copyright © Houghton Mifflin Company. All rights reserved.

chapter

3 Decimals

This woman is the owner and operator of a wholesale groceries store. As such, she is a self-employed person. Anyone who operates a trade, business, or profession is self-employed. There are over 10 million self-employed people in the United States. Among them are authors, musicians, computer technicians, contractors, landscape designers, lawyers, psychologists, sales agents, and dentists. The table associated with **Exercises 28 to 30 on page 136** lists the numbers of self-employed persons in the United States by annual earnings, from less than $5000 per year to $50,000 or more.

OBJECTIVES

Section 3.1

A To write decimals in standard form and in words

B To round a decimal to a given place value

Section 3.2

A To add decimals

B To solve application problems

Section 3.3

A To subtract decimals

B To solve application problems

Section 3.4

A To multiply decimals

B To solve application problems

Section 3.5

A To divide decimals

B To solve application problems

Section 3.6

A To convert fractions to decimals

B To convert decimals to fractions

C To identify the order relation between two decimals or between a decimal and a fraction

Need help? For online student resources, such as section quizzes, visit this textbook's website at **math.college.hmco.com/students.**

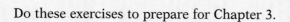
Do these exercises to prepare for Chapter 3.

1. Express the shaded portion of the rectangle as a fraction.

    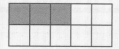

2. Round 36,852 to the nearest hundred.

3. Write 4791 in words.

4. Write six thousand eight hundred forty-two in standard form.

For Exercises 5 to 8, add, subtract, multiply, or divide.

5. 37 + 8892 + 465

6. 2403 − 765

7. 844 × 91

8. 23)‾6412

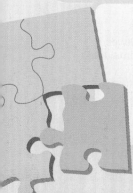

GO FIGURE • • •

Maria and Pedro are siblings. Pedro has as many brothers as sisters. Maria has twice as many brothers as sisters. How many children are in the family?

Copyright © Houghton Mifflin Company. All rights reserved.

Copyright © Houghton Mifflin Company. All rights reserved.

3.1 Introduction to Decimals

Objective A   To write decimals in standard form and in words

TAKE NOTE

In decimal notation, that part of the number that appears to the left of the decimal is the **whole-number part.** That part of the number that appears to the right of the decimal point is the **decimal part.** The **decimal point** separates the whole-number part from the decimal part.

The price tag on a sweater reads $61.88. The number 61.88 is in **decimal notation.** A number written in decimal notation is often called simply a **decimal.**

A number written in decimal notation has three parts.

61	.	88
Whole-number part	**Decimal point**	**Decimal part**

The decimal part of the number represents a number less than 1. For example, $.88 is less than $1. The decimal point (.) separates the whole-number part from the decimal part.

The position of a digit in a decimal determines the digit's place value. The place-value chart is extended to the right to show the place value of digits to the right of a decimal point.

Point of Interest

The idea that all fractions should be represented in tenths, hundredths, and thousandths was presented in 1585 in Simon Stevin's publication *De Thiende* and its French translation, *La Disme*, which was widely read and accepted by the French. This may help to explain why the French accepted the metric system so easily two hundred years later.

In *De Thiende*, Stevin argued in favor of his notation by including examples for astronomers, tapestry makers, surveyors, tailors, and the like. He stated that using decimals would enable calculations to be "performed . . . with as much ease as counterreckoning."

In the decimal 458.302719, the position of the digit 7 determines that its place value is ten-thousandths.

Note the relationship between fractions and numbers written in decimal notation.

Seven tenths	Seven hundredths	Seven thousandths
$\dfrac{7}{10} = 0.7$	$\dfrac{7}{100} = 0.07$	$\dfrac{7}{1000} = 0.007$
1 zero in 10	2 zeros in 100	3 zeros in 1000
1 decimal place in 0.7	2 decimal places in 0.07	3 decimal places in 0.007

To write a decimal in words, write the decimal part of the number as though it were a whole number, and then name the place value of the last digit.

0.9684   Nine thousand six hundred eighty-four ten-thousandths

The decimal point in a decimal is read as "and."

372.516   Three hundred seventy-two and five hundred sixteen thousandths

Point of Interest

The decimal point did not make its appearance until the early 1600s. Stevin's notation used subscripts with circles around them after each digit: 0 for ones, 1 for tenths (which he called "primes"), 2 for hundredths (called "seconds"), 3 for thousandths ("thirds"), and so on. For example, 1.375 would have been written

1   3   7   5
⓪   ①   ②   ③

To write a decimal in standard form when it is written in words, write the whole-number part, replace the word *and* with a decimal point, and write the decimal part so that the last digit is in the given place-value position.

Four and twenty-three <u>hundredths</u>

3 is in the hundredths place.   4.2<u>3</u>

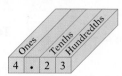

When writing a decimal in standard form, you may need to insert zeros after the decimal point so that the last digit is in the given place-value position.

Ninety-one and eight <u>thousandths</u>

8 is in the thousandths place.
Insert two zeros so that the 8 is in   91.00<u>8</u>
the thousandths place.

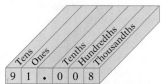

Sixty-five <u>ten-thousandths</u>

5 is in the ten-thousandths place.
Insert two zeros so that the 5 is in   0.006<u>5</u>
the ten-thousandths place.

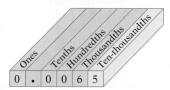

Example 1 Name the place value of the digit 8 in the number 45.687.

Solution The digit 8 is in the hundredths place.

You Try It 1 Name the place value of the digit 4 in the number 907.1342.

Your solution

Example 2 Write $\frac{43}{100}$ as a decimal.

Solution $\frac{43}{100} = 0.43$ • Forty-three hundredths

You Try It 2 Write $\frac{501}{1000}$ as a decimal.

Your solution

Example 3 Write 0.289 as a fraction.

Solution $0.289 = \frac{289}{1000}$ • 289 thousandths

You Try It 3 Write 0.67 as a fraction.

Your solution

Example 4 Write 293.50816 in words.

Solution Two hundred ninety-three and fifty thousand eight hundred sixteen hundred-thousandths

You Try It 4 Write 55.6083 in words.

Your solution

Solutions on p. S8

Copyright © Houghton Mifflin Company. All rights reserved.

Example 5	Write twenty-three and two hundred forty-seven millionths in standard form.	**You Try It 5**	Write eight hundred six and four hundred ninety-one hundred-thousandths in standard form.
Solution	23.000247 • **7 is in the millionths place.**	**Your solution**	

Solution on p. S8

Objective B

To round a decimal to a given place value

Study Tip

Have you considered joining a study group? Getting together regularly with other students in the class to go over material and quiz each other can be very beneficial. See *AIM for Success* at the front of the book.

In general, rounding decimals is similar to rounding whole numbers except that the digits to the right of the given place value are dropped instead of being replaced by zeros.

If the digit to the right of the given place value is less than 5, that digit and all digits to the right are dropped.

Round 6.9237 to the nearest hundredth.

┌── Given place value (hundredths)

6.9237

  └── 3 < 5   Drop the digits 3 and 7.

6.9237 rounded to the nearest hundredth is 6.92.

If the digit to the right of the given place value is greater or equal to 5, increase the digit in the given place value by 1, and drop all digits to its right.

Round 12.385 to the nearest tenth.

┌── Given place value (tenths)

12.385

  └── 8 > 5   Increase 3 by 1 and drop all digits to the right of 3.

12.385 rounded to the nearest tenth is 12.4.

TAKE NOTE

In the example at the right, the zero in the given place value is not dropped. This indicates that the number is rounded to the nearest thousandth. If we dropped the zero and wrote 0.47, it would indicate that the number was rounded to the nearest hundredth.

HOW TO   Round 0.46972 to the nearest thousandth.

┌── Given place value (thousandths)

0.46972

  └── 7 > 5   Round up by adding 1 to the 9 (9 + 1 = 10). Carry the 1 to the hundredths place (6 + 1 = 7).

0.46972 rounded to the nearest thousandth is 0.470.

Copyright © Houghton Mifflin Company. All rights reserved.

Example 6 Round 0.9375 to the nearest thousandth.

Solution

```
         ┌─── Given place value
  0.9375
       └── 5 = 5
```

0.9375 rounded to the nearest thousandth is 0.938.

You Try It 6 Round 3.675849 to the nearest ten-thousandth.

Your solution

Example 7 Round 2.5963 to the nearest hundredth.

Solution

```
        ┌─── Given place value
  2.5963
      └── 6 > 5
```

2.5963 rounded to the nearest hundredth is 2.60.

You Try It 7 Round 48.907 to the nearest tenth.

Your solution

Example 8 Round 72.416 to the nearest whole number.

Solution

```
        ┌─── Given place value
  72.416
      └── 4 < 5
```

72.416 rounded to the nearest whole number is 72.

You Try It 8 Round 31.8652 to the nearest whole number.

Your solution

Example 9

On average, an American goes to the movies 4.56 times per year. To the nearest whole number, how many times per year does an American go to the movies?

Solution

4.56 rounded to the nearest whole number is 5. An American goes to the movies about 5 times per year.

You Try It 9

One of the driest cities in the Southwest is Yuma, Arizona, with an average annual precipitation of 2.65 inch. To the nearest inch, what is the average annual precipitation in Yuma?

Your solution

Solutions on pp. S8–S9

Copyright © Houghton Mifflin Company. All rights reserved.

3.1 Exercises

Objective A   **To write decimals in standard form and in words**

For Exercises 1 to 6, name the place value of the digit 5.

1. 76.31587

2. 291.508

3. 432.09157

4. 0.0006512

5. 38.2591

hundredth

6. 0.0000853

For Exercises 7 to 14, write the fraction as a decimal.

7. $\dfrac{3}{10}$

8. $\dfrac{9}{10}$

9. $\dfrac{21}{100}$

10. $\dfrac{87}{100}$

11. $\dfrac{461}{1000}$

12. $\dfrac{853}{1000}$

13. $\dfrac{93}{1000}$

14. $\dfrac{61}{1000}$

For Exercises 15 to 22, write the decimal as a fraction.

15. 0.1

16. 0.3 $\dfrac{3}{10}$

17. 0.47

18. 0.59

19. 0.289

20. 0.601

21. 0.09

22. 0.013

For Exercises 23 to 31, write the number in words.

23. 0.37

24. 25.6

25. 9.4

26. 1.004

27. 0.0053

28. 41.108

29. 0.045

30. 3.157

31. 26.04

For Exercises 32 to 39, write the number in standard form.

32. Six hundred seventy-two thousandths

33. Three and eight hundred six ten-thousandths

34. Nine and four hundred seven ten-thousandths

35. Four hundred seven and three hundredths

Copyright © Houghton Mifflin Company. All rights reserved.

36. Six hundred twelve and seven hundred four thousandths

37. Two hundred forty-six and twenty-four thousandths

38. Two thousand sixty-seven and nine thousand two ten-thousandths

39. Seventy-three and two thousand six hundred eighty-four hundred-thousandths

Objective B   **To round a decimal to a given place value**

For Exercises 40 to 54, round the number to the given place value.

40. 6.249   Tenths

41. 5.398   Tenths

42. 21.007   Tenths

43. 30.0092   Tenths

44. 18.40937   Hundredths

45. 413.5972   Hundredths

46. 72.4983   Hundredths

47. 6.061745   Thousandths

48. 936.2905   Thousandths

49. 96.8027   Whole number

50. 47.3192   Whole number

51. 5439.83   Whole number

52. 7014.96   Whole number

53. 0.023591   Ten-thousandths

54. 2.975268   Hundred-thousandths

55. **Measurement**   A nickel weighs about 0.1763668 ounce. Find the weight of a nickel to the nearest hundredth of an ounce.

0.18

56. **Business**   The total cost of a parka, including sales tax, is $124.1093. Round the total cost to the nearest cent to find the amount a customer pays for the parka.

57. **Sports**   Runners in the Boston Marathon run a distance of 26.21875 miles. To the nearest tenth of a mile, find the distance that an entrant who completes the Boston Marathon runs.

APPLYING THE CONCEPTS

58. Indicate which digits of the number, if any, need not be entered on a calculator.

   a. 1.500   **b.** 0.908   **c.** 60.07   **d.** 0.0032

59. **a.** Find a number between 0.1 and 0.2.   **b.** Find a number between 1 and 1.1   **c.** Find a number between 0 and 0.005.

Copyright © Houghton Mifflin Company. All rights reserved.

3.2 Addition of Decimals

Objective A | **To add decimals**

To add decimals, write the numbers so that the decimal points are on a vertical line. Add as for whole numbers, and write the decimal point in the sum directly below the decimal points in the addends.

HOW TO Add: 0.237 + 4.9 + 27.32

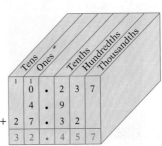

Note that by placing the decimal points on a vertical line, we make sure that digits of the same place value are added.

Example 1 Find the sum of 42.3, 162.903, and 65.0729.

Solution
```
  1 1 1
   42.3
  162.903
+  65.0729
  270.2759
```
• Place the decimal points on a vertical line.

You Try It 1 Find the sum of 4.62, 27.9, and 0.62054.

Your solution

Example 2 Add: 0.83 + 7.942 + 15

Solution
```
 1 1
   0.83
   7.942
+ 15.
  23.772
```

You Try It 2 Add: 6.05 + 12 + 0.374

Your solution

Solutions on p. S9

● E S T I M A T I O N ●

Estimating the Sum of Two or More Decimals

Calculate 23.037 + 16.7892. Then use estimation to determine whether the sum is reasonable.

Add to find the exact sum.

$$23.037 + 16.7892 = 39.8262$$

To estimate the sum, round each number to the same place value. Here we have rounded to the nearest whole number. Then add. The estimated answer is 40, which is very close to the exact sum, 39.8262.

$$
\begin{array}{r}
23.037 \approx 23 \\
+16.7892 \approx +17 \\
\hline
40
\end{array}
$$

Copyright © Houghton Mifflin Company. All rights reserved.

Objective B **To solve application problems**

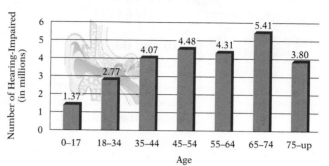

The graph at the right shows the breakdown by age group of Americans who are hearing-impaired. Use this graph for Example 3 and You Try It 3.

Breakdown by Age Group of Americans Who Are Hearing-Impaired

Source: American Speech-Language-Hearing Association

Example 3

Determine the number of Americans under the age of 45 who are hearing-impaired.

Strategy

To determine the number, add the numbers of hearing impaired ages 0 to 17, 18 to 34, and 35 to 44.

Solution

$$\begin{array}{r} 1.37 \\ 2.77 \\ +4.07 \\ \hline 8.21 \end{array}$$

8.21 million Americans under the age of 45 are hearing-impaired.

You Try It 3

Determine the number of Americans ages 45 and older who are hearing-impaired.

Your strategy

Your solution

Example 4

Dan Burhoe earned a salary of $138.50 for working 3 days this week as a food server. He also received $42.75, $35.80, and $59.25 in tips during the 3 days. Find his total income for the 3 days of work.

Strategy

To find the total income, add the tips (42.75, 35.80, and 59.25) to the salary (138.50).

Solution

$138.50 + 42.75 + 35.80 + 59.25 = 276.30$

Dan's total income for the 3 days of work was $276.30.

You Try It 4

Anita Khavari, an insurance executive, earns a salary of $875 every 4 weeks. During the past 4-week period, she received commissions of $985.80, $791.46, $829.75, and $635.42. Find her total income for the past 4-week period.

Your strategy

Your solution

Solutions on p. S9

Copyright © Houghton Mifflin Company. All rights reserved.

3.2 Exercises

Objective A   **To add decimals**

For Exercises 1 to 17, add.

1. 16.008 + 2.0385 + 132.06

2. 17.32 + 1.0579 + 16.5

3. 1.792 + 67 + 27.0526

4. 8.772 + 1.09 + 26.5027

5. 3.02 + 62.7 + 3.924

6. 9.06 + 4.976 + 59.6

7. 82.006 + 9.95 + 0.927

8. 0.826 + 8.76 + 79.005

9. 4.307 + 99.82 + 9.078

10.   0.3
     + 0.07

11.   0.29
     + 0.4

12.   1.007
     + 2.1

13.   7.3
     + 9.005

14.   4.9257
      27.05
     + 9.0063

15.   8.72
      99.073
     + 2.9736

16.   62.4
      9.827
     + 692.44

17.   8
      89.43
     + 7.0659

For Exercises 18 to 21, use a calculator to add. Then round the numbers to the nearest whole number and use estimation to determine whether the sum you calculated is reasonable.

18.   342.42
      89.625
     + 176.2

19.   219.9
      0.872
     + 13.42

20.   823.9
      82.65
     + 46.923

21.   678.92
      97.6
     + 5.423

Objective B   **To solve application problems**

22. Finances   A family has a mortgage payment of $814.72, a Visa bill of $216.40, and an electric bill of $87.32.
   a. Calculate the exact amount of the three payments.
   b. Round the numbers to the nearest hundred and use estimation to determine whether the sum you calculated is reasonable.

23. Mechanics   Find the length of the shaft.

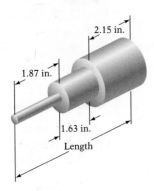

2.15 in.

1.87 in.

1.63 in.

Length

24. Mechanics   Find the length of the shaft.

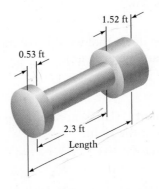

1.52 ft

0.53 ft

2.3 ft

Length

Copyright © Houghton Mifflin Company. All rights reserved.

25. Banking You have $2143.57 in your checking account. You make deposits of $210.98, $45.32, $1236.34, and $27.99. Find the amount in your checking account after you have made the deposits if no money has been withdrawn.

26. Geometry The perimeter of a triangle is the sum of the lengths of the three sides of the triangle. Find the perimeter of a triangle that has sides that measure 4.9 meters, 6.1 meters, and 7.5 meters.

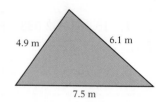

4.9 m   6.1 m

7.5 m

27. **Demography** The world's population in 2050 is expected to be 8.9 billion people. It is projected that in that year, Asia's population will be 5.3 billion and Africa's population will be 1.8 billion. What are the combined populations of Asia and Africa expected to be in 2050? (*Source:* United Nations Population Division, World Population Prospects)

Employment The table at the right shows the number of self-employed people, in millions, in the United States. They are listed by annual earnings. Use this table for Exercises 28 to 30.

Annual Earnings	Number of Self-Employed (in millions)
Less than $5000	5.1
$5000 – $24,999	3.9
$25,000 – $49,999	1.6
$50,000 or more	1.0

Source: U.S. Small Business Administration

28. How many self-employed people earn less than $50,000?

29. How many self-employed people earn $5000 or more?

30. How many people in the United States are self-employed?

APPLYING THE CONCEPTS

Consumerism The table at the right gives the prices for selected products in a grocery store. Use this table for Exercises 31 and 32.

31. Does a customer with $10 have enough money to purchase raisin bran, bread, milk, and butter?

32. Name three items that would cost more than $7 but less than $8. (There is more than one answer.)

Product	Cost
Raisin bran	$3.29
Butter	$2.79
Bread	$1.49
Popcorn	$1.19
Potatoes	$2.49
Cola (6-pack)	$1.99
Mayonnaise	$2.99
Lunch meat	$3.39
Milk	$2.59
Toothpaste	$2.69

33. Measurement Can a piece of rope 4 feet long be wrapped around the box shown at the right?

1.4 ft   1.4 ft   1.4 ft

Copyright © Houghton Mifflin Company. All rights reserved.

3.3 Subtraction of Decimals

Objective A To subtract decimals

To subtract decimals, write the numbers so that the decimal points are on a vertical line. Subtract as for whole numbers, and write the decimal point in the difference directly below the decimal point in the subtrahend.

HOW TO Subtract $21.532 - 9.875$ and check.

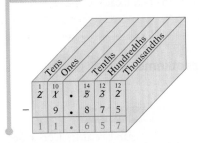

Placing the decimal points on a vertical line ensures that digits of the same place value are subtracted.

		$\overset{1\ 1\ \ 11}{}$
Check:	Subtrahend	9.875
	+ Difference	+ 11.657
	= Minuend	21.532

HOW TO Subtract $4.3 - 1.7942$ and check.

$$\begin{array}{r} \overset{3\ \ \ 12\ 9\ 9\ 10}{\cancel{4}.3\,\cancel{0}\,\cancel{0}\,\cancel{0}} \\ -\ 1.7942 \\ \hline 2.5058 \end{array}$$

If necessary, insert zeros in the minuend before subtracting.

	$\overset{1\ \ 1\,1\,1}{}$
Check:	1.7942
	+ 2.5058
	4.3000

Example 1 Subtract $39.047 - 7.96$ and check.

Solution

$$\begin{array}{r} \overset{8\ \ 9\ 14}{3\cancel{9}.\cancel{0}\cancel{4}\,7} \\ -\ \ 7.96 \\ \hline 31.087 \end{array} \qquad Check: \begin{array}{r} \overset{1\ 1}{7.96} \\ +\ 31.087 \\ \hline 39.047 \end{array}$$

You Try It 1 Subtract $72.039 - 8.47$ and check.

Your solution

Example 2 Find 9.23 less than 29 and check.

Solution

$$\begin{array}{r} \overset{1\ 18\ \ 9\ 10}{\cancel{2}\cancel{9}.\cancel{0}\,\cancel{0}} \\ -\ \ 9.23 \\ \hline 19.77 \end{array} \qquad Check: \begin{array}{r} \overset{1\ 1\ 1}{9.23} \\ +\ 19.77 \\ \hline 29.00 \end{array}$$

You Try It 2 Subtract $35 - 9.67$ and check.

Your solution

Example 3 Subtract $1.2 - 0.8235$ and check.

Solution

$$\begin{array}{r} \overset{0\ \ 11\ 9\ 9\ 10}{\cancel{1}.2\,\cancel{0}\,\cancel{0}\,\cancel{0}} \\ -\ 0.8235 \\ \hline 0.3765 \end{array} \qquad Check: \begin{array}{r} \overset{1\ \ 1\,1\,1}{0.8235} \\ +\ 0.3765 \\ \hline 1.2000 \end{array}$$

You Try It 3 Subtract $3.7 - 1.9715$ and check.

Your solution

Solutions on p. S9

Copyright © Houghton Mifflin Company. All rights reserved.

● E S T I M A T I O N ●

Estimating the Difference Between Two Decimals

Calculate 820.23 − 475.748. Then use estimation to determine whether the difference is reasonable.

Subtract to find the exact difference.

820.23 − 475.748 = 344.482

To estimate the difference, round each number to the same place value. Here we have rounded to the nearest ten. Then subtract. The estimated answer is 340, which is very close to the exact difference, 344.482.

$$
\begin{array}{r}
820.23 \approx 820 \\
-475.748 \approx -480 \\
\hline
340
\end{array}
$$

Objective B   **To solve application problems**

Example 4

You bought a book for $15.87. How much change did you receive from a $20.00 bill?

Strategy
To find the amount of change, subtract the cost of the book (15.87) from $20.00.

Solution

$$
\begin{array}{r}
20.00 \\
-15.87 \\
\hline
4.13
\end{array}
$$

You received $4.13 in change.

You Try It 4

Your breakfast cost $6.85. How much change did you receive from a $10.00 bill?

Your strategy

Your solution

Example 5

You had a balance of $87.93 on your student debit card. You then used the card, deducting $15.99 for a CD, $6.85 for lunch, and $18.50 for a ticket to the football game. What is your new student debit card balance?

Strategy
To find your new debit card balance:
• Add to find the total of the three deductions (15.99 + 6.85 + 18.50).
• Subtract the total of the three deductions from the old balance (87.93).

Solution

$$
\begin{array}{r}
15.99 \\
6.85 \\
+18.50 \\
\hline
41.34 \text{ total of deductions}
\end{array}
\qquad
\begin{array}{r}
87.93 \\
-41.34 \\
\hline
46.59
\end{array}
$$

Your new debit card balance is $46.59.

You Try It 5

You had a balance of $2472.69 in your checking account. You then wrote checks for $1025.60, $79.85, and $162.47. Find the new balance in your checking account.

Your strategy

Your solution

Solutions on p. S9

Copyright © Houghton Mifflin Company. All rights reserved.

3.3 Exercises

Objective A   **To subtract decimals**

For Exercises 1 to 24, subtract and check.

1. 24.037 − 18.41   **2.** 26.029 − 19.31   **3.** 123.07 − 9.4273   **4.** 214 − 7.143

5. 16.5 − 9.7902   **6.** 13.2 − 8.6205   **7.** 235.79 − 20.093   **8.** 463.27 − 40.095

9. 63.005 − 9.1274   **10.** 23.004 − 7.2175   **11.** 92 − 19.2909   **12.** 41.2405 − 25.2709

13.   0.32   **14.**   0.78   **15.**   3.005   **16.**   6.007
       − 0.0058   − 0.0073   − 1.982   − 2.734

17.   352.16   **18.**   872   **19.**   724.32   **20.**   625.46
       − 90.994   − 80.753   − 69   − 77.509

21.   362.394   **22.**   421.385   **23.**   19   **24.**   23.4
       − 19.4672   − 17.5293   − 10.372   − 0.921

For Exercises 25 to 27, use the relationship between addition and subtraction to complete the statement.

25. _____ + 2.325 = 7.01   **26.** 5.392 + _____ = 8.07   **27.** _____ + 8.967 = 19.35

For Exercises 28 to 31, use a calculator to subtract. Then round the numbers to the nearest whole number and use estimation to determine whether the difference you calculated is reasonable.

28.   93.079256   **29.**   3.7529   **30.**   76.53902   **31.**   9.07325
       − 66.09249   − 1.00784   − 45.73005   − 1.924

Copyright © Houghton Mifflin Company. All rights reserved.

Objective B **To solve application problems**

32. Business The manager of the Edgewater Cafe takes a reading of the cash register tape each hour. At 1:00 P.M. the tape read $967.54. At 2:00 P.M. the tape read $1437.15. Find the amount of sales between 1:00 P.M. and 2:00 P.M.

33. Mechanics Find the missing dimension.

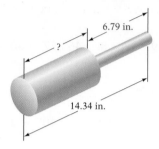

6.79 in.

?

14.34 in.

34. Mechanics Find the missing dimension.

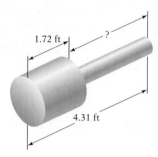

1.72 ft

?

4.31 ft

35. Banking You had a balance of $1029.74 in your checking account. You then wrote checks for $67.92, $43.10, and $496.34.
a. Find the total amount of the checks written.
b. Find the new balance in your checking account.

36. Consumerism The price of gasoline is $1.82 per gallon after the price rose $.07 one month and $.12 the next month. Find the price of gasoline before these increases in price.

37. **Population Growth** The annual number of births in the United States is projected to rise from 3.98 million in 2003 to 4.37 million in 2012. (*Sources:* U.S. Census Bureau; National Center for Health Statistics) Find the increase in the annual number of births from 2003 to 2012.

38. Investments Grace Herrera owned 357.448 shares of a mutual fund on January 1. On December 31 of the same year, she had 439.917 shares. What was the increase in the number of shares Grace owned during that year?

E-Commerce The table at the right shows the actual and projected number of households shopping online, in millions. Use the table for Exercises 39 and 40.

39. Calculate the growth in online shopping from 2000 to 2003.

40. Calculate the growth in online shopping from 1998 to 2003.

Year	Number of Households Shopping Online (in millions)
1998	8.7
1999	13.1
2000	17.7
2001	23.1
2002	30.3
2003	40.3

Source: Forrester Research Inc.

APPLYING THE CONCEPTS

41. Find the largest amount by which the estimate of the sum of two decimals rounded to the given place value could differ from the exact sum.
a. Tenths   **b.** Hundredths   **c.** Thousandths

Copyright © Houghton Mifflin Company. All rights reserved.

3.4 Multiplication of Decimals

Objective A   ## To multiply decimals

Point of Interest

Benjamin Banneker (1731–1806) was the first African American to earn distinction as a mathematician and scientist. He was on the survey team that determined the boundaries of Washington, D.C. The mathematics of surveying requires extensive use of decimals.

Decimals are multiplied as though they were whole numbers. Then the decimal point is placed in the product. Writing the decimals as fractions shows where to write the decimal point in the product.

$$0.\underline{3} \times 5 = \frac{3}{10} \times \frac{5}{1} = \frac{15}{10} = 1.\underline{5}$$

1 decimal place   1 decimal place

$$0.\underline{3} \times 0.\underline{5} = \frac{3}{10} \times \frac{5}{10} = \frac{15}{100} = 0.\underline{15}$$

1 decimal place   1 decimal place   2 decimal places

$$0.\underline{3} \times 0.\underline{05} = \frac{3}{10} \times \frac{5}{100} = \frac{15}{1000} = 0.\underline{015}$$

1 decimal place   2 decimal places   3 decimal places

To multiply decimals, multiply the numbers as with whole numbers. Write the decimal point in the product so that the number of decimal places in the product is the sum of the decimal places in the factors.

Integrating Technology

Scientific calculators have a floating decimal point. This means that the decimal point is automatically placed in the answer. For example, for the product at the right, enter

21 · 4 × 0 · 36 =

The display reads 7.704, with the decimal point in the correct position.

HOW TO   Multiply: 21.4×0.36

$$
\begin{array}{r}
21.4 \\
\times\ 0.36 \\
\hline
1284 \\
642 \\
\hline
7.704
\end{array}
$$

1 decimal place
2 decimal places

3 decimal places

HOW TO   Multiply: 0.037×0.08

$$
\begin{array}{r}
0.037 \\
\times\ 0.08 \\
\hline
0.00296
\end{array}
$$

3 decimal places
2 decimal places
5 decimal places

• Two zeros must be inserted between the 2 and the decimal point so that there are 5 decimal places in the product.

To multiply a decimal by a power of 10 (10, 100, 1000, . . .), move the decimal point to the right the same number of places as there are zeros in the power of 10.

$3.8925 \times 1\underline{0} \qquad = 38.925$

1 zero   1 decimal place

$3.8925 \times 1\underline{00} \qquad = 389.25$

2 zeros   2 decimal places

$3.8925 \times 1\underline{000} \qquad = 3892.5$

3 zeros   3 decimal places

$3.8925 \times 1\underline{0,000} \qquad = 38,925.$

4 zeros   4 decimal places

$3.8925 \times 1\underline{00,000} = 389,250.$

5 zeros   5 decimal places

• Note that a zero must be inserted before the decimal point.

Copyright © Houghton Mifflin Company. All rights reserved.

Note that if the power of 10 is written in exponential notation, the exponent indicates how many places to move the decimal point.

$3.8925 \times 10^1 = 38.925$

1 decimal place

$3.8925 \times 10^2 = 389.25$

2 decimal places

$3.8925 \times 10^3 = 3892.5$

3 decimal places

$3.8925 \times 10^4 = 38,925.$

4 decimal places

$3.8925 \times 10^5 = 389,250.$

5 decimal places

Example 1 Multiply: 920×3.7

Solution

$$
\begin{array}{r}
920 \\
\times\ \ 3.7 \\
\hline
644\ 0 \\
2760 \\
\hline
3404.0
\end{array}
$$

• 1 decimal place

• 1 decimal place

You Try It 1 Multiply: 870×4.6

Your solution

Example 2 Find 0.00079 multiplied by 0.025.

Solution

$$
\begin{array}{r}
0.00079 \\
\times\ \ \ \ 0.025 \\
\hline
395 \\
158 \\
\hline
0.00001975
\end{array}
$$

• 5 decimal places

• 3 decimal places

• 8 decimal places

You Try It 2 Find 0.000086 multiplied by 0.057.

Your solution

Example 3 Find the product of 3.69 and 2.07.

Solution

$$
\begin{array}{r}
3.69 \\
\times\ 2.07 \\
\hline
2583 \\
7380 \\
\hline
7.6383
\end{array}
$$

• 2 decimal places

• 2 decimal places

• 4 decimal places

You Try It 3 Find the product of 4.68 and 6.03.

Your solution

Example 4 Multiply: $42.07 \times 10,000$

Solution $42.07 \times 10,000 = 420,700$

You Try It 4 Multiply: 6.9×1000

Your solution

Example 5 Find 3.01 times 10^3.

Solution $3.01 \times 10^3 = 3010$

You Try It 5 Find 4.0273 times 10^2.

Your solution

Solutions on pp. S9–S10

Copyright © Houghton Mifflin Company. All rights reserved.

● E S T I M A T I O N ●

Estimating the Product of Two Decimals

Calculate 28.259×0.029. Then use estimation to determine whether the product is reasonable.

Multiply to find the exact product. $28.259 \times 0.029 = 0.819511$

To estimate the product, round each number so there is one nonzero digit. Then multiply. The estimated answer is 0.90, which is very close to the exact product, 0.819511.

$$28.259 \approx 30$$
$$\times\ 0.029 \approx \times 0.03$$
$$0.90$$

Objective B **To solve application problems**

The tables that follow list water rates and meter fees for a city. These tables are used for Example 6 and You Try It 6.

Water Charges	
Commercial	$1.39/1000 gal
Comm Restaurant	$1.39/1000 gal
Industrial	$1.39/1000 gal
Institutional	$1.39/1000 gal
Res—No Sewer	
Residential—SF	
>0 <200 gal per day	$1.15/1000 gal
>200 <1500 gal per day	$1.39/1000 gal
>1500 gal per day	$1.54/1000 gal

Meter Charges	
Meter	Meter Fee
5/8" & 3/4"	$13.50
1"	$21.80
1-1/2"	$42.50
2"	$67.20
3"	$133.70
4"	$208.20
6"	$415.10
8"	$663.70

Example 6

Find the total bill for an industrial water user with a 6-inch meter that used 152,000 gallons of water for July and August.

Strategy

To find the total cost of water:

- Find the cost of water by multiplying the cost per 1000 gallons (1.39) by the number of 1000-gallon units used.
- Add the cost of the water to the meter fee (415.10).

Solution

Cost of water $= \dfrac{152,000}{1000} \cdot 1.39 = 211.28$

Total cost $= 211.28 + 415.10 = 626.38$

The total cost is $626.38.

You Try It 6

Find the total bill for a commercial user that used 5000 gallons of water per day for July and August. The user has a 3-inch meter.

Your strategy

Your solution

Solution on p. S10

Copyright © Houghton Mifflin Company. All rights reserved.

Example 7

It costs $.036 an hour to operate an electric motor. How much does it cost to operate the motor for 120 hours?

Strategy

To find the cost of running the motor for 120 hours, multiply the hourly cost (0.036) by the number of hours the motor is run (120).

Solution

$$
\begin{array}{r}
0.036 \\
\times\ \ 120 \\
\hline
720 \\
36\ \ \ \\
\hline
4.320
\end{array}
$$

The cost of running the motor for 120 hours is $4.32.

You Try It 7

The cost of electricity to run a freezer for 1 hour is $.035. This month the freezer has run for 210 hours. Find the total cost of running the freezer this month.

Your strategy

Your solution

Example 8

Jason Ng earns a salary of $440 for a 40-hour workweek. This week he worked 12 hours of overtime at a rate of $16.50 for each hour of overtime worked. Find his total income for the week.

Strategy

To find Jason's total income for the week:

- Find the overtime pay by multiplying the hourly overtime rate (16.50) by the number of hours of overtime worked (12).
- Add the overtime pay to the weekly salary (440).

Solution

$$
\begin{array}{r}
16.50 \\
\times\ \ \ 12 \\
\hline
33\ 00 \\
165\ 0\ \ \\
\hline
198.00
\end{array}
\qquad
\begin{array}{r}
440.00 \\
+\ 198.00 \\
\hline
638.00
\end{array}
$$

198.00 Overtime pay

Jason's total income for this week is $638.00.

You Try It 8

You make a down payment of $175 on a stereo and agree to make payments of $37.18 a month for the next 18 months to repay the remaining balance. Find the total cost of the stereo.

Your strategy

Your solution

Solutions on p. S10

Copyright © Houghton Mifflin Company. All rights reserved.

3.4 Exercises

Objective A   To multiply decimals

For Exercises 1 to 94, multiply.

1. 0.9
       $\times$ 0.4

2. 0.7
       $\times$ 0.9

3. 0.5
       $\times$ 0.6

4. 0.3
       $\times$ 0.7

5. 0.5
       $\times$ 0.5

6. 0.7
       $\times$ 0.7

7. 0.9
       $\times$ 0.5

8. 0.2
       $\times$ 0.6

9. 7.7
       $\times$ 0.9

10. 3.4
        $\times$ 0.4

11. 9.2
        $\times$ 0.2

12. 2.6
        $\times$ 0.7

13. 7.2
        $\times$ 0.6

14. 6.8
        $\times$ 0.4

15. 7.4
        $\times$ 0.1

16. 3.8
        $\times$ 0.1

17. 7.9
        $\times$ 5

18. 9.3
        $\times$ 7

19. 0.68
        $\times$ 4

20. 0.83
        $\times$ 9

21. 0.67
        $\times$ 0.9

22. 0.84
        $\times$ 0.3

23. 0.16
        $\times$ 0.6

24. 0.47
        $\times$ 0.8

25. 2.5
        $\times$ 5.4

26. 3.9
        $\times$ 1.9

27. 8.4
        $\times$ 9.5

28. 7.6
        $\times$ 5.8

29. 0.83
        $\times$ 5.2

30. 0.24
        $\times$ 2.7

31. 0.46
        $\times$ 3.9

32. 0.78
        $\times$ 6.8

33. 0.2
        $\times$ 0.3

34. 0.3
        $\times$ 0.3

35. 0.24
        $\times$ 0.3

36. 0.17
        $\times$ 0.5

37. 1.47
        $\times$ 0.09

38. 6.37
        $\times$ 0.05

39. 8.92
        $\times$ 0.004

40. 6.75
        $\times$ 0.007

Copyright © Houghton Mifflin Company. All rights reserved.

41. $\begin{array}{r} 0.49 \\ \times\ 0.16 \\ \hline \end{array}$

42. $\begin{array}{r} 0.38 \\ \times\ 0.21 \\ \hline \end{array}$

43. $\begin{array}{r} 7.6 \\ \times\ 0.01 \\ \hline \end{array}$

44. $\begin{array}{r} 5.1 \\ \times\ 0.01 \\ \hline \end{array}$

45. $\begin{array}{r} 8.62 \\ \times\ 4 \\ \hline \end{array}$

46. $\begin{array}{r} 5.83 \\ \times\ 7 \\ \hline \end{array}$

47. $\begin{array}{r} 64.5 \\ \times\ 9 \\ \hline \end{array}$

48. $\begin{array}{r} 37.8 \\ \times\ 8 \\ \hline \end{array}$

49. $\begin{array}{r} 2.19 \\ \times\ 9.2 \\ \hline \end{array}$

50. $\begin{array}{r} 1.25 \\ \times\ 5.6 \\ \hline \end{array}$

51. $\begin{array}{r} 1.85 \\ \times\ 0.023 \\ \hline \end{array}$

52. $\begin{array}{r} 37.8 \\ \times\ 0.052 \\ \hline \end{array}$

53. $\begin{array}{r} 0.478 \\ \times\ 0.37 \\ \hline \end{array}$

54. $\begin{array}{r} 0.526 \\ \times\ 0.22 \\ \hline \end{array}$

55. $\begin{array}{r} 48.3 \\ \times\ 0.0041 \\ \hline \end{array}$

56. $\begin{array}{r} 67.2 \\ \times\ 0.0086 \\ \hline \end{array}$

57. $\begin{array}{r} 2.437 \\ \times\ 6.1 \\ \hline \end{array}$

58. $\begin{array}{r} 4.237 \\ \times\ 0.54 \\ \hline \end{array}$

59. $\begin{array}{r} 0.413 \\ \times\ 0.0016 \\ \hline \end{array}$

60. $\begin{array}{r} 0.517 \\ \times\ 0.0029 \\ \hline \end{array}$

61. $\begin{array}{r} 94.73 \\ \times\ 0.57 \\ \hline \end{array}$

62. $\begin{array}{r} 89.23 \\ \times\ 0.62 \\ \hline \end{array}$

63. $\begin{array}{r} 8.005 \\ \times\ 0.067 \\ \hline \end{array}$

64. $\begin{array}{r} 9.032 \\ \times\ 0.019 \\ \hline \end{array}$

65. 4.29×0.1

66. 6.78×0.1

67. 5.29×0.4

68. 6.78×0.5

69. 0.68×0.7

70. 0.56×0.9

71. 1.4×0.73

72. 6.3×0.37

73. 5.2×7.3

74. 7.4×2.9

75. 3.8×0.61

76. 7.2×0.72

77. 0.32×10

78. 6.93×10

79. 0.065×100

80. 0.039×100

81. 6.2856×1000

82. 3.2954×1000

83. 3.2×1000

84. $0.006 \times 10,000$

85. $3.57 \times 10,000$

Copyright © Houghton Mifflin Company. All rights reserved.

86. 8.52×10^1 **87.** 0.63×10^1 **88.** 82.9×10^2

89. 0.039×10^2 **90.** 6.8×10^3 **91.** 4.9×10^4

92. 6.83×10^4 **93.** 0.067×10^2 **94.** 0.052×10^2

95. Find the product of 0.0035 and 3.45.   **96.** Find the product of 237 and 0.34.

97. Multiply 3.005 by 0.00392.   **98.** Multiply 20.34 by 1.008.

99. Multiply 1.348 by 0.23.   **100.** Multiply 0.000358 by 3.56.

101. Find the product of 23.67 and 0.0035.   **102.** Find the product of 0.00346 and 23.1.

103. Find the product of 5, 0.45, and 2.3.   **104.** Find the product of 0.03, 23, and 9.45.

For Exercises 105 to 120, use a calculator to multiply. Then use estimation to determine whether the product you calculated is reasonable.

105.
$$\begin{array}{r} 28.5 \\ \times\ \ 3.2 \\ \hline \end{array}$$

106.
$$\begin{array}{r} 86.3 \\ \times\ \ 4.4 \\ \hline \end{array}$$

107.
$$\begin{array}{r} 2.38 \\ \times 0.44 \\ \hline \end{array}$$

108.
$$\begin{array}{r} 9.82 \\ \times 0.77 \\ \hline \end{array}$$

109.
$$\begin{array}{r} 0.866 \\ \times\ \ \ 4.5 \\ \hline \end{array}$$

110.
$$\begin{array}{r} 0.239 \\ \times\ \ \ 8.2 \\ \hline \end{array}$$

111.
$$\begin{array}{r} 4.34 \\ \times 2.59 \\ \hline \end{array}$$

112.
$$\begin{array}{r} 6.87 \\ \times 9.98 \\ \hline \end{array}$$

113.
$$\begin{array}{r} 8.434 \\ \times 0.044 \\ \hline \end{array}$$

114.
$$\begin{array}{r} 7.037 \\ \times 0.094 \\ \hline \end{array}$$

115.
$$\begin{array}{r} 28.44 \\ \times\ \ 1.12 \\ \hline \end{array}$$

116.
$$\begin{array}{r} 86.57 \\ \times\ \ 7.33 \\ \hline \end{array}$$

117.
$$\begin{array}{r} 49.6854 \\ \times 39.0672 \\ \hline \end{array}$$

118.
$$\begin{array}{r} 2.00547 \\ \times\ \ \ 9.672 \\ \hline \end{array}$$

119.
$$\begin{array}{r} 0.00456 \\ \times 0.009542 \\ \hline \end{array}$$

120.
$$\begin{array}{r} 7.00637 \\ \times\ \ 0.0128 \\ \hline \end{array}$$

Copyright © Houghton Mifflin Company. All rights reserved.

To solve application problems

121. Consumerism An electric motor costing $315.45 has an operating cost of $.027 for 1 hour of operation. Find the cost to run the motor for 56 hours. Round to the nearest cent.

122. Recycling Four hundred empty soft drink cans weigh 18.75 pounds. A recycling center pays $.75 per pound for the cans. Find the amount received for the 400 cans. Round to the nearest cent.

123. Recycling A recycling center pays $.045 per pound for newspapers.
a. Estimate the payment for recycling 520 pounds of newspapers.
b. Find the actual amount received from recycling the newspapers.

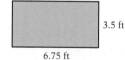

124. Geometry The perimeter of a square is equal to four times the length of a side of the square. Find the perimeter of a square whose side measures 2.8 meters.

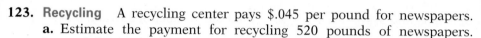
2.8 m

125. Geometry The area of a rectangle is equal to the product of the length of the rectangle times its width. Find the area of a rectangle that has a length of 6.75 feet and a width of 3.5 feet. The area will be in square feet.

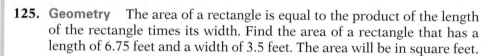
3.5 ft
6.75 ft

126. Finance You bought a car for $5000 down and made payments of $399.50 each month for 36 months.
a. Find the amount of the payments over the 36 months.
b. Find the total cost of the car.

127. Compensation A nurse earns a salary of $1156 for a 40-hour week. This week the nurse worked 15 hours of overtime at a rate of $43.35 for each hour of overtime worked.
a. Find the nurse's overtime pay.
b. Find the nurse's total income for the week.

128. Consumerism Bay Area Rental Cars charges $15 a day and $.15 per mile for renting a car. You rented a car for 3 days and drove 235 miles. Find the total cost of renting the car.

129. 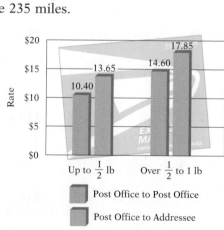 **Shipping** The graph at the right shows United States Postal Service rates for express mail. How much would it cost a company to mail 25 express mail packages, each weighing $\frac{1}{4}$ pound, post office to addressee?

130. Transportation A taxi costs $2.50 and $.20 for each $\frac{1}{8}$ mile driven. Find the cost of hiring a taxi to get from the airport to the hotel—a distance of $5\frac{1}{2}$ miles.

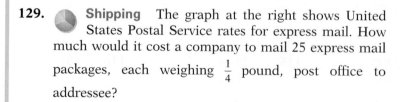
$20
$15
$10
$5
$0
17.85
13.65
14.60
10.40
Rate
Up to $\frac{1}{2}$ lb   Over $\frac{1}{2}$ to 1 lb

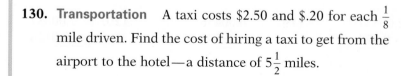
Post Office to Post Office
Post Office to Addressee
U.S. Postal Service Rates for Express Mail

Copyright © Houghton Mifflin Company. All rights reserved.

131. Business   The table at the right lists three pieces of steel required for a repair project.
   a. Find the total cost of grade 1.
   b. Find the total cost of grade 2.
   c. Find the total cost of grade 3.
   d. Find the total cost of the three pieces of steel.

Grade of Steel	Weight (pounds per foot)	Required Number of Feet	Cost per Pound
1	2.2	8	$1.20
2	3.4	6.5	$1.35
3	6.75	15.4	$1.94

132. Business   A confectioner ships holiday packs of candy and nuts anywhere in the United States. At the right is a price list for nuts and candy, and below is a table of shipping charges to zones in the United States. For any fraction of a pound, use the next higher weight. (16 oz = 1 lb)

Code	Description	Price
112	Almonds 16 oz	$4.75
116	Cashews 8 oz	$2.90
117	Cashews 16 oz	$5.50
130	Macadamias 7 oz	$5.25
131	Macadamias 16 oz	$9.95
149	Pecan halves 8 oz	$6.25
155	Mixed nuts 8 oz	$4.80
160	Cashew brittle 8 oz	$1.95
182	Pecan roll 8 oz	$3.70
199	Chocolate peanuts 8 oz	$1.90

Pounds	Zone 1	Zone 2	Zone 3	Zone 4
1–3	$6.55	$6.85	$7.25	$7.75
4–6	$7.10	$7.40	$7.80	$8.30
7–9	$7.50	$7.80	$8.20	$8.70
10–12	$7.90	$8.20	$8.60	$9.10

Find the cost of sending the following orders to the given mail zone.

a. Code	Quantity	b. Code	Quantity	c. Code	Quantity
116	2	112	1	117	3
130	1	117	4	131	1
149	3	131	2	155	2
182	4	160	3	160	4
Mail to zone 4.		182	5	182	1
		Mail to zone 3.		199	3
				Mail to zone 2.	

133. Air Pollution   An emissions test for cars requires that of the total engine exhaust, less than 1 part per thousand $\left(\dfrac{1}{1000} = 0.001\right)$ be hydrocarbon emissions. Using this figure, determine which of the cars in the table below would fail the emissions test.

Car	Total Engine Exhaust	Hydrocarbon Emission
1	367,921	360
2	401,346	420
3	298,773	210
4	330,045	320
5	432,989	450

Copyright © Houghton Mifflin Company. All rights reserved.

APPLYING THE CONCEPTS

134. Automotive Repair   Chris works at B & W Garage as an auto mechanic and has just completed an engine overhaul for a customer. To determine the cost of the repair job, Chris keeps a list of times worked and parts used. A parts list and a list of the times worked are shown below.

Parts Used		Time Spent	
Item	Quantity	Day	Hours
Gasket set	1	Monday	7.0
Ring set	1	Tuesday	7.5
Valves	8	Wednesday	6.5
Wrist pins	8	Thursday	8.5
Valve springs	16	Friday	9.0
Rod bearings	8		
Main bearings	5		
Valve seals	16		
Timing chain	1		

Price List		
Item Number	Description	Unit Price
27345	Valve spring	$9.25
41257	Main bearing	$17.49
54678	Valve	$16.99
29753	Ring set	$169.99
45837	Gasket set	$174.90
23751	Timing chain	$50.49
23765	Fuel pump	$429.99
28632	Wrist pin	$13.55
34922	Rod bearing	$4.69
2871	Valve seal	$1.69

a. Organize a table of data showing the parts used, the unit price for each, and the price of the quantity used. *Hint:* Use the following headings for the table.

> Quantity   Item Number   Description   Unit Price   Total

b. Add up the numbers in the "Total" column to find the total cost of the parts.

c. If the charge for labor is $46.75 per hour, compute the cost of labor.

d. What is the total cost for parts and labor?

135. Explain how the decimal point is placed when a number is multiplied by 10, 100, 1000, 10,000, etc.

136. Explain how the decimal point is placed in the product of two decimals.

137. Show how the decimal is placed in the product of 1.3×2.31 by first writing each number as a fraction and then multiplying. Then change the product back to decimal notation.

Copyright © Houghton Mifflin Company. All rights reserved.

3.5 Division of Decimals

Objective A **To divide decimals**

To divide decimals, move the decimal point in the divisor to the right to make the divisor a whole number. Move the decimal point in the dividend the same number of places to the right. Place the decimal point in the quotient directly over the decimal point in the dividend, and then divide as with whole numbers.

HOW TO Divide: $3.25\overline{)15.275}$

$$3.25.\overline{)15.27.5}$$

- Move the decimal point 2 places to the right in the divisor and then in the dividend. Place the decimal point in the quotient.

$$
\begin{array}{r}
4.7 \\
325.\overline{)\ 1527.5} \\
-1300 \\
\hline
227\ 5 \\
-227\ 5 \\
\hline
0
\end{array}
$$

- Divide as with whole numbers.

Study Tip

To learn mathematics, you must be an active participant. Listening and watching your professor do mathematics are not enough. Take notes in class, mentally think through every question your instructor asks, and try to answer it even if you are not called on to answer it verbally. Ask questions when you have them. See *AIM for Success* at the front of the book for other ways to be an active learner.

Moving the decimal point the same number of decimal places in the divisor and dividend does not change the value of the quotient, because this process is the same as multiplying the numerator and denominator of a fraction by the same number. In the example above,

$$3.25\overline{)15.275} = \frac{15.275}{3.25} = \frac{15.275 \times 100}{3.25 \times 100} = \frac{1527.5}{325} = 325\overline{)1527.5}$$

When dividing decimals, we usually round the quotient off to a specified place value, rather than writing the quotient with a remainder.

HOW TO Divide: $0.3\overline{)0.56}$
Round to the nearest hundredth.

$$
\begin{array}{r}
1.866 \approx 1.87 \\
0.3.\overline{)\ 0.5.600} \\
-\ 3 \\
\hline
2\ 6 \\
-2\ 4 \\
\hline
20 \\
-18 \\
\hline
20 \\
-18
\end{array}
$$

We must carry the division to the thousandths place to round the quotient to the nearest hundredth. Therefore, zeros must be inserted in the dividend so that the quotient has a digit in the thousandths place.

Copyright © Houghton Mifflin Company. All rights reserved.

Copyright © Houghton Mifflin Company. All rights reserved.

Integrating Technology

A calculator displays the quotient to the limit of the calculator's display. Enter

57 · 93 ÷ 3 · 24 =

to determine the number of places your calculator displays.

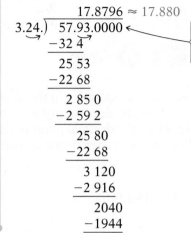

HOW TO Divide 57.93 by 3.24. Round to the nearest thousandth.

$$
\begin{array}{r}
17.8796 \approx 17.880 \\
3.24\overline{)\ 57.93.0000} \\
-32\ 4 \\
\hline
25\ 53 \\
-22\ 68 \\
\hline
2\ 85\ 0 \\
-2\ 59\ 2 \\
\hline
25\ 80 \\
-22\ 68 \\
\hline
3\ 120 \\
-2\ 916 \\
\hline
2040 \\
-1944 \\
\hline
\end{array}
$$

Zeros must be inserted in the dividend so that the quotient has a digit in the ten-thousandths place.

To divide a decimal by a power of 10 (10, 100, 1000, . . .), move the decimal point to the left the same number of places as there are zeros in the power of 10.

$34.65 \div \underline{10} = 3.465$
      1 zero   1 decimal place

$34.65 \div \underline{100} = 0.3465$
      2 zeros   2 decimal places

$34.65 \div \underline{1000} = 0.03465$
      3 zeros   3 decimal places

- Note that a zero must be inserted between the 3 and the decimal point.

$34.65 \div \underline{10,000} = 0.003465$
      4 zeros   4 decimal places

- Note that two zeros must be inserted between the 3 and the decimal point.

If the power of 10 is written in exponential notation, the exponent indicates how many places to move the decimal point.

$34.65 \div 10^1 = 3.465$ 1 decimal place
$34.65 \div 10^2 = 0.3465$ 2 decimal places
$34.65 \div 10^3 = 0.03465$ 3 decimal places
$34.65 \div 10^4 = 0.003465$ 4 decimal places

Example 1   Divide: $0.1344 \div 0.032$

Solution

$$
\begin{array}{r}
4.2 \\
0.032.\overline{)0.134.4} \\
-128 \\
\hline
6\ 4 \\
-6\ 4 \\
\hline
0
\end{array}
$$

- Move the decimal point 3 places to the right in the divisor and the dividend.

You Try It 1   Divide: $0.1404 \div 0.052$

Your solution

Solution on p. S10

Example 2   Divide: 58.092 ÷ 82
Round to the nearest
thousandth.

Solution

$$
\begin{array}{r}
0.7084 \approx 0.708 \\
82\overline{)\ 58.0920} \\
-57\ 4 \\
\hline
69 \\
-\ \ 0 \\
\hline
692 \\
-656 \\
\hline
360 \\
-328 \\
\end{array}
$$

You Try It 2   Divide: 37.042 ÷ 76
Round to the nearest
thousandth.

Your solution

Example 3   Divide: 420.9 ÷ 7.06
Round to the nearest tenth.

Solution

$$
\begin{array}{r}
59.61 \approx 59.6 \\
7.06\overline{)\ 420.90.00} \\
-353\ 0 \\
\hline
67\ 90 \\
-63\ 54 \\
\hline
4\ 36\ 0 \\
-4\ 23\ 6 \\
\hline
12\ 40 \\
-\ 7\ 06 \\
\end{array}
$$

You Try It 3   Divide: 370.2 ÷ 5.09
Round to the nearest tenth.

Your solution

Example 4   Divide: 402.75 ÷ 1000

Solution   402.75 ÷ 1000 = 0.40275

You Try It 4   Divide: 309.21 ÷ 10,000

Your solution

Example 5   What is 0.625 divided by 10^2?

Solution $0.625 \div 10^2 = 0.00625$

You Try It 5   What is 42.93 divided by 10^4?

Your solution

Solutions on p. S10

● E S T I M A T I O N ●

Estimating the Quotient of Two Decimals

Calculate 282.18 ÷ 0.48. Then use estimation to determine whether the quotient is reasonable.

Divide to find the exact quotient.   282.18 ÷ 0.48 = 587.875

To estimate the quotient, round each   282.18 ÷ 0.48 ≈ 300 ÷ 0.5
number so there is one nonzero digit.   = 600
Then divide. The estimated answer is 600,
which is very close to the exact quotient,
587.875.

Copyright © Houghton Mifflin Company. All rights reserved.

Objective B **To solve application problems**

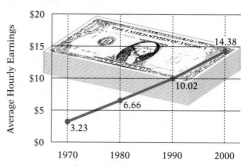

The graph at the right shows average hourly earnings in the United States. Use this table for Example 6 and You Try It 6.

Average Hourly Earnings in the United States

Sources: Statistical Abstract of the United States; Bureau of Labor Statistics

---

Example 6

How many times greater were the average hourly earnings in 2000 than in 1970? Round to the nearest whole number.

Strategy

To find how many times greater the average hourly earnings were, divide the 2000 average hourly earnings (14.38) by the 1970 average hourly earnings (3.23).

Solution

$14.38 \div 3.23 \approx 4$

The average hourly earnings in 2000 were about 4 times greater than in 1970.

You Try It 6

How many times greater were the average hourly earnings in 1990 than in 1970? Round to the nearest tenth.

Your strategy

Your solution

Example 7

A 1-year subscription to a monthly magazine costs $90. The price of each issue at the newsstand is $9.80. How much would you save per issue by buying a year's subscription rather than buying each issue at the newsstand?

Strategy

To find the amount saved:

• Find the subscription price per issue by dividing the cost of the subscription (90) by the number of issues (12).

• Subtract the subscription price per issue from the newsstand price (9.80).

Solution

$90 \div 12 = 7.50$
$9.80 - 7.50 = 2.30$

The savings would be $2.30 per issue.

You Try It 7

A Nielsen survey of the number of people (in millions) who watch television each day of the week is given in the table below.

Mon.	Tues.	Wed.	Thu.	Fri.	Sat.	Sun.
91.9	89.8	90.6	93.9	78.0	77.1	87.7

Find the average number of people who watch television per day.

Your strategy

Your solution

Solutions on pp. S10–S11

Copyright © Houghton Mifflin Company. All rights reserved.

3.5 Exercises

Objective A   **To divide decimals**

For Exercises 1 to 20, divide.

1. $3\overline{)2.46}$

2. $7\overline{)3.71}$

3. $0.8\overline{)3.84}$

4. $0.9\overline{)6.93}$

5. $0.7\overline{)62.3}$

6. $0.4\overline{)52.8}$

7. $0.4\overline{)24}$

8. $0.5\overline{)65}$

9. $0.7\overline{)59.01}$

10. $0.9\overline{)8.721}$

11. $0.5\overline{)16.15}$

12. $0.8\overline{)77.6}$

13. $0.7\overline{)3.542}$

14. $0.6\overline{)2.436}$

15. $6.3\overline{)8.19}$

16. $3.2\overline{)7.04}$

17. $3.6\overline{)0.396}$

18. $2.7\overline{)0.648}$

19. $6.9\overline{)26.22}$

20. $1.7\overline{)84.66}$

For Exercises 21 to 29, divide. Round to the nearest tenth.

21. $55.62 \div 8.8$

22. $25.43 \div 5.4$

23. $5.427 \div 9.5$

24. $1.837 \div 1.4$

25. $18.4 \div 7.3$

26. $52.9 \div 8.1$

27. $0.183 \div 0.17$

28. $0.381 \div 0.47$

29. $6.924 \div 0.053$

For Exercises 30 to 38, divide. Round to the nearest hundredth.

30. $4.817 \div 16$

31. $6.467 \div 8$

32. $0.0418 \div 0.53$

33. $0.0647 \div 0.72$

34. $7 \div 0.55$

35. $38.665 \div 0.95$

36. $13.97 \div 25.4$

37. $27.738 \div 60.8$

38. $3.171 \div 45.6$

Copyright © Houghton Mifflin Company. All rights reserved.

For Exercises 39 to 47, divide. Round to the nearest thousandth.

39. 1.028 ÷ 54

40. 6.729 ÷ 27

41. 0.0437 ÷ 0.5

42. 75.469 ÷ 77.8

43. 34.31 ÷ 95.3

44. 0.2695 ÷ 2.67

45. 0.4871 ÷ 4.72

46. 0.1142 ÷ 17.2

47. 0.2307 ÷ 26.7

For Exercises 48 to 56, divide. Round to the nearest whole number.

48. 16.5 ÷ 4

49. 89.76 ÷ 90

50. 1.94 ÷ 0.3

51. 1.0478 ÷ 0.413

52. 2.148 ÷ 0.519

53. 0.79 ÷ 0.778

54. 3.092 ÷ 0.075

55. 392 ÷ 6.9

56. 8.729 ÷ 0.075

For Exercises 57 to 74, divide.

57. 4.07 ÷ 10

58. 0.039 ÷ 10

59. 42.67 ÷ 10

60. 389.7 ÷ 100

61. 1.037 ÷ 100

62. 237.835 ÷ 100

63. 8.295 ÷ 1000

64. 82,547 ÷ 1000

65. 825.37 ÷ 1000

66. $8.35 \div 10^1$

67. $0.32 \div 10^1$

68. $87.65 \div 10^1$

69. $23.627 \div 10^2$

70. $2.954 \div 10^2$

71. $0.0053 \div 10^2$

72. $289.32 \div 10^3$

73. $1.8932 \div 10^3$

74. $0.139 \div 10^3$

Copyright © Houghton Mifflin Company. All rights reserved.

75. Divide 44.208 by 2.4.

76. Divide 0.04664 by 0.44.

77. Find the quotient of 723.15 and 45.

78. Find the quotient of 3.3463 and 3.07.

79. Divide 13.5 by 10^3.

80. Divide 0.045 by 10^5.

81. Find the quotient of 23.678 and 1000.

82. Find the quotient of 7.005 and 10,000.

83. What is 0.0056 divided by 0.05?

84. What is 123.8 divided by 0.02?

For Exercises 85 to 96, use a calculator to divide. Round to the nearest ten-thousandth. Then use estimation to determine whether the quotient you calculated is reasonable.

85. $42.42 \div 3.8$

86. $69.8 \div 7.2$

87. $389 \div 0.44$

88. $642 \div 0.83$

89. $6.394 \div 3.5$

90. $8.429 \div 4.2$

91. $1.235 \div 0.021$

92. $7.456 \div 0.072$

93. $95.443 \div 1.32$

94. $423.0925 \div 4.0927$

95. $1.000523 \div 429.07$

96. $0.03629 \div 0.00054$

Objective B   **To solve application problems**

97. Sports   Ramon, a high school football player, gained 162 yards on 26 carries in a high school football game. Find the average number of yards gained per carry. Round to the nearest hundredth.

98. Fuel Efficiency   A car with an odometer reading of 17,814.2 is filled with 9.4 gallons of gas. At an odometer reading of 18,130.4, the tank is empty and the car is filled with 12.4 gallons of gas. How many miles does the car travel on 1 gallon of gasoline?

99. Consumerism   A case of diet cola costs $6.79. If there are 24 cans in a case, find the cost per can. Round to the nearest cent.

100. Carpentry   Anne is building bookcases that are 3.4 feet long. How many complete shelves can be cut from a 12-foot board?

101. ● **Travel**   When the Massachusetts Turnpike opened, the toll for a passenger car that traveled the entire 136 miles of it was $5.60. Calculate the cost per mile. Round to the nearest cent.

Copyright © Houghton Mifflin Company. All rights reserved.

102. Investments An oil company has issued 3,541,221,500 shares of stock. The company paid $6,090,990,120 in dividends. Find the dividend for each share of stock. Round to the nearest cent.

103. Insurance Earl is 52 years old and is buying $70,000 of life insurance for an annual premium of $703.80. If he pays each annual premium in 12 equal installments, how much is each monthly payment?

104. Email The graph at the right shows the growth in the number of spam messages sent daily worldwide. Figures given are in billions. How many times greater is the number of spam messages sent daily in 2004 than the number sent in 2000? Round to the nearest tenth.

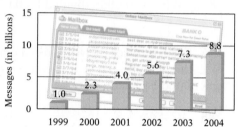

Spam Messages Sent World Wide

Source: IDC

APPLYING THE CONCEPTS

105. Education According to the National Center for Education Statistics, 10.2 million women and 7.3 million men will be enrolled at institutions of higher learning in 2010. How many more women than men are expected to be attending institutions of higher learning in 2010?

The Military The table at the right shows the advertising budgets of four branches of the U.S. armed services in a recent year. Use this table for Exercises 106 to 108.

Service	Advertising Budget
Army	$85.3 million
Air Force	$41.1 million
Navy	$20.5 million
Marines	$15.9 million

Source: CMR/TNS Media Intelligence

106. Find the difference between the Army's advertising budget and the Marines' advertising budget.

107. How many times greater was the Army's advertising budget than the Navy's advertising budget? Round to the nearest tenth.

108. What was the total of the advertising budgets for the four branches of the service?

109. Population Growth The U.S. population of people ages 85 and over is expected to grow from 4.2 million in 2000 to 8.9 million in 2030. How many times greater is the population of this segment expected to be in 2030 than in 2000? Round to the nearest tenth.

Copyright © Houghton Mifflin Company. All rights reserved.

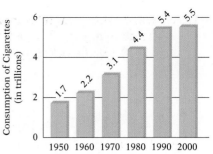

Health   The graph at the right shows that the worldwide consumption of cigarettes has been increasing. Use this table for Exercises 110 to 112.

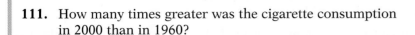

World Wide Consumption of Cigarettes

Source: The Tobacco Atlas; U.S. Department of Agriculture

110. Find the increase in cigarette consumption from 1950 to 1990.

111. How many times greater was the cigarette consumption in 2000 than in 1960?

112. **a.** During which 10-year period was the increase in cigarette consumption greatest?
  b. During which 10-year period was the increase in cigarette consumption the least?

Forestry   The table at the right shows the number of acres burned in wildfires through July 24 of each year listed. Use this table for Exercises 113 and 114.

Year	Acres Burned
2000	2.9 million
2001	1.5 million
2002	3.8 million
2003	1.5 million

Source: National Interagency Fire Center

113. Find the total number of acres burned in the 4 years listed.

114. How many more acres were burned in 2002 and 2003 than in 2000 and 2001?

115. Explain how the decimal point is moved when a number is divided by 10, 100, 1000, 10,000, etc.

116. **Sports**   Explain how baseball batting averages are determined. Then find Chicago Cubs shortstop Nomar Garciaparra's batting average with 190 hits out of 532 at bats. Round to the nearest thousandth.

117. Explain how the decimal point is placed in the quotient when a number is divided by a decimal.

For Exercises 118 to 123, insert $+$, $-$, $\times$, or $\div$ into the square so that the statement is true.

118. $3.45 \,\square\, 0.5 = 6.9$

119. $3.46 \,\square\, 0.24 = 0.8304$

120. $6.009 \,\square\, 4.68 = 1.329$

121. $0.064 \,\square\, 1.6 = 0.1024$

122. $9.876 \,\square\, 23.12 = 32.996$

123. $3.0381 \,\square\, 1.23 = 2.47$

For Exercises 124 to 126, fill in the square to make a true statement.

124. $6.47 - \square = 1.253$

125. $6.47 + \square = 9$

126. $0.009 \div \square = 0.36$

Copyright © Houghton Mifflin Company. All rights reserved.

3.6 Comparing and Converting Fractions and Decimals

Copyright © Houghton Mifflin Company. All rights reserved.

Objective A To convert fractions to decimals

Every fraction can be written as a decimal. To write a fraction as a decimal, divide the numerator of the fraction by the denominator. The quotient can be rounded to the desired place value.

HOW TO Convert $\frac{3}{7}$ to a decimal.

$$7)\overline{3.00000} \quad 0.42857$$

$\frac{3}{7}$ rounded to the nearest hundredth is 0.43.

$\frac{3}{7}$ rounded to the nearest thousandth is 0.429.

$\frac{3}{7}$ rounded to the nearest ten-thousandth is 0.4286.

HOW TO Convert $3\frac{2}{9}$ to a decimal. Round to the nearest thousandth.

$$3\frac{2}{9} = \frac{29}{9} \qquad 9)\overline{29.0000} \quad 3.2222$$

$3\frac{2}{9}$ rounded to the nearest thousandth is 3.222.

Example 1 Convert $\frac{3}{8}$ to a decimal. Round to the nearest hundredth.

Solution $8)\overline{3.000} \quad 0.375 \approx 0.38$

Example 2 Convert $2\frac{3}{4}$ to a decimal. Round to the nearest tenth.

Solution $2\frac{3}{4} = \frac{11}{4} \qquad 4)\overline{11.00} \quad 2.75 \approx 2.8$

You Try It 1 Convert $\frac{9}{16}$ to a decimal. Round to the nearest tenth.

Your solution

You Try It 2 Convert $4\frac{1}{6}$ to a decimal. Round to the nearest hundredth.

Your solution

Solutions on p. S11

Objective B To convert decimals to fractions

To convert a decimal to a fraction, remove the decimal point and place the decimal part over a denominator equal to the place value of the last digit in the decimal.

$$0.47 = \frac{47}{100} \text{ hundredths}$$

$$7.45 = 7\frac{45}{100} = 7\frac{9}{20} \text{ hundredths}$$

$$0.275 = \frac{275}{1000} = \frac{11}{40} \text{ thousandths}$$

$$0.16\frac{2}{3} = \frac{16\frac{2}{3}}{100} = 16\frac{2}{3} \div 100 = \frac{50}{3} \times \frac{1}{100} = \frac{1}{6} \text{ hundredths}$$

Example 3 Convert 0.82 and 4.75 to fractions.

Solution $0.82 = \dfrac{82}{100} = \dfrac{41}{50}$

$4.75 = 4\dfrac{75}{100} = 4\dfrac{3}{4}$

You Try It 3 Convert 0.56 and 5.35 to fractions.

Your solution

Example 4 Convert $0.15\dfrac{2}{3}$ to a fraction.

Solution $0.15\dfrac{2}{3} = \dfrac{15\dfrac{2}{3}}{100} = 15\dfrac{2}{3} \div 100$

$= \dfrac{47}{3} \times \dfrac{1}{100} = \dfrac{47}{300}$

You Try It 4 Convert $0.12\dfrac{7}{8}$ to a fraction.

Your solution

Solutions on p. S11

Objective C **To identify the order relation between two decimals or between a decimal and a fraction**

Decimals, like whole numbers and fractions, can be graphed as points on the number line. The number line can be used to show the order of decimals. A decimal that appears to the right of a given number is greater than the given number. A decimal that appears to the left of a given number is less than the given number.

3.00 3.05 3.10 3.15 3.20 3.25 3.30 3.35 3.40

Note that 3, 3.0, and 3.00 represent the same number.

HOW TO Find the order relation between $\dfrac{3}{8}$ and 0.38.

$\dfrac{3}{8} = 0.375$ $0.38 = 0.380$ • Convert the fraction $\dfrac{3}{8}$ to a decimal.

$0.375 < 0.380$ • Compare the two decimals.

$\dfrac{3}{8} < 0.38$ • Convert 0.375 back to a fraction.

Example 5 Place the correct symbol, < or >, between the numbers.

$\dfrac{5}{16}$ 0.32

Solution $\dfrac{5}{16} \approx 0.313$ • Convert $\dfrac{5}{16}$ to a decimal.

$0.313 < 0.32$ • Compare the two decimals.

$\dfrac{5}{16} < 0.32$ • Convert 0.313 back to a fraction.

You Try It 5 Place the correct symbol, < or >, between the numbers.

0.63 $\dfrac{5}{8}$

Your solution

Solution on p. S11

Copyright © Houghton Mifflin Company. All rights reserved.

3.6 Exercises   Do odds for homework

To convert fractions to decimals

For Exercises 1 to 24, convert the fraction to a decimal. Round to the nearest thousandth.

1. $\dfrac{5}{8}$ 2. $\dfrac{7}{12}$ 3. $\dfrac{2}{3}$ 4. $\dfrac{5}{6}$ 5. $\dfrac{1}{6}$ 6. $\dfrac{7}{8}$

7. $\dfrac{5}{12}$ 8. $\dfrac{9}{16}$ 9. $\dfrac{7}{4}$ 10. $\dfrac{5}{3}$ 11. $1\dfrac{1}{2}$ 12. $2\dfrac{1}{3}$

13. $\dfrac{16}{4}$ 14. $\dfrac{36}{9}$ 15. $\dfrac{3}{1000}$ 16. $\dfrac{5}{10}$ 17. $7\dfrac{2}{25}$ 18. $16\dfrac{7}{9}$

19. $37\dfrac{1}{2}$ 20. $\dfrac{5}{24}$ 21. $\dfrac{4}{25}$ 22. $3\dfrac{1}{3}$ 23. $8\dfrac{2}{5}$ 24. $5\dfrac{4}{9}$

To convert decimals to fractions

For Exercises 25 to 49, convert the decimal to a fraction.

25. 0.8 26. 0.4 27. 0.32 28. 0.48 29. 0.125

30. 0.485 31. 1.25 32. 3.75 33. 16.9 34. 17.5

35. 8.4 36. 10.7 37. 8.437 38. 9.279 39. 2.25

40. 7.75 41. $0.15\dfrac{1}{3}$ 42. $0.17\dfrac{2}{3}$ 43. $0.87\dfrac{7}{8}$ 44. $0.12\dfrac{5}{9}$

45. 7.38 46. 0.33 47. 0.57 48. $0.33\dfrac{1}{3}$ 49. $0.66\dfrac{2}{3}$

Copyright © Houghton Mifflin Company. All rights reserved.

> **Objective C**   **To identify the order relation between two decimals or between a decimal and a fraction**

For Exercises 50 to 69, place the correct symbol, $<$ or $>$, between the numbers.

50. 0.15   0.5

51. 0.6   0.45

52. 6.65   6.56

53. 3.89   3.98

54. 2.504   2.054

55. 0.025   0.105

56. $\dfrac{3}{8}$ 0.365

57. $\dfrac{4}{5}$ 0.802

58. $\dfrac{2}{3}$ 0.65

59. 0.85 $\dfrac{7}{8}$

60. $\dfrac{5}{9}$ 0.55

61. $\dfrac{7}{12}$ 0.58

62. 0.62 $\dfrac{7}{15}$

63. $\dfrac{11}{12}$ 0.92

64. 0.161 $\dfrac{1}{7}$

65. 0.623   0.6023

66. 0.86   0.855

67. 0.87   0.087

68. 1.005   0.5

69. 0.033   0.3

APPLYING THE CONCEPTS

Demography   The graph at the right shows the U.S. population according to the 2000 census. Use this table for Exercises 70 to 72.

70. Are there more individuals ages 0 to 39 or ages 40 and over in the United States?

71. Is the population ages 0 to 19 more than or less than $\dfrac{1}{4}$ of the total population?

72. Is the population ages 0 to 39 more than or less than $\dfrac{1}{2}$ of the total population?

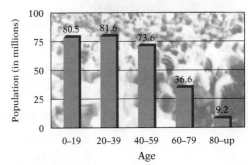

U.S. Population
Source: U.S. Census Bureau

73. Is $\dfrac{7}{13}$ in decimal form a repeating decimal? Why or why not? (*Hint:* See Projects and Group Activities, page 165.)

74. If a number is rounded to the nearest thousandth, is it always greater than if it were rounded to the nearest hundredth? Give examples to support your answer.

75. Explain how terminating, repeating, and nonrepeating decimals differ. Give an example of each kind of decimal.

Copyright © Houghton Mifflin Company. All rights reserved.

Focus on Problem Solving

Relevant Information

Problems in mathematics or real life involve a question or a need and information or circumstances related to that question or need. Solving problems in the sciences usually involves a question, an observation, and measurements of some kind.

One of the challenges of problem solving in the sciences is to separate the information that is relevant to the problem from other information. Following is an example from the physical sciences in which some relevant information was omitted.

Hooke's Law states that the distance that a weight will stretch a spring is directly proportional to the weight on the spring. That is, $d = kF$, where d is the distance the spring is stretched and F is the force. In an experiment to verify this law, some physics students were continually getting inconsistent results. Finally, the instructor discovered that the heat produced when the lights were turned on was affecting the experiment. In this case, relevant information was omitted—namely, that the temperature of the spring can affect the distance it will stretch.

A lawyer drove 8 miles to the train station. After a 35-minute ride of 18 miles, the lawyer walked 10 minutes to the office. Find the total time it took the lawyer to get to work.

From this situation, answer the following before reading on.

a. What is asked for?

b. Is there enough information to answer the question?

c. Is information given that is not needed?

Here are the answers.

a. We want the total time for the lawyer to get to work.

b. No. We do not know the time it takes the lawyer to get to the train station.

c. Yes. Neither the distance to the train station nor the distance of the train ride is necessary to answer the question.

For each of the following problems, answer the questions printed in red above.

1. A customer bought 6 boxes of strawberries and paid with a $20 bill. What was the change?

2. A board is cut into two pieces. One piece is 3 feet longer than the other piece. What is the length of the original board?

3. A family rented a car for their vacation and drove 680 miles. The cost of the rental car was $21 per day with 150 free miles per day and $.15 for each mile driven above the number of free miles allowed. How many miles did the family drive per day?

4. An investor bought 8 acres of land for $80,000. One and one-half acres were set aside for a park, and the remainder was developed into one-half-acre lots. How many lots were available for sale?

5. You wrote checks of $43.67, $122.88, and $432.22 after making a deposit of $768.55. How much do you have left in your checking account?

Copyright © Houghton Mifflin Company. All rights reserved.

Projects and Group Activities

Fractions as Terminating or Repeating Decimals

The fraction $\frac{3}{4}$ is equivalent to 0.75. The decimal 0.75 is a terminating decimal because there is a remainder of zero when 3 is divided by 4. The fraction $\frac{1}{3}$ is equivalent to 0.333 . . . . The three dots mean the pattern continues on and on. 0.333 . . . is a repeating decimal. To determine whether a fraction can be written as a terminating decimal, first write the fraction in simplest form. Then look at the denominator of the fraction. If it contains prime factors of only 2s and/or 5s, then it can be expressed as a terminating decimal. If it contains prime factors other than 2s or 5s, it represents a repeating decimal.

> **TAKE NOTE**
>
> If the denominator of a fraction in simplest form is 20, then it can be written as a terminating decimal because $20 = 2 \cdot 2 \cdot 5$ (only prime factors of 2 and 5). If the denominator of a fraction in simplest form is 6, it represents a repeating decimal because it contains the prime factor 3 (a number other than 2 or 5).

1. Assume that each of the following numbers is the denominator of a fraction written in simplest form. Does the fraction represent a terminating or repeating decimal?

   a. 4   **b.** 5   **c.** 7   **d.** 9   **e.** 10   **f.** 12   **g.** 15
   h. 16   **i.** 18   **j.** 21   **k.** 24   **l.** 25   **m.** 28   **n.** 40

2. Write two other numbers that, as denominators of fractions in simplest form, represent terminating decimals, and write two other numbers that, as denominators of fractions in simplest form, represent repeating decimals.

Chapter 3 Summary

Key Words

A number written in *decimal notation* has three parts: a *whole-number part*, a *decimal point*, and a *decimal part*. The decimal part of a number represents a number less than 1. A number written in decimal notation is often simply called a *decimal*. [3.1A, p. 127]

Examples

For the decimal 31.25, 31 is the whole-number part and 25 is the decimal part.

Essential Rules and Procedures

To write a decimal in words, write the decimal part as if it were a whole number. Then name the place value of the last digit. The decimal point is read as "and." [3.1A, p. 127]

Examples

The decimal 12.875 is written in words as twelve and eight hundred seventy-five thousandths.

To write a decimal in standard form when it is written in words, write the whole-number part, replace the word *and* with a decimal point, and write the decimal part so that the last digit is in the given place-value position. [3.1A, p. 128]

The decimal forty-nine and sixty-three thousandths is written in standard form as 49.063.

To round a decimal to a given place value, use the same rules used with whole numbers, except drop the digits to the right of the given place value instead of replacing them with zeros. [3.1B, p. 129]

2.7134 rounded to the nearest tenth is 2.7.
0.4687 rounded to the nearest hundredth is 0.47.

Copyright © Houghton Mifflin Company. All rights reserved.

To add decimals, write the decimals so that the decimal points are on a vertical line. Add as you would with whole numbers. Then write the decimal point in the sum directly below the decimal points in the addends. [3.2A, p. 133]

$$
\begin{array}{r}
{\scriptstyle 1\ 1} \\
1.35 \\
20.8 \\
+\ \ 0.76 \\
\hline
22.91
\end{array}
$$

To subtract decimals, write the decimals so that the decimal points are on a vertical line. Subtract as you would with whole numbers. Then write the decimal point in the difference directly below the decimal point in the subtrahend. [3.3A, p. 137]

$$
\begin{array}{r}
{\scriptstyle 2\ 15\quad 6\ 10} \\
\cancel{3}\ \cancel{5}.\cancel{8}\ \cancel{7}\ \cancel{0} \\
-\ \ 9.6\ 4\ 1 \\
\hline
2\ 6.2\ 2\ 9
\end{array}
$$

To multiply decimals, multiply the numbers as you would whole numbers. Then write the decimal point in the product so that the number of decimal places in the product is the sum of the decimal places in the factors. [3.4A, p. 141]

$$
\begin{array}{rl}
26.83 & \text{2 decimal places} \\
\times\ \ \ \ 0.45 & \text{2 decimal places} \\
\hline
13415 & \\
10732\ \ \ & \\
\hline
12.0735 & \text{4 decimal places}
\end{array}
$$

To multiply a decimal by a power of 10, move the decimal point to the right the same number of places as there are zeros in the power of 10. If the power of 10 is written in exponential notation, the exponent indicates how many places to move the decimal point. [3.4A, pp. 141, 142]

$3.97 \cdot 10{,}000 = 39{,}700$
$0.641 \cdot 10^5 = 64{,}100$

To divide decimals, move the decimal point in the divisor to the right so that it is a whole number. Move the decimal point in the dividend the same number of places to the right. Place the decimal point in the quotient directly above the decimal point in the dividend. Then divide as you would with whole numbers. [3.5A, p. 151]

$$
\begin{array}{r}
6.2 \\
0.39.\overline{)2.41.8} \\
-2\ 34 \\
\hline
7\ 8 \\
-7\ 8 \\
\hline
0
\end{array}
$$

To divide a decimal by a power of 10, move the decimal point to the left the same number of places as there are zeros in the power of 10. If the power of 10 is written in exponential notation, the exponent indicates how many places to move the decimal point. [3.5A, p. 152]

$972.8 \div 1000 = 0.9728$
$61.305 \div 10^4 = 0.0061305$

To convert a fraction to a decimal, divide the numerator of the fraction by the denominator. [3.6A, p. 160]

$\dfrac{7}{8} = 7 \div 8 = 0.875$

To convert a decimal to a fraction, remove the decimal point and place the decimal part over a denominator equal to the place value of the last digit in the decimal. [3.6B, p. 160]

0.85 is eighty-five <u>hundredths</u>.
$0.85 = \dfrac{85}{100} = \dfrac{17}{20}$

To find the order relation between a decimal and a fraction, first rewrite the fraction as a decimal. Then compare the two decimals. [3.6C, p. 161]

Because $\dfrac{3}{11} \approx 0.273$, and $0.273 > 0.26$, $\dfrac{3}{11} > 0.26$.

Copyright © Houghton Mifflin Company. All rights reserved.

Chapter 3 Review Exercises

1. Find the quotient of 3.6515 and 0.067.

2. Find the sum of 369.41, 88.3, 9.774, and 366.474.

3. Place the correct symbol, $<$ or $>$, between the two numbers.
0.055   0.1

4. Write 22.0092 in words.

5. Round 0.05678235 to the nearest hundred-thousandth.

6. Convert $2\frac{1}{3}$ to a decimal. Round to the nearest hundredth.

7. Convert 0.375 to a fraction.

8. Add: $3.42 + 0.794 + 32.5$

9. Write thirty-four and twenty-five thousandths in standard form.

10. Place the correct symbol, $<$ or $>$, between the two numbers.
$\frac{5}{8}$ 0.62

11. Convert $\frac{7}{9}$ to a decimal. Round to the nearest thousandth.

12. Convert 0.66 to a fraction.

13. Subtract: $27.31 - 4.4465$

14. Round 7.93704 to the nearest hundredth.

Copyright © Houghton Mifflin Company. All rights reserved.

15. Find the product of 3.08 and 2.9.

16. Write 342.37 in words.

17. Write three and six thousand seven hundred fifty-three hundred-thousandths in standard form.

18. Multiply: 34.79
$\times$ 0.74

19. Divide: $0.053\overline{)0.349482}$

20. What is 7.796 decreased by 2.9175?

21. **Banking**   You had a balance of $895.68 in your checking account. You then wrote checks of $145.72 and $88.45. Find the new balance in your checking account.

Education   The graph at the right shows where American children in grades K–12 are being educated. Figures are in millions of children. Use this graph for Exercises 22 and 23.

22. Find the total number of American children in grades K–12.

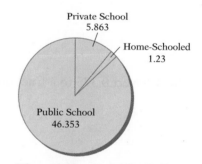

Private School
5.863

Home-Schooled
1.23

Public School
46.353

Where Children in Grades K–12 Are
Being Educated in America
(in millions)

Sources: U.S. Department of Education;
Home School Legal Defense Association

23. How many more children are being educated in public school than in private school?

24. **Nutrition**   According to the American School Food Service Association, 1.9 million gallons of milk are served in school cafeterias every day. How many gallons of milk are served in school cafeterias during a 5-day school week?

25. **Travel**   In a recent year, 30.6 million Americans drove to their destinations over Thanksgiving, and 4.8 million Americans traveled by plane. (*Source:* AAA) How many times greater is the number who drove than the number who flew? Round to the nearest tenth.

Copyright © Houghton Mifflin Company. All rights reserved.

Chapter 3 Test

1. Place the correct symbol, $<$ or $>$, between the two numbers.
0.66   0.666

2. Subtract: 13.027
    $-$ 8.94

3. Write 45.0302 in words.

4. Convert $\frac{9}{13}$ to a decimal. Round to the nearest thousandth.

5. Convert 0.825 to a fraction.

6. Round 0.07395 to the nearest ten-thousandth.

7. Find 0.0569 divided by 0.037. Round to the nearest thousandth.

8. Find 9.23674 less than 37.003.

9. Round 7.0954625 to the nearest thousandth.

10. Divide: $0.006\overline{)1.392}$

11. Add: 270.93
        97.
        1.976
    $+$ 88.675

12. **Mechanics**   Find the missing dimension.

4.86 in.

?

6.23 in.

Copyright © Houghton Mifflin Company. All rights reserved.

13. Multiply: 1.37
               $\times$ 0.004

14. What is the total of 62.3, 4.007, and 189.65?

15. Write two hundred nine and seven thousand eighty-six hundred-thousandths in standard form.

16. Finances A car was bought for $16,734.40, with a down payment of $2500. The balance was paid in 36 monthly payments. Find the amount of each monthly payment.

17. Compensation You received a salary of $727.50, a commission of $1909.64, and a bonus of $450. Find your total income.

18. Consumerism A long-distance telephone call costs $.85 for the first 3 minutes and $.42 for each additional minute. Find the cost of a 12-minute long-distance telephone call.

Computers The table at the right shows the average number of hours per week that students use a computer. Use this table for Exercises 19 and 20.

19. On average, how many hours per year does a 10th-grade student use a computer? Use a 52-week year.

20. On average, how many more hours per year does a 2nd-grade student use a computer than a 5th-grade student? Use a 52-week year.

Grade Level	Average Number of Hours of Computer Use per Week
Prekindergarten – kindergarten	3.9
1st – 3rd	4.9
4th – 6th	4.2
7th – 8th	6.9
9th – 12th	6.7

Source: Find/SVP American Learning Household Survey

Copyright © Houghton Mifflin Company. All rights reserved.

Cumulative Review Exercises

1. Divide: $89\overline{)20,932}$

2. Simplify: $2^3 \cdot 4^2$

3. Simplify: $2^2 - (7 - 3) \div 2 + 1$

4. Find the LCM of 9, 12, and 24.

5. Write $\frac{22}{5}$ as a mixed number.

6. Write $4\frac{5}{8}$ as an improper fraction.

7. Write an equivalent fraction with the given denominator.
$$\frac{5}{12} = \frac{}{60}$$

8. Add: $\frac{3}{8} + \frac{5}{12} + \frac{9}{16}$

9. What is $5\frac{7}{12}$ increased by $3\frac{7}{18}$?

10. Subtract: $9\frac{5}{9} - 3\frac{11}{12}$

11. Multiply: $\frac{9}{16} \times \frac{4}{27}$

12. Find the product of $2\frac{1}{8}$ and $4\frac{5}{17}$.

13. Divide: $\frac{11}{12} \div \frac{3}{4}$

14. What is $2\frac{3}{8}$ divided by $2\frac{1}{2}$?

15. Simplify: $\left(\frac{2}{3}\right)^2 \cdot \left(\frac{3}{4}\right)^3$

16. Simplify: $\left(\frac{2}{3}\right)^2 - \left(\frac{2}{3} - \frac{1}{2}\right) + 2$

17. Write 65.0309 in words.

18. Add:
$$
\begin{array}{r}
379.006 \\
27.523 \\
9.8707 \\
+ \ 88.2994 \\
\end{array}
$$

Copyright © Houghton Mifflin Company. All rights reserved.

19. What is 29.005 decreased by 7.9286?

20. Multiply: 9.074
  $\times$ 6.09

21. Divide: $8.09\overline{)17.42963}$.
Round to the nearest thousandth.

22. Convert $\frac{11}{15}$ to a decimal. Round to the nearest thousandth.

23. Convert $0.16\frac{2}{3}$ to a fraction.

24. Place the correct symbol, $<$ or $>$, between the two numbers.
$\frac{8}{9}$ 0.98

25. **Vacation** The graph at the right shows the number of vacation days per year that are legally mandated in several countries. How many more vacation days does Sweden mandate than Germany?

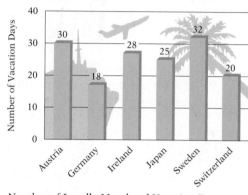

Number of Legally Mandated Vacation Days

Sources: Economic Policy Institute; *World Almanac*

26. **Health** A patient is put on a diet to lose 24 pounds in 3 months. The patient loses $9\frac{1}{2}$ pounds the first month and $6\frac{3}{4}$ pounds the second month. How much weight must this patient lose the third month to achieve the goal?

27. **Banking** You have a checking account balance of $814.35. You then write checks for $42.98, $16.43, and $137.56. Find your checking account balance after you write the checks.

28. **Mechanics** A machine lathe takes 0.017 inch from a brass bushing that is 1.412 inches thick. Find the resulting thickness of the bushing.

29. **Taxes** The state income tax on your business is $820 plus 0.08 times your profit. You made a profit of $64,860 last year. Find the amount of income tax you paid last year.

30. **Finances** You bought a camera costing $210.96. The down payment was $20, and the balance is to be paid in 8 equal monthly payments. Find the monthly payment.

Copyright © Houghton Mifflin Company. All rights reserved.

4 Ratio and Proportion

Egypt is known around the world for its enormous pyramids, built thousands of years ago by the pharaohs. This photo shows the pyramids at Giza. The largest pyramid at Giza, called the Great Pyramid, dates to approximately 2600 B.C. and is the oldest of the Seven Ancient Wonders of the World. It is also the only ancient wonder that still survives today. Some historians believe that some of the pyramids of Egypt incorporate a special ratio called the golden ratio. This ratio, which is described in the **project on page 192,** has been used extensively in art and architecture for centuries. The ratio of the slant height to a side of the base of the pyramid reflects the golden ratio.

OBJECTIVES

Section 4.1
A To write the ratio of two quantities in simplest form
B To solve application problems

Section 4.2
A To write rates
B To write unit rates
C To solve application problems

Section 4.3
A To determine whether a proportion is true
B To solve proportions
C To solve application problems

Copyright © Houghton Mifflin Company. All rights reserved.

Need help? For online student resources, such as section quizzes, visit this textbook's website at **math.college.hmco.com/students.**

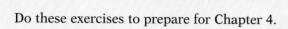

Do these exercises to prepare for Chapter 4.

1. Simplify: $\dfrac{8}{10}$

2. Simplify: $\dfrac{450}{650 + 250}$

3. Write as a decimal: $\dfrac{372}{15}$

4. Which is greater, 4×33 or 62×2?

5. Complete: $? \times 5 = 20$

GO FIGURE • • •

Luis, Kim, Reggie, and Dave are standing in line. Dave is not first. Kim is between Luis and Reggie. Luis is between Dave and Kim. Give the order in which the men are standing.

Copyright © Houghton Mifflin Company. All rights reserved.

4.1 Ratio

Objective A **To write the ratio of two quantities in simplest form**

Copyright © Houghton Mifflin Company. All rights reserved.

Point of Interest

In the 1990s, the major-league pitchers with the best strikeout-to-walk ratios (having pitched a minimum of 100 innings) were

Dennis Eckersley	6.46:1
Shane Reynolds	4.13:1
Greg Maddux	4:1
Bret Saberhagen	3.92:1
Rod Beck	3.81:1

The best single-season strikeout-to-walk ratio for starting pitchers in the same period was that of Bret Saberhagen, 11:1. (*Source:* Elias Sports Bureau)

Quantities such as 3 feet, 12 cents, and 9 cars are number quantities written with units.

$$\begin{matrix} 3 \text{ feet} \\ 12 \text{ cents} \\ 9 \text{ cars} \\ \uparrow \\ \text{units} \end{matrix}$$

These are some examples of units. Shirts, dollars, trees, miles, and gallons are further examples.

A **ratio** is a comparison of two quantities that have the *same* units. This comparison can be written three different ways:

1. As a fraction
2. As two numbers separated by a colon (:)
3. As two numbers separated by the word *to*

The ratio of the lengths of two boards, one 8 feet long and the other 10 feet long, can be written as

1. $\dfrac{8 \text{ feet}}{10 \text{ feet}} = \dfrac{8}{10} = \dfrac{4}{5}$
2. $8 \text{ feet} : 10 \text{ feet} = 8 : 10 = 4 : 5$
3. $8 \text{ feet to } 10 \text{ feet} = 8 \text{ to } 10 = 4 \text{ to } 5$

Writing the **simplest form of a ratio** means writing it so that the two numbers have no common factor other than 1.

This ratio means that the smaller board is $\dfrac{4}{5}$ the length of the longer board.

Example 1

Write the comparison $6 to $8 as a ratio in simplest form using a fraction, a colon, and the word *to*.

Solution $\dfrac{\$6}{\$8} = \dfrac{6}{8} = \dfrac{3}{4}$

$\$6 : \$8 = 6 : 8 = 3 : 4$

$\$6 \text{ to } \$8 = 6 \text{ to } 8 = 3 \text{ to } 4$

You Try It 1

Write the comparison 20 pounds to 24 pounds as a ratio in simplest form using a fraction, a colon, and the word *to*.

Your solution

Example 2

Write the comparison 18 quarts to 6 quarts as a ratio in simplest form using a fraction, a colon, and the word *to*.

Solution $\dfrac{18 \text{ quarts}}{6 \text{ quarts}} = \dfrac{18}{6} = \dfrac{3}{1}$

$18 \text{ quarts} : 6 \text{ quarts} =$
$18 : 6 = 3 : 1$

$18 \text{ quarts to } 6 \text{ quarts} =$
$18 \text{ to } 6 = 3 \text{ to } 1$

You Try It 2

Write the comparison 64 miles to 8 miles as a ratio in simplest form using a fraction, a colon, and the word *to*.

Your solution

Solutions on p. S11

Objective B   To solve application problems

Use the table below for Example 3 and You Try It 3.

Board Feet of Wood at a Lumber Store			
Pine	Ash	Oak	Cedar
20,000	18,000	10,000	12,000

Example 3

Find, as a fraction in simplest form, the ratio of the number of board feet of pine to the number of board feet of oak.

Strategy
To find the ratio, write the ratio of board feet of pine (20,000) to board feet of oak (10,000) in simplest form.

Solution
$$\frac{20,000}{10,000} = \frac{2}{1}$$

The ratio is $\frac{2}{1}$.

You Try It 3

Find, as a fraction in simplest form, the ratio of the number of board feet of cedar to the number of board feet of ash.

Your strategy

Your solution

Example 4

The cost of building a patio cover was $500 for labor and $700 for materials. What, as a fraction in simplest form, is the ratio of the cost of materials to the total cost for labor and materials?

Strategy
To find the ratio, write the ratio of the cost of materials ($700) to the total cost ($500 + $700) in simplest form.

Solution
$$\frac{\$700}{\$500 + \$700} = \frac{700}{1200} = \frac{7}{12}$$

The ratio is $\frac{7}{12}$.

You Try It 4

A company spends $60,000 a month for television advertising and $45,000 a month for radio advertising. What, as a fraction in simplest form, is the ratio of the cost of radio advertising to the total cost of radio and television advertising?

Your strategy

Your solution

Solutions on p. S11

Copyright © Houghton Mifflin Company. All rights reserved.

4.1 Exercises

odds

Objective A   **To write the ratio of two quantities in simplest form**

For Exercises 1 to 24, write the comparison as a ratio in simplest form using a fraction, a colon (:), and the word *to*.

1. 3 pints to 15 pints

2. 6 pounds to 8 pounds

3. $40 to $20

4. 10 feet to 2 feet

5. 3 miles to 8 miles

6. 2 hours to 3 hours

7. 37 hours to 24 hours

8. 29 inches to 12 inches

9. 6 minutes to 6 minutes

10. 8 days to 12 days

11. 35 cents to 50 cents

12. 28 inches to 36 inches

13. 30 minutes to 60 minutes

14. 25 cents to 100 cents

15. 32 ounces to 16 ounces

16. 12 quarts to 4 quarts

17. 3 cups to 4 cups

18. 6 years to 7 years

19. $5 to $3

20. 30 yards to 12 yards

21. 12 quarts to 18 quarts

22. 20 gallons to 28 gallons

23. 14 days to 7 days

24. 9 feet to 3 feet

Objective B   **To solve application problems**

For Exercises 25 to 28, write ratios in simplest form using a fraction.

Family Budget						
Housing	Food	Transportation	Taxes	Utilities	Miscellaneous	Total
$1600	$800	$600	$700	$300	$800	$4800

25. Budgets   Use the table to find the ratio of housing cost to total expenses.

26. Budgets   Use the table to find the ratio of food cost to total expenses.

27. Budgets   Use the table to find the ratio of utilities cost to food cost.

28. Budgets   Use the table to find the ratio of transportation cost to housing cost.

Copyright © Houghton Mifflin Company. All rights reserved.

29. 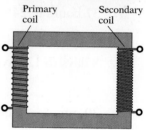 **Sports**   National Collegiate Athletic Association (NCAA) statistics show that for every 154,000 high school seniors playing basketball, only 4000 will play college basketball as first-year students. Write the ratio of the number of first-year students playing college basketball to the number of high school seniors playing basketball.

30. 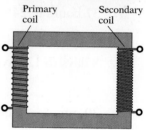 **Sports**   NCAA statistics show that for every 2800 college seniors playing college basketball, only 50 will play as rookies in the National Basketball Association. Write the ratio of the number of National Basketball Association rookies to the number of college seniors playing basketball.

31. **Electricity**   A transformer has 40 turns in the primary coil and 480 turns in the secondary coil. State the ratio of the number of turns in the primary coil to the number of turns in the secondary coil.

Primary coil   Secondary coil

32. **Consumerism**   Rita Sterling bought a computer system for $2400. Five years later she sold the computer for $900. Find the ratio of the amount she received for the computer to the cost of the computer.

33. **Real Estate**   A house with an original value of $90,000 increased in value to $110,000 in 5 years.
  a. Find the increase in the value of the house.
  b. What is the ratio of the increase in value to the original value of the house?

34. **Energy Prices**   The price of gasoline jumped from $1.35 to $1.62 in 1 year.
  a. What was the increase in the price per gallon?
  b. What is the ratio of the increase in price to the original price?

APPLYING THE CONCEPTS

Banking   A bank uses the ratio of a borrower's total monthly debts to the borrower's total monthly income to determine the maximum monthly payment for a potential homeowner. This ratio is called the debt–income ratio. Use the homeowner's debt–income table at the right for Exercises 35 and 36.

Income	Debt
$5500	$1200
$450	$300
$250	$450
	$250

35. Compute the debt–income ratio for the potential homeowner.

36. Central Trust Bank will make a loan to a customer whose debt–income ratio is less than $\frac{1}{3}$. Will the potential homeowner qualify? Explain your answer.

37. **Banking**   To make a home loan, First National Bank requires a debt–income ratio that is less than $\frac{2}{5}$. Would the home-owner whose debt–income table is given at the right qualify for a loan using these standards? Explain.

Income		Debt	
Salary	$3400	Mortgage	$1800
Interest	$83	Property tax	$104
Rent	$650	Insurance	$35
Dividends	$34	Liabilities	$120
		Credit card	$234
		Car loan	$197

38. 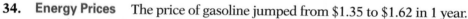 Is the value of a ratio always less than 1? Explain.

Copyright © Houghton Mifflin Company. All rights reserved.

4.2 Rates

Objective A   **To write rates**

Point of Interest

Listed below are rates at which some crimes are committed in our nation.

Crime	Every
Larceny	4 seconds
Burglary	14 seconds
Robbery	60 seconds
Rape	6 minutes
Murder	31 minutes

A **rate** is a comparison of two quantities that have *different* units. A rate is written as a fraction.

A distance runner ran 26 miles in 4 hours. The distance-to-time rate is written

$$\frac{26 \text{ miles}}{4 \text{ hours}} = \frac{13 \text{ miles}}{2 \text{ hours}}$$

Writing the **simplest form of a rate** means writing it so that the two numbers that form the rate have no common factor other than 1.

Example 1   Write "6 roof supports for every 9 feet" as a rate in simplest form.

You Try It 1   Write "15 pounds of fertilizer for 12 trees" as a rate in simplest form.

Solution $\dfrac{6 \text{ supports}}{9 \text{ feet}} = \dfrac{2 \text{ supports}}{3 \text{ feet}}$

Your solution

Solution on p. S11

Objective B   **To write unit rates**

Point of Interest

According to a Gallup Poll, women see doctors more often than men do. On average, men visit the doctor 3.8 times per year, whereas women go to the doctor 5.8 times per year.

A **unit rate** is a rate in which the number in the denominator is 1.

$$\frac{\$3.25}{1 \text{ pound}}$$ or $3.25/pound is read "$3.25 per pound."

To find unit rates, divide the number in the numerator of the rate by the number in the denominator of the rate.

A car traveled 344 miles on 16 gallons of gasoline. To find the miles per gallon (unit rate), divide the numerator of the rate by the denominator of the rate.

$$\frac{344 \text{ miles}}{16 \text{ gallons}}$$ is the rate.

$$\begin{array}{r} 21.5 \\ 16\overline{)344.0} \end{array}$$ 21.5 miles/gallon is the unit rate.

Example 2   Write "300 feet in 8 seconds" as a unit rate.

You Try It 2   Write "260 miles in 8 hours" as a unit rate.

Solution $\dfrac{300 \text{ feet}}{8 \text{ seconds}}$ $8\overline{)300.0}^{\,37.5}$

37.5 feet/second

Your solution

Solution on p. S11

Copyright © Houghton Mifflin Company. All rights reserved.

Objective C **To solve application problems**

HOW TO The table at the right shows air fares for some routes in the continental United States. Find the cost per mile for the four routes in order to determine the most expensive route and the least expensive route on the basis of mileage flown.

Long Routes	Miles	Fare
New York–Los Angeles	2475	$683
San Francisco–Dallas	1464	$536
Denver–Pittsburgh	1302	$525
Minneapolis–Hartford	1050	$483

Strategy
To find the cost per mile, divide the fare by the miles flown for each route. Compare the costs per mile to determine the most expensive and least expensive routes per mile.

Solution   New York–Los Angeles $\dfrac{683}{2475} \approx 0.28$

San Francisco–Dallas $\dfrac{536}{1464} \approx 0.37$

Denver–Pittsburgh $\dfrac{525}{1302} \approx 0.40$

Minneapolis–Hartford $\dfrac{483}{1050} = 0.46$

$0.28 < 0.37 < 0.40 < 0.46$

The Minneapolis–Hartford route is the most expensive per mile, and the New York–Los Angeles route is the least expensive per mile.

Denver Airport

Integrating Technology

To calculate the costs per mile using a calculator, perform four divisions:
683 ÷ 2475 =
536 ÷ 1464 =
525 ÷ 1302 =
483 ÷ 1050 =
In each case, round the number in the display to the nearest hundredth.

Example 3

As an investor, Jung Ho purchased 100 shares of stock for $1500. One year later, Jung sold the 100 shares for $1800. What was his profit per share?

Strategy
To find Jung's profit per share:
• Find the total profit by subtracting the original cost ($1500) from the selling price ($1800).
• Find the profit per share (unit rate) by dividing the total profit by the number of shares of stock (100).

Solution
$1800 - 1500 = 300$

$300 \div 100 = 3$

Jung Ho's profit was $3/share.

You Try It 3

Erik Peltier, a jeweler, purchased 5 ounces of gold for $1625. Later, he sold the 5 ounces for $1720. What was Erik's profit per ounce?

Your strategy

Your solution

Solution on p. S11

Copyright © Houghton Mifflin Company. All rights reserved.

4.2 Exercises

Objective A   **To write rates**

For Exercises 1 to 10, write each phrase as a rate in simplest form.

1. 3 pounds of meat for 4 people

2. 30 ounces in 24 glasses

3. $80 for 12 boards

4. 84 cents for 3 bars of soap

5. 300 miles on 15 gallons

6. 88 feet in 8 seconds

7. 20 children in 8 families

8. 48 leaves on 9 plants

9. 16 gallons in 2 hours

10. 25 ounces in 5 minutes

Objective B   **To write unit rates**

For Exercises 11 to 24, write each phrase as a unit rate.   *odds*

11. 10 feet in 4 seconds

12. 816 miles in 6 days

13. $3900 earned in 4 weeks

14. $51,000 earned in 12 months

15. 1100 trees planted on 10 acres

16. 3750 words on 15 pages

17. $131.88 earned in 7 hours

18. $315.70 earned in 22 hours

19. 628.8 miles in 12 hours

20. 388.8 miles in 8 hours

21. 344.4 miles on 12.3 gallons of gasoline

22. 409.4 miles on 11.5 gallons of gasoline

23. $349.80 for 212 pounds

24. $11.05 for 3.4 pounds

Objective C   **To solve application problems**

25. **Fuel Efficiency**   An automobile was driven 326.6 miles on 11.5 gallons of gas. Find the number of miles driven per gallon of gas.

26. **Travel**   You drive 246.6 miles in 4.5 hours. Find the number of miles you drove per hour.

27.   **Fuel Efficiency**   The Saturn-5 rocket uses 534,000 gallons of fuel in 2.5 minutes. How much fuel does the rocket use per minute?

Copyright © Houghton Mifflin Company. All rights reserved.

28. Manufacturing Assume that Regency Computer produced 5000 zip disks for $26,536.32. Of the disks made, 122 did not meet company standards.
   a. How many disks did meet company standards?
   b. What was the cost per disk for those disks that met company standards?

29. Consumerism The Pierre family purchased a 250-pound side of beef for $365.75 and had it packaged. During the packaging, 75 pounds of beef were discarded as waste.
   a. How many pounds of beef were packaged?
   b. What was the cost per pound for the packaged beef?

30. Advertising In 2003, the average price of a 30-second commercial during the sitcom *Friends* was $455,700. (*Source: Time*) Find the price per second.

31. The Film Industry During filming, an IMAX camera uses 65-mm film at a rate of 5.6 feet per second.
   a. At what rate per minute does the camera go through film?
   b. At what rate does it use a 500-foot roll of 65-mm film? Round to the nearest second.

32. Demography The table at the right shows the population and area of three countries. The population density of a country is the number of people per square mile.
   a. Which country has the least population density?
   b. How many more people per square mile are there in India than in the United States? Round to the nearest whole number.

Country	Population	Area (in square miles)
Australia	19,547,000	2,968,000
India	1,045,845,000	1,269,000
United States	291,929,000	3,619,000

Another application of rates is in the area of international trade. Suppose a company in Canada purchases a shipment of sneakers from an American company. The Canadian company must exchange Canadian dollars for U.S. dollars in order to pay for the order. The number of Canadian dollars that are equivalent to 1 U.S. dollar is called the **exchange rate.**

33. Exchange Rates The table at the right shows the exchange rates per U.S. dollar for three foreign countries and for the euro at the time of this writing.
   a. How many euros would be paid for an order of American computer hardware costing $120,000?
   b. Calculate the cost, in Japanese yen, of an American car costing $34,000.

Exchange Rates per U.S. Dollar	
Australian Dollar	1.545
Canadian Dollar	1.386
Japanese Yen	117.000
The Euro	0.9103

APPLYING THE CONCEPTS

34. Compensation You have a choice of receiving a wage of $34,000 per year, $2840 per month, $650 per week, or $18 per hour. Which pay choice would you take? Assume a 40-hour week with 52 weeks per year.

35. The price–earnings ratio of a company's stock is one measure used by stock market analysts to assess the financial well-being of the company. Explain the meaning of the price–earnings ratio.

Copyright © Houghton Mifflin Company. All rights reserved.

Proportions

Copyright © Houghton Mifflin Company. All rights reserved.

Objective A   **To determine whether a proportion is true**

A **proportion** is an expression of the equality of two ratios or rates.

Point of Interest

Proportions were studied by the earliest mathematicians. Clay tablets uncovered by archaeologists show evidence of proportions in Egyptian and Babylonian cultures dating from 1800 B.C.

$$\frac{50 \text{ miles}}{4 \text{ gallons}} = \frac{25 \text{ miles}}{2 \text{ gallons}}$$

Note that the units of the numerators are the same and the units of the denominators are the same.

$$\frac{3}{6} = \frac{1}{2}$$

This is the equality of two ratios.

A proportion is **true** if the fractions are equal when written in lowest terms.

In any true proportion, the **cross products** are equal.

HOW TO   Is $\frac{2}{3} = \frac{8}{12}$ a true proportion?

$$\frac{2}{3} \bowtie \frac{8}{12} \quad \begin{array}{l} 3 \times 8 = 24 \\ 2 \times 12 = 24 \end{array}$$

The cross products *are* equal.
$\frac{2}{3} = \frac{8}{12}$ is a true proportion.

A proportion is **not true** if the fractions are not equal when reduced to lowest terms.

If the cross products are not equal, then the proportion is not true.

HOW TO   Is $\frac{4}{5} = \frac{8}{9}$ a true proportion?

$$\frac{4}{5} \bowtie \frac{8}{9} \quad \begin{array}{l} 5 \times 8 = 40 \\ 4 \times 9 = 36 \end{array}$$

The cross products *are not* equal.
$\frac{4}{5} = \frac{8}{9}$ is not a true proportion.

Example 1   Is $\frac{5}{8} = \frac{10}{16}$ a true proportion?

Solution
$$\frac{5}{8} \bowtie \frac{10}{16} \quad \begin{array}{l} 8 \times 10 = 80 \\ 5 \times 16 = 80 \end{array}$$

The cross products are equal.
The proportion is true.

You Try It 1   Is $\frac{6}{10} = \frac{9}{15}$ a true proportion?

Your solution

Example 2   Is $\frac{62 \text{ miles}}{4 \text{ gallons}} = \frac{33 \text{ miles}}{2 \text{ gallons}}$ a true proportion?

Solution
$$\frac{62}{4} \bowtie \frac{33}{2} \quad \begin{array}{l} 4 \times 33 = 132 \\ 62 \times 2 = 124 \end{array}$$

The cross products are not equal.
The proportion is not true.

You Try It 2   Is $\frac{\$32}{6 \text{ hours}} = \frac{\$90}{8 \text{ hours}}$ a true proportion?

Your solution

Solutions on p. S11

Objective B   **To solve proportions**

S t u d y   T i p

An important element of success is practice. We cannot do anything well if we do not practice it repeatedly. Practice is crucial to success in mathematics. In this objective you are learning a new skill: how to solve a proportion. You will need to practice this skill over and over again in order to be successful at it.

Sometimes one of the numbers in a proportion is unknown. In this case, it is necessary to *solve* the proportion.

To **solve a proportion**, find a number to replace the unknown so that the proportion is true.

HOW TO   Solve: $\dfrac{9}{6} = \dfrac{3}{n}$

$$\frac{9}{6} = \frac{3}{n}$$

$9 \times n = 6 \times 3$ • Find the cross products.

$9 \times n = 18$

$n = 18 \div 9$ • Think of $9 \times n = 18$ as $9\overline{)18}$.

$n = 2$

Check:

$\dfrac{9}{6} \bowtie \dfrac{3}{2} \rightarrow 6 \times 3 = 18$
$\phantom{\dfrac{9}{6} \bowtie \dfrac{3}{2} } \rightarrow 9 \times 2 = 18$

Example 3   Solve $\dfrac{n}{12} = \dfrac{25}{60}$ and check.

Solution

$n \times 60 = 12 \times 25$ • Find the cross
$n \times 60 = 300$ products. Then
$n = 300 \div 60$ solve for n.
$n = 5$

Check:

$\dfrac{5}{12} \bowtie \dfrac{25}{60} \rightarrow 12 \times 25 = 300$
$\phantom{\dfrac{5}{12} \bowtie \dfrac{25}{60}} \rightarrow 5 \times 60 = 300$

You Try It 3   Solve $\dfrac{n}{14} = \dfrac{3}{7}$ and check.

Your solution

Example 4   Solve $\dfrac{4}{9} = \dfrac{n}{16}$. Round to the nearest tenth.

Solution

$4 \times 16 = 9 \times n$ • Find the cross
$64 = 9 \times n$ products. Then
$64 \div 9 = n$ solve for n.
$7.1 \approx n$

Note: A rounded answer is an approximation. Therefore, the answer to a check will not be exact.

You Try It 4   Solve $\dfrac{5}{7} = \dfrac{n}{20}$. Round to the nearest tenth.

Your solution

Solutions on p. S12

Copyright © Houghton Mifflin Company. All rights reserved.

Example 5   Solve $\frac{28}{52} = \frac{7}{n}$ and check.

Solution

$28 \times n = 52 \times 7$ • **Find the cross**
$28 \times n = 364$ **products. Then**
$n = 364 \div 28$ **solve for *n*.**
$n = 13$

Check:

$\frac{28}{52} \diagtimes \frac{7}{13} \rightarrow \begin{matrix} 52 \times 7 = 364 \\ 28 \times 13 = 364 \end{matrix}$

You Try It 5   Solve $\frac{15}{20} = \frac{12}{n}$ and check.

Your solution

Example 6   Solve $\frac{15}{n} = \frac{8}{3}$. Round to the nearest hundredth.

Solution

$15 \times 3 = n \times 8$
$45 = n \times 8$
$45 \div 8 = n$
$5.63 \approx n$

You Try It 6   Solve $\frac{12}{n} = \frac{7}{4}$. Round to the nearest hundredth.

Your solution

Example 7   Solve $\frac{n}{9} = \frac{3}{1}$ and check.

Solution

$n \times 1 = 9 \times 3$
$n \times 1 = 27$
$n = 27 \div 1$
$n = 27$

Check:

$\frac{27}{9} \diagtimes \frac{3}{1} \rightarrow \begin{matrix} 9 \times 3 = 27 \\ 27 \times 1 = 27 \end{matrix}$

You Try It 7   Solve $\frac{n}{12} = \frac{4}{1}$ and check.

Your solution

Solutions on p. S12

Objective C   **To solve application problems**

The application problems in this objective require you to write and solve a proportion. When setting up a proportion, remember to keep the same units in the numerator and the same units in the denominator.

Copyright © Houghton Mifflin Company. All rights reserved.

Example 8

The dosage of a certain medication is 2 ounces for every 50 pounds of body weight. How many ounces of this medication are required for a person who weighs 175 pounds?

Strategy

To find the number of ounces of medication for a person weighing 175 pounds, write and solve a proportion using n to represent the number of ounces of medication for a 175-pound person.

Solution

$$\frac{2 \text{ ounces}}{50 \text{ pounds}} = \frac{n \text{ ounces}}{175 \text{ pounds}}$$

- The unit "ounces" is in the numerator. The unit "pounds" is in the denominator.

$$2 \times 175 = 50 \times n$$
$$350 = 50 \times n$$
$$350 \div 50 = n$$
$$7 = n$$

A 175-pound person requires 7 ounces of medication.

You Try It 8

Three tablespoons of a liquid plant fertilizer are to be added to every 4 gallons of water. How many tablespoons of fertilizer are required for 10 gallons of water?

Your strategy

Your solution

Example 9

A mason determines that 9 cement blocks are required for a retaining wall 2 feet long. At this rate, how many cement blocks are required for a retaining wall that is 24 feet long?

Strategy

To find the number of cement blocks for a retaining wall 24 feet long, write and solve a proportion using n to represent the number of blocks required.

Solution

$$\frac{9 \text{ cement blocks}}{2 \text{ feet}} = \frac{n \text{ cement blocks}}{24 \text{ feet}}$$

$$9 \times 24 = 2 \times n$$
$$216 = 2 \times n$$
$$216 \div 2 = n$$
$$108 = n$$

A 24-foot retaining wall requires 108 cement blocks.

You Try It 9

Twenty-four jars can be packed in 6 identical boxes. At this rate, how many jars can be packed in 15 boxes?

Your strategy

Your solution

Solutions on p. S12

Copyright © Houghton Mifflin Company. All rights reserved.

4.3 Exercises   *Odds*

For Exercises 1 to 24, determine whether the proportion is true or not true.

1. $\dfrac{4}{8} \bcancel{=} \dfrac{10}{20}$ T $4 \cdot 20 = 80$ $8 \cdot 10 = 80$

2. $\dfrac{39}{48} = \dfrac{13}{16}$

3. $\dfrac{7}{8} = \dfrac{11}{12}$

4. $\dfrac{15}{7} = \dfrac{17}{8}$

5. $\dfrac{27}{8} = \dfrac{9}{4}$

6. $\dfrac{3}{18} = \dfrac{4}{19}$

7. $\dfrac{45}{135} = \dfrac{3}{9}$

8. $\dfrac{3}{4} = \dfrac{54}{72}$

9. $\dfrac{16}{3} = \dfrac{48}{9}$

10. $\dfrac{15}{5} = \dfrac{3}{1}$

11. $\dfrac{7}{40} = \dfrac{7}{8}$

12. $\dfrac{9}{7} = \dfrac{6}{5}$

13. $\dfrac{50 \text{ miles}}{2 \text{ gallons}} = \dfrac{25 \text{ miles}}{1 \text{ gallon}}$

14. $\dfrac{16 \text{ feet}}{10 \text{ seconds}} = \dfrac{24 \text{ feet}}{15 \text{ seconds}}$

15. $\dfrac{6 \text{ minutes}}{5 \text{ cents}} = \dfrac{30 \text{ minutes}}{25 \text{ cents}}$

16. $\dfrac{16 \text{ pounds}}{12 \text{ days}} = \dfrac{20 \text{ pounds}}{14 \text{ days}}$

17. $\dfrac{\$15}{4 \text{ pounds}} = \dfrac{\$45}{12 \text{ pounds}}$

18. $\dfrac{270 \text{ trees}}{6 \text{ acres}} = \dfrac{90 \text{ trees}}{2 \text{ acres}}$

19. $\dfrac{300 \text{ feet}}{4 \text{ rolls}} = \dfrac{450 \text{ feet}}{7 \text{ rolls}}$

20. $\dfrac{1 \text{ gallon}}{4 \text{ quarts}} = \dfrac{7 \text{ gallons}}{28 \text{ quarts}}$

21. $\dfrac{\$65}{5 \text{ days}} = \dfrac{\$26}{2 \text{ days}}$

22. $\dfrac{80 \text{ miles}}{2 \text{ hours}} = \dfrac{110 \text{ miles}}{3 \text{ hours}}$

23. $\dfrac{7 \text{ tiles}}{4 \text{ feet}} = \dfrac{42 \text{ tiles}}{20 \text{ feet}}$

24. $\dfrac{15 \text{ feet}}{3 \text{ yards}} = \dfrac{90 \text{ feet}}{18 \text{ yards}}$

Copyright © Houghton Mifflin Company. All rights reserved.

Objective B   **To solve proportions**

For Exercises 25 to 52, solve. Round to the nearest hundredth, if necessary.

25. $\dfrac{n}{4} = \dfrac{6}{8}$

26. $\dfrac{n}{7} = \dfrac{9}{21}$

27. $\dfrac{12}{18} = \dfrac{n}{9}$

28. $\dfrac{7}{21} = \dfrac{35}{n}$

29. $\dfrac{6}{n} = \dfrac{24}{36}$

30. $\dfrac{3}{n} = \dfrac{15}{10}$

31. $\dfrac{n}{45} = \dfrac{17}{135}$

32. $\dfrac{9}{4} = \dfrac{18}{n}$

33. $\dfrac{n}{6} = \dfrac{2}{3}$

34. $\dfrac{5}{12} = \dfrac{n}{144}$

35. $\dfrac{n}{5} = \dfrac{7}{8}$

36. $\dfrac{4}{n} = \dfrac{9}{5}$

37. $\dfrac{n}{11} = \dfrac{32}{4}$

38. $\dfrac{3}{4} = \dfrac{8}{n}$

39. $\dfrac{5}{12} = \dfrac{n}{8}$

40. $\dfrac{36}{20} = \dfrac{12}{n}$

41. $\dfrac{n}{15} = \dfrac{21}{12}$

42. $\dfrac{40}{n} = \dfrac{15}{8}$

43. $\dfrac{32}{n} = \dfrac{1}{3}$

44. $\dfrac{5}{8} = \dfrac{42}{n}$

45. $\dfrac{18}{11} = \dfrac{16}{n}$

46. $\dfrac{25}{4} = \dfrac{n}{12}$

47. $\dfrac{28}{8} = \dfrac{12}{n}$

48. $\dfrac{n}{30} = \dfrac{65}{120}$

49. $\dfrac{0.3}{5.6} = \dfrac{n}{25}$

50. $\dfrac{1.3}{16} = \dfrac{n}{30}$

51. $\dfrac{0.7}{9.8} = \dfrac{3.6}{n}$

52. $\dfrac{1.9}{7} = \dfrac{13}{n}$

Objective C   **To solve application problems**

For Exercises 53 to 71, solve. Round to the nearest hundredth.

53.   **Nutrition**   A 6-ounce package of Puffed Wheat contains 600 calories. How many calories are in a 0.5-ounce serving of the cereal?

54.   **Fuel Efficiency**   A car travels 70.5 miles on 3 gallons of gas. Find the distance that the car can travel on 14 gallons of gas.

Copyright © Houghton Mifflin Company. All rights reserved.

55. **Landscaping**   Ron Stokes uses 2 pounds of fertilizer for every 100 square feet of lawn for landscape maintenance. At this rate, how many pounds of fertilizer did he use on a lawn that measures 3500 square feet?

56. **Gardening**   A nursery prepares a liquid plant food by adding 1 gallon of water for each 2 ounces of plant food. At this rate, how many gallons of water are required for 25 ounces of plant food?

57. **Manufacturing**   A manufacturer of baseball equipment makes 4 aluminum bats for every 15 bats made from wood. On a day when 100 aluminum bats are made, how many wooden bats are produced?

58. **Masonry**   A brick wall 20 feet in length contains 1040 bricks. At the same rate, how many bricks would it take to build a wall 48 feet in length?

59. **Cartography**   The scale on the map at the right is "1.25 inches equals 10 miles." Find the distance between Carlsbad and Del Mar, which are 2 inches apart on the map.

60. ~~Architecture   The scale on the plans~~ for a new house is "1 inch equals ~~3 feet." Find the~~ width and the length of a room that measures 5 inches by 8 inches on the drawing.

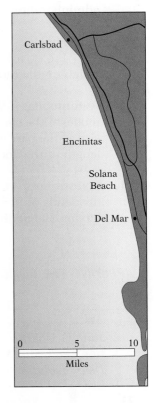

61. **Medicine**   The dosage for a medication is $\frac{1}{3}$ ounce for every 40 pounds of body weight. At this rate, how many ounces of medication should a physician prescribe for a patient who weighs 150 pounds? Write the answer as a decimal.

62. **Banking**   A bank requires a monthly payment of $33.45 on a $2500 loan. At the same rate, find the monthly payment on a $10,000 loan.

63. **Elections**   A pre-election survey showed that 2 out of every 3 eligible voters would cast ballots in the county election. At this rate, how many people in a county of 240,000 eligible voters would vote in the election?

64. **Interior Design**   A paint manufacturer suggests using 1 gallon of paint for every 400 square feet of a wall. At this rate, how many gallons of paint would be required for a room that has 1400 square feet of wall?

65. **Insurance**   A 60-year-old male can obtain $10,000 of life insurance for $35.35 per month. At this rate, what is the monthly cost of $50,000 of life insurance?

66. **Manufacturing**   Suppose a computer chip manufacturer knows from experience that in an average production run of 2000 circuit boards, 60 will be defective. How many defective circuit boards can be expected from a run of 25,000 circuit boards?

Copyright © Houghton Mifflin Company. All rights reserved.

67. **Investments**   You own 240 shares of stock in a computer company. The company declares a stock split of 5 shares for every 3 owned. How many shares of stock will you own after the stock split?

68. **Computers**   The director of data processing at a college estimates that the ratio of student time to administrative time on a certain computer is 3:2. During a month in which the computer was used 200 hours for administration, how many hours was it used by students?

69. **Physics**   The ratio of weight on the moon to weight on Earth is 1:6. If a bowling ball weighs 16 pounds on Earth, what would it weigh on the moon?

70. **Automobiles**   When engineers designed a new car, they first built a model of the car. The ratio of the size of a part on the model to the actual size of the part is 2:5. If a door is 1.3 feet long on the model, what is the length of the door on the car?

71. **Investments**   Carlos Capasso owns 50 shares of Texas Utilities that pay dividends of $153. At this rate, what dividend would Carlos receive after buying 300 additional shares of Texas Utilities?

APPLYING THE CONCEPTS

72. **Publishing**   In January 2003, *USA Today* reported that for every 100 copies of *Dr. Atkins' New Diet Revolution* sold, John Grisham's *The Summons* sold 7.3 copies. Explain how a proportion can be used to determine the number of copies of *Dr. Atkins' New Diet Revolution* sold given the number of copies of *The Summons* sold.

73. **Social Security**   According to the Social Security Administration, the numbers of workers per retiree in the future are expected to be as given in the table below.

Year	2010	2020	2030	2040
Number of workers per retiree	3.1	2.5	2.1	2.0

Why is the shrinking number of workers per retiree of importance to the Social Security Administration?

74. **Compensation**   In June 2002, *Time* magazine reported, "In 1980 the average CEO made 40 times the pay of the average factory worker; by 2000 the ratio had climbed to 531 to 1." What information would you need to know in order to determine the average pay of a CEO in 2000? With that information, how would you calculate the average pay of a CEO in 2000?

75. **Elections**   A survey of voters in a city claimed that 2 people of every 5 who voted cast a ballot in favor of city amendment A and that 3 people of every 4 who voted cast a ballot against amendment A. Is this possible? Explain your answer.

76. Write a word problem that requires solving a proportion to find the answer.

Copyright © Houghton Mifflin Company. All rights reserved.

Focus on Problem Solving

Looking for a Pattern

A very useful problem-solving strategy is looking for a pattern.

Problem   A legend says that a peasant invented the game of chess and gave it to a very rich king as a present. The king so enjoyed the game that he gave the peasant the choice of anything in the kingdom. The peasant's request was simple: "Place one grain of wheat on the first square, 2 grains on the second square, 4 grains on the third square, 8 on the fourth square, and continue doubling the number of grains until the last square of the chessboard is reached." How many grains of wheat must the king give the peasant?

Solution   A chessboard consists of 64 squares. To find the total number of grains of wheat on the 64 squares, we begin by looking at the amount of wheat on the first few squares.

Square 1	Square 2	Square 3	Square 4	Square 5	Square 6	Square 7	Square 8
1	2	4	8	16	32	64	128
1	3	7	15	31	63	127	255

The bottom row of numbers represents the sum of the number of grains of wheat up to and including that square. For instance, the number of grains of wheat on the first 7 squares is $1 + 2 + 4 + 8 + 16 + 32 + 64 = 127$.

One pattern to observe is that the number of grains of wheat on a square can be expressed as a power of 2.

The number of grains on square $n = 2^{n-1}$.

For example, the number of grains on square $7 = 2^{7-1} = 2^6 = 64$.

A second pattern of interest is that **the number *below* a square** (the total number of grains up to and including that square) **is 1 less than the number of grains of wheat *on the next* square.** For example, the number *below* square 7 is 1 less than the number *on* square 8 ($128 - 1 = 127$). From this observation, the number of grains of wheat on the first 8 squares is the number on square 8 (128) plus 1 less than the number on square 8 (127): The total number of grains of wheat on the first 8 squares is $128 + 127 = 255$.

From this observation,

$$\begin{array}{c} \text{Number of grains of} \\ \text{wheat on the chessboard} \end{array} = \begin{array}{c} \text{number of grains} \\ \text{on square 64} \end{array} + \begin{array}{c} \text{1 less than the number} \\ \text{of grains on square 64} \end{array}$$

$$= 2^{64-1} + (2^{64-1} - 1)$$

$$= 2^{63} + 2^{63} - 1 \approx 18{,}000{,}000{,}000{,}000{,}000{,}000$$

To give you an idea of the magnitude of this number, this is more wheat than has been produced in the world since chess was invented.

Copyright © Houghton Mifflin Company. All rights reserved.

The same king decided to have a banquet in the long banquet room of the palace to celebrate the invention of chess. The king had 50 square tables, and each table could seat only one person on each side. The king pushed the tables together to form one long banquet table. How many people could sit at this table? *Hint:* Try constructing a pattern by using 2 tables, 3 tables, and 4 tables.

Projects and Group Activities

The Golden Ratio   There are certain designs that have been repeated over and over in both art and architecture. One of these involves the **golden rectangle.**

A golden rectangle is drawn at the right. Begin with a square that measures, say, 2 inches on a side. Let *A* be the midpoint of a side (halfway between two corners). Now measure the distance from *A* to *B*. Place this length along the bottom of the square, starting at *A*. The resulting rectangle is a golden rectangle.

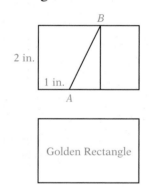

The **golden ratio** is the ratio of the length of the golden rectangle to its width. If you have drawn the rectangle following the procedure above, you will find that the golden ratio is approximately 1.6 to 1.

The golden ratio appears in many different situations. Some historians claim that some of the great pyramids of Egypt are based on the golden ratio. The drawing at the right shows the Pyramid of Giza, which dates from approximately 2600 B.C. The ratio of the height to a side of the base is approximately 1.6 to 1.

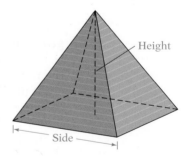

1. There are instances of the golden rectangle in the Mona Lisa painted by Leonardo da Vinci. Do some research on this painting and write a few paragraphs summarizing your findings.
2. What do 3 × 5 and 5 × 8 index cards have to do with the golden rectangle?
3. What does the United Nations Building in New York City have to do with the golden rectangle?
4. When was the Parthenon in Athens, Greece, built? What does the front of that building have to do with the golden rectangle?

Copyright © Houghton Mifflin Company. All rights reserved.

Drawing the Floor Plans for a Building

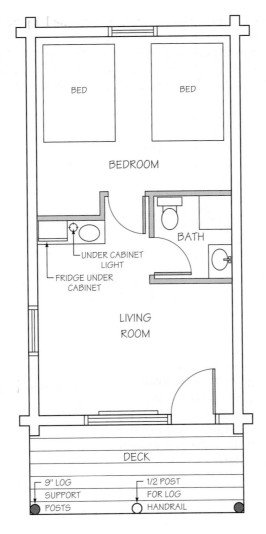

BED

BED

BEDROOM

BATH

UNDER CABINET
LIGHT

FRIDGE UNDER
CABINET

LIVING
ROOM

DECK

9" LOG
SUPPORT
POSTS

1/2 POST
FOR LOG
HANDRAIL

The drawing at the left is a sketch of the floor plan for a cabin at a resort in the mountains of Utah. The measurements are missing. Assume that you are the architect and will finish the drawing. You will have to decide the size of the rooms and put in the measurements to scale.

Design a cabin that you would like to own. Select a scale and draw all the rooms to scale.

If you are interested in architecture, visit an architect who is using CAD (computer-aided design) to create a floor plan. Computer technology has revolutionized the field of architectural design.

The U.S. House of Representatives

The framers of the Constitution decided to use a ratio to determine the number of representatives from each state. It was determined that each state would have one representative for every 30,000 citizens, with a minimum of one representative. Congress has changed this ratio over the years, and we now have 435 representatives.

Find the number of representatives from your state. Determine the ratio of citizens to representatives. Also do this for the most populous state and for the least populous state.

 You might consider getting information on the number of representatives for each state and the populations of different states via the Internet.

Copyright © Houghton Mifflin Company. All rights reserved.

Chapter 4 Summary

Key Words	Examples
A *ratio* is the comparison of two quantities with the same units. A ratio can be written in three ways: as a fraction, as two numbers separated by a colon (:), or as two numbers separated by the word *to*. A ratio is in *simplest form* when the two numbers do not have a common factor. [4.1A, p. 175]	The comparison 16 to 24 ounces can be written as a ratio in simplest form as $\frac{2}{3}$, 2:3, or 2 to 3.
A *rate* is the comparison of two quantities with different units. A rate is written as a fraction. A rate is in *simplest form* when the numbers that form the ratio do not have a common factor. [4.2A, p. 179]	You earned \$63 for working 6 hours. The rate is written in simplest form as $\frac{\$21}{2\ \text{hours}}$.
A *unit rate* is a rate in which the number in the denominator is 1. [4.2B, p. 179]	You traveled 144 miles in 3 hours. The unit rate is 48 miles per hour.
A *proportion* is an expression of the equality of two ratios or rates. A proportion is true if the fractions are equal when written in lowest terms; in any true proportion, the *cross products* are equal. A proportion is not true if the fractions are not equal when written in lowest terms; if the cross products are not equal, the proportion is not true. [4.3A, p. 183]	The proportion $\frac{3}{5} = \frac{12}{20}$ is true because the cross products are equal: $3 \times 20 = 5 \times 12$. The proportion $\frac{3}{4} = \frac{12}{20}$ is not true because the cross products are not equal: $3 \times 20 \neq 4 \times 12$.

Essential Rules and Procedures	Examples
To find a unit rate, divide the number in the numerator of the rate by the number in the denominator of the rate. [4.2B, p. 179]	You earned \$41 for working 4 hours. $$41 \div 4 = 10.25$$ The unit rate is \$10.25/hour.
To solve a proportion, find a number to replace the unknown so that the proportion is true. [4.3B, p. 184]	$\frac{6}{24} = \frac{9}{n}$ $6 \times n = 24 \times 9$ • Find the cross products. $6 \times n = 216$ $n = 216 \div 6$ $n = 36$
To set up a proportion, keep the same units in the numerator and the same units in the denominator. [4.3C, p. 185]	Three machines fill 5 cereal boxes per minute. How many boxes can 8 machines fill per minute? $$\frac{3\ \text{machines}}{5\ \text{cereal boxes}} = \frac{8\ \text{machines}}{n\ \text{cereal boxes}}$$

Copyright © Houghton Mifflin Company. All rights reserved.

Chapter 4 Review Exercises

1. Determine whether the proportion is true or not true.
$$\frac{2}{9} = \frac{10}{45}$$

2. Write the comparison 32 dollars to 80 dollars as a ratio in simplest form using a fraction, a colon (:), and the word *to*.

3. Write "250 miles in 4 hours" as a unit rate.

4. Determine whether the proportion is true or not true.
$$\frac{8}{15} = \frac{32}{60}$$

5. Solve the proportion.
$$\frac{16}{n} = \frac{4}{17}$$

6. Write "$300 earned in 40 hours" as a unit rate.

7. Write "$8.75 for 5 pounds" as a unit rate.

8. Write the comparison 8 feet to 28 feet as a ratio in simplest form using a fraction, a colon (:), and the word *to*.

9. Solve the proportion.
$$\frac{n}{8} = \frac{9}{2}$$

10. Solve the proportion. Round to the nearest hundredth.
$$\frac{18}{35} = \frac{10}{n}$$

11. Write the comparison 6 inches to 15 inches as a ratio in simplest form using a fraction, a colon (:), and the word *to*.

12. Determine whether the proportion is true or not true.
$$\frac{3}{8} = \frac{10}{24}$$

13. Write "$15 in 4 hours" as a rate in simplest form.

14. Write "326.4 miles on 12 gallons" as a unit rate.

15. Write the comparison 12 days to 12 days as a ratio in simplest form using a fraction, a colon (:), and the word *to*.

16. Determine whether the proportion is true or not true.
$$\frac{5}{7} = \frac{25}{35}$$

Copyright © Houghton Mifflin Company. All rights reserved.

17. Solve the proportion. Round to the nearest hundredth.

$$\frac{24}{11} = \frac{n}{30}$$

18. Write "100 miles in 3 hours" as a rate in simplest form.

19. **Business** In 5 years, the price of a calculator went from $40 to $24. What is the ratio, as a fraction in simplest form, of the decrease in price to the original price?

20. **Taxes** The property tax on a $245,000 home is $4900. At the same rate, what is the property tax on a home valued at $320,000?

21. **Meteorology** The high temperature during a 24-hour period was 84 degrees, and the low temperature was 42 degrees. Write the ratio, as a fraction in simplest form, of the high temperature to the low temperature for the 24-hour period.

22. **Manufacturing** The total cost of manufacturing 1000 cordless phones was $36,600. Of the phones made, 24 did not pass inspection. What is the cost per phone of the phones that *did* pass inspection?

23. **Masonry** A brick wall 40 feet in length contains 448 concrete blocks. At the same rate, how many blocks would it take to build a wall that is 120 feet in length?

24. **Advertising** A retail computer store spends $30,000 a year on radio advertising and $12,000 on newspaper advertising. Find the ratio, as a fraction in simplest form, of radio advertising to newspaper advertising.

25. **Consumerism** A 15-pound turkey costs $13.95. What is the cost per pound?

26. **Travel** Mahesh drove 198.8 miles in 3.5 hours. Find the average number of miles he drove per hour.

27. **Insurance** An insurance policy costs $9.87 for every $1000 of insurance. At this rate, what is the cost of $50,000 of insurance?

28. **Investments** Pascal Hollis purchased 80 shares of stock for $3580. What was the cost per share?

29. **Landscaping** Monique uses 1.5 pounds of fertilizer for every 200 square feet of lawn. How many pounds of fertilizer will she have to use on a lawn that measures 3000 square feet?

30. **Real Estate** A house had an original value of $80,000, but its value increased to $120,000 in 2 years. Find the ratio, as a fraction in simplest form, of the increase to the original value.

Copyright © Houghton Mifflin Company. All rights reserved.

Chapter 4 Test

1. Write "$46,036.80 earned in 12 months" as a unit rate.

2. Write the comparison 40 miles to 240 miles as a ratio in simplest form using a fraction, a colon (:), and the word *to*.

3. Write "18 supports for every 8 feet" as a rate in simplest form.

4. Determine whether the proportion is true or not true.
$$\frac{40}{125} = \frac{5}{25}$$

5. Write the comparison 12 days to 8 days as a ratio in simplest form using a fraction, a colon (:), and the word *to*.

6. Solve the proportion.
$$\frac{5}{12} = \frac{60}{n}$$

7. Write "256.2 miles on 8.4 gallons of gas" as a unit rate.

8. Write the comparison 27 dollars to 81 dollars as a ratio in simplest form using a fraction, a colon (:), and the word *to*.

9. Determine whether the proportion is true or not true.
$$\frac{5}{14} = \frac{25}{70}$$

10. Solve the proportion.
$$\frac{n}{18} = \frac{9}{4}$$

11. Write "$81 for 12 boards" as a rate in simplest form.

12. Write the comparison 18 feet to 30 feet as a ratio in simplest form using a fraction, a colon (:), and the word *to*.

Copyright © Houghton Mifflin Company. All rights reserved.

13. **Investments**   Fifty shares of a utility stock pay a dividend of $62.50. At the same rate, what is the dividend paid on 500 shares of the utility stock?

14. **Meteorology**   The average summer temperature in a California desert is 112 degrees. In a city 100 miles away, the average summer temperature is 86 degrees. Write the ratio, as a fraction in simplest form, of the average city temperature to the average desert temperature.

15. **Travel**   A plane travels 2421 miles in 4.5 hours. Find the plane's speed in miles per hour.

16. **Physiology**   A research scientist estimates that the human body contains 88 pounds of water for every 100 pounds of body weight. At this rate, estimate the number of pounds of water in a college student who weighs 150 pounds.

17. **Business**   If 40 feet of lumber costs $69.20, what is the per-foot cost of the lumber?

18. **Medicine**   The dosage of a certain medication is $\frac{1}{4}$ ounce for every 50 pounds of body weight. How many ounces of this medication are required for a person who weighs 175 pounds? Write the answer as a decimal.

19. **Sports**   A basketball team won 20 games and lost 5 games during the season. Write, as a fraction in simplest form, the ratio of the number of games won to the total number of games played.

20. **Manufacturing**   A computer manufacturer discovers through experience that an average of 3 defective hard drives are found in every 100 hard drives manufactured. How many defective hard drives are expected to be found in the production of 1200 hard drives?

Copyright © Houghton Mifflin Company. All rights reserved.

Cumulative Review Exercises

1. Subtract: $\begin{array}{r} 20,095 \\ -\ 10,937 \\ \hline \end{array}$

2. Write $2 \cdot 2 \cdot 2 \cdot 2 \cdot 3 \cdot 3 \cdot 3$ in exponential notation.

3. Simplify: $4 - (5 - 2)^2 \div 3 + 2$

4. Find the prime factorization of 160.

5. Find the LCM of 9, 12, and 18.

6. Find the GCF of 28 and 42.

7. Write $\frac{40}{64}$ in simplest form.

8. Find $4\frac{7}{15}$ more than $3\frac{5}{6}$.

9. What is $4\frac{5}{9}$ less than $10\frac{1}{6}$?

10. Multiply: $\frac{11}{12} \times 3\frac{1}{11}$

11. Find the quotient of $3\frac{1}{3}$ and $\frac{5}{7}$.

12. Simplify: $\left(\frac{2}{5} + \frac{3}{4}\right) \div \frac{3}{2}$

13. Write 4.0709 in words.

14. Round 2.09762 to the nearest hundredth.

15. Divide: $8.09\overline{)16.0976}$
Round to the nearest thousandth.

16. Convert $0.06\frac{2}{3}$ to a fraction.

Copyright © Houghton Mifflin Company. All rights reserved.

17. Write the comparison 25 miles to 200 miles as a ratio in simplest form using a fraction.

18. Write "87 cents for 6 pencils" as a rate in simplest form.

19. Write "250.5 miles on 7.5 gallons of gas" as a unit rate.

20. Solve $\dfrac{40}{n} = \dfrac{160}{17}$.

21. **Travel**   A car traveled 457.6 miles in 8 hours. Find the car's speed in miles per hour.

22. Solve the proportion.
$$\dfrac{12}{5} = \dfrac{n}{15}$$

23. **Banking**   You had $1024 in your checking account. You then wrote checks for $192 and $88. What is your new checking account balance?

24. **Finance**   Malek Khatri buys a tractor for $32,360. A down payment of $5000 is required. The balance remaining is paid in 48 equal monthly installments. What is the monthly payment?

25. **Homework Assignments**   Yuko is assigned to read a book containing 175 pages. She reads $\frac{2}{5}$ of the book during Thanksgiving vacation. How many pages of the assignment remain to be read?

26. **Real Estate**   A building contractor bought $2\frac{1}{3}$ acres of land for $84,000. What was the cost of each acre?

27. **Consumerism**   Benjamin Eli bought a shirt for $22.79 and a tie for $9.59. He used a $50 bill to pay for the purchases. Find the amount of change.

28. **Compensation**   If you earn an annual salary of $41,619, what is your monthly salary?

29. **Erosion**   A soil conservationist estimates that a river bank is eroding at the rate of 3 inches every 6 months. At this rate, how many inches will be eroded in 50 months?

30. **Medicine**   The dosage of a certain medication is $\frac{1}{2}$ ounce for every 50 pounds of body weight. How many ounces of this medication are required for a person who weighs 160 pounds? Write the answer as a decimal.

Copyright © Houghton Mifflin Company. All rights reserved.

5 Percents

Everyone knows that good health depends on eating right, watching your weight, not smoking, and exercising regularly. In order to get the most out of your workout, the American College of Sports Medicine (ACSM) recommends that you know how to determine your target heart rate. Your target heart rate is the rate at which your heart should beat during any aerobic exercise, such as running, fast walking, or bicycling. Your target heart rate depends on how fit you are, so athletes have higher target heart rates than people who are more sedentary. According to the ACSM, you should reach and then maintain your target heart rate for 20 minutes or more during a workout to achieve cardiovascular fitness. The **Projects and Group Activities on page 224** explain how you can calculate your target heart rate.

Copyright © Houghton Mifflin Company. All rights reserved.

OBJECTIVES

Section 5.1
A To write a percent as a fraction or a decimal
B To write a fraction or a decimal as a percent

Section 5.2
A To find the amount when the percent and the base are given
B To solve application problems

Section 5.3
A To find the percent when the base and amount are given
B To solve application problems

Section 5.4
A To find the base when the percent and amount are given
B To solve application problems

Section 5.5
A To solve percent problems using proportions
B To solve application problems

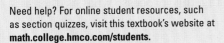 Need help? For online student resources, such as section quizzes, visit this textbook's website at **math.college.hmco.com/students.**

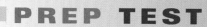

Do these exercises to prepare for Chapter 5.

For Exercises 1 to 6, multiply or divide.

1. $19 \times \dfrac{1}{100}$

2. 23×0.01

3. 0.47×100

4. $0.06 \times 47,500$

5. $60 \div 0.015$

6. $8 \div \dfrac{1}{4}$

7. Multiply $\dfrac{5}{8} \times 100$. Write the answer as a decimal.

8. Write $\dfrac{200}{3}$ as a mixed number.

9. Divide $28 \div 16$. Write the answer as a decimal.

GO FIGURE • • •

A whole number that remains unchanged when its digits are written in reverse order is a palindrome. For example, 818 is a palindrome.
a. Find the smallest three-digit multiple of 6 that is a palindrome.
b. Many nonpalindrome numbers can be converted into palindromes: reverse the digits of the number and add the result to the original; continue until a palindrome is achieved. Using this process, what is the palindrome created from 874?

Copyright © Houghton Mifflin Company. All rights reserved.

5.1 Introduction to Percents

Copyright © Houghton Mifflin Company. All rights reserved.

Objective A   **To write a percent as a fraction or a decimal**

Percent means "parts of 100." In the figure at the right, there are 100 parts. Because 13 of the 100 parts are shaded, 13% of the figure is shaded. The symbol % is the **percent sign.**

In most applied problems involving percents, it is necessary either to rewrite a percent as a fraction or a decimal or to rewrite a fraction or a decimal as a percent.

> **TAKE NOTE**
>
> Recall that division is defined as multiplication by the reciprocal. Therefore, multiplying by $\frac{1}{100}$ is equivalent to dividing by 100.

To write a percent as a fraction, remove the percent sign and multiply by $\frac{1}{100}$.

$$13\% = 13 \times \frac{1}{100} = \frac{13}{100}$$

To write a percent as a decimal, remove the percent sign and multiply by 0.01.

$$13\% \quad = \quad 13 \times 0.01 \quad = \quad 0.13$$

> Move the decimal point two places to the left. Then remove the percent sign.

Example 1   **a.** Write 120% as a fraction.
b. Write 120% as a decimal.

Solution   **a.** $120\% = 120 \times \frac{1}{100} = \frac{120}{100}$

$$= 1\frac{1}{5}$$

b. $120\% = 120 \times 0.01 = 1.2$

Note that percents larger than 100 are greater than 1.

You Try It 1   **a.** Write 125% as a fraction.
b. Write 125% as a decimal.

Your solution

Example 2   Write $16\frac{2}{3}\%$ as a fraction.

Solution $16\frac{2}{3}\% = 16\frac{2}{3} \times \frac{1}{100}$

$$= \frac{50}{3} \times \frac{1}{100} = \frac{50}{300} = \frac{1}{6}$$

You Try It 2   Write $33\frac{1}{3}\%$ as a fraction.

Your solution

Example 3   Write 0.5% as a decimal.

Solution $0.5\% = 0.5 \times 0.01 = 0.005$

You Try It 3   Write 0.25% as a decimal.

Your solution

Solutions on pp. S12–S13

Objective B   **To write a fraction or a decimal as a percent**

A fraction or a decimal can be written as a percent by multiplying by 100%.

> **HOW TO**   Write $\frac{3}{8}$ as a percent.
>
> $$\frac{3}{8} = \frac{3}{8} \times 100\% = \frac{3}{8} \times \frac{100}{1}\% = \frac{300}{8}\% = 37\frac{1}{2}\% \text{ or } 37.5\%$$
>
> **HOW TO**   Write 0.37 as a percent.
>
> $$0.37 \quad = \quad 0.37 \times 100\% \quad = \quad 37\%$$
>
> Move the decimal point two places to the right. Then write the percent sign.

Example 4   Write 0.015 as a percent.

Solution
$$0.015 = 0.015 \times 100\%$$
$$= 1.5\%$$

You Try It 4   Write 0.048 as a percent.

Your solution

Example 5   Write 2.15 as a percent.

Solution
$$2.15 = 2.15 \times 100\% = 215\%$$

You Try It 5   Write 3.67 as a percent.

Your solution

Example 6   Write $0.33\frac{1}{3}$ as a percent.

Solution
$$0.33\frac{1}{3} = 0.33\frac{1}{3} \times 100\%$$
$$= 33\frac{1}{3}\%$$

You Try It 6   Write $0.62\frac{1}{2}$ as a percent.

Your solution

Example 7   Write $\frac{2}{3}$ as a percent.
Write the remainder in fractional form.

Solution
$$\frac{2}{3} = \frac{2}{3} \times 100\% = \frac{200}{3}\%$$
$$= 66\frac{2}{3}\%$$

You Try It 7   Write $\frac{5}{6}$ as a percent.
Write the remainder in fractional form.

Your solution

Example 8   Write $2\frac{2}{7}$ as a percent.
Round to the nearest tenth.

Solution
$$2\frac{2}{7} = \frac{16}{7} = \frac{16}{7} \times 100\%$$
$$= \frac{1600}{7}\% \approx 228.6\%$$

You Try It 8   Write $1\frac{4}{9}$ as a percent.
Round to the nearest tenth.

Your solution

Solutions on p. S13

Copyright © Houghton Mifflin Company. All rights reserved.

5.1 Exercises

Objective A   **To write a percent as a fraction or a decimal**

For Exercises 1 to 16, write as a fraction and as a decimal.

1. 25%

2. 40%

3. 130%

4. 150%

5. 100%

6. 87%

7. 73%

8. 45%

9. 383%

10. 425%

11. 70%

12. 55%

13. 88%

14. 64%

15. 32%

16. 18%

For Exercises 17 to 28, write as a fraction.

17. $66\frac{2}{3}\%$

18. $12\frac{1}{2}\%$

19. $83\frac{1}{3}\%$

20. $3\frac{1}{8}\%$

21. $11\frac{1}{9}\%$

22. $\frac{3}{8}\%$

23. $45\frac{5}{11}\%$

24. $15\frac{3}{8}\%$

25. $4\frac{2}{7}\%$

26. $5\frac{3}{4}\%$

27. $6\frac{2}{3}\%$

28. $8\frac{2}{3}\%$

For Exercises 29 to 43, write as a decimal.

29. 6.5%

30. 9.4%

31. 12.3%

32. 16.7%

33. 0.55%

34. 0.45%

35. 8.25%

36. 6.75%

37. 5.05%

38. 3.08%

39. 2%

40. 7%

41. 80.4%

42. 36.2%

43. 4.9%

Objective B   **To write a fraction or a decimal as a percent**

For Exercises 44 to 55, write as a percent.

44. 0.16

45. 0.73

46. 0.05

47. 0.01

48. 1.07

49. 2.94

50. 0.004

51. 0.006

52. 1.012

53. 3.106

54. 0.8

55. 0.7

Copyright © Houghton Mifflin Company. All rights reserved.

For Exercises 56 to 67, write as a percent. Round to the nearest tenth of a percent.

56. $\dfrac{27}{50}$ **57.** $\dfrac{37}{100}$ **58.** $\dfrac{1}{3}$ **59.** $\dfrac{2}{5}$

60. $\dfrac{5}{8}$ **61.** $\dfrac{1}{8}$ **62.** $\dfrac{1}{6}$ **63.** $1\dfrac{1}{2}$

64. $\dfrac{7}{40}$ **65.** $1\dfrac{2}{3}$ **66.** $1\dfrac{7}{9}$ **67.** $\dfrac{7}{8}$

For Exercises 68 to 75, write as a percent. Write the remainder in fractional form.

68. $\dfrac{15}{50}$ **69.** $\dfrac{12}{25}$ **70.** $\dfrac{7}{30}$ **71.** $\dfrac{1}{3}$

72. $2\dfrac{3}{8}$ **73.** $1\dfrac{2}{3}$ **74.** $2\dfrac{1}{6}$ **75.** $\dfrac{7}{8}$

76. Write the part of the square that is shaded as a fraction, as a decimal, and as a percent. Write the part of the square that is not shaded as a fraction, as a decimal, and as a percent.

APPLYING THE CONCEPTS

77. **The Food Industry**   In a survey conducted by Opinion Research Corp. for Lloyd's Barbeque Co., people were asked to name their favorite barbeque side dishes. 38% named corn on the cob, 35% named cole slaw, 11% named corn bread, and 10% named fries. What percent of those surveyed named something other than corn on the cob, cole slaw, corn bread, or fries?

78. **Consumerism**   A sale on computers advertised $\frac{1}{3}$ off the regular price. What percent of the regular price does this represent?

79. **Consumerism**   A suit was priced at 50% off the regular price. What fraction of the regular price does this represent?

80. **Elections**   If $\frac{2}{5}$ of the population voted in an election, what percent of the population did not vote?

81. a. Is the statement "Multiplying a number by a percent always decreases the number" true or false?
   b. If it is false, give an example to show that the statement is false.

Copyright © Houghton Mifflin Company. All rights reserved.

5.2 Percent Equations: Part I

Objective A   **To find the amount when the percent and the base are given**

A real estate broker receives a payment that is 4% of a $285,000 sale. To find the amount the broker receives requires answering the question "4% of $285,000 is what?"

This sentence can be written using mathematical symbols and then solved for the unknown number.

4%	of	$285,000	is	what?
↓	↓	↓	↓	↓

$$\boxed{\begin{array}{c}\text{Percent}\\4\%\end{array}} \times \boxed{\begin{array}{c}\text{base}\\285{,}000\end{array}} = \boxed{\begin{array}{c}\text{amount}\\n\end{array}}$$

of   is written as × (times)
is   is written as = (equals)
what is written as n (the unknown number)

$$0.04 \times 285{,}000 = n$$
$$11{,}400 = n$$

Note that the percent is written as a decimal.

The broker receives a payment of $11,400.

The solution was found by solving the **basic percent equation** for amount.

> **The Basic Percent Equation**
>
> $$\boxed{\text{Percent}} \times \boxed{\text{base}} = \boxed{\text{amount}}$$

In most cases, the percent is written as a decimal before the basic percent equation is solved. However, some percents are more easily written as a fraction than as a decimal. For example,

$$33\frac{1}{3}\% = \frac{1}{3} \qquad 66\frac{2}{3}\% = \frac{2}{3} \qquad 16\frac{2}{3}\% = \frac{1}{6} \qquad 83\frac{1}{3}\% = \frac{5}{6}$$

Example 1   Find 5.7% of 160.

Solution

Percent × base = amount   • The word *Find* is
    $0.057 \times 160 = n$ used instead of the
        $9.12 = n$ words *what is*.

You Try It 1   Find 6.3% of 150.

Your solution

Example 2   What is $33\frac{1}{3}\%$ of 90?

Solution   Percent × base = amount
    $\frac{1}{3} \times 90 = n$ • $33\frac{1}{3}\% = \frac{1}{3}$
        $30 = n$

You Try It 2   What is $16\frac{2}{3}\%$ of 66?

Your solution

Solutions on p. S13

Copyright © Houghton Mifflin Company. All rights reserved.

Objective B **To solve application problems**

Solving percent problems requires identifying the three elements of the basic percent equation. Recall that these three parts are the *percent,* the *base,* and the *amount.* Usually the base follows the phrase "percent of."

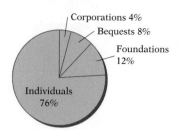

Corporations 4%
Bequests 8%
Foundations 12%
Individuals 76%

Charitable Giving
Sources: American Association of Fundraising Counsel; AP

During a recent year, Americans gave $212 billion to charities. The circle graph at the right shows where that money came from. Use these data for Example 3 and You Try It 3.

Example 3

How much of the amount given to charities came from individuals?

Strategy

To determine the amount that came from individuals, write and solve the basic percent equation using n to represent the amount. The percent is 76%. The base is $212 billion.

Solution   Percent × base = amount
           76% × 212 = n
          0.76 × 212 = n
               161.12 = n

Individuals gave $161.12 billion to charities.

You Try It 3

How much of the amount given to charities was given by corporations?

Your strategy

Your solution

Example 4

A quality control inspector found that 1.2% of 2500 telephones inspected were defective. How many telephones inspected were not defective?

Strategy

To find the number of nondefective phones:

• Find the number of defective phones. Write and solve the basic percent equation using n to represent the number of defective phones (amount). The percent is 1.2% and the base is 2500.

• Subtract the number of defective phones from the number of phones inspected (2500).

Solution   1.2% × 2500 = n
          0.012 × 2500 = n
                  30 = n defective phones

          2500 − 30 = 2470

2470 telephones were not defective.

You Try It 4

An electrician's hourly wage was $33.50 before an 8% raise. What is the new hourly wage?

Your strategy

Your solution

Solutions on p. S13

Copyright © Houghton Mifflin Company. All rights reserved.

5.2 Exercises

Objective A   **To find the amount when the percent and the base are given**

1.   8% of 100 is what?

2.   16% of 50 is what?

3.   27% of 40 is what?

4.   52% of 95 is what?

5.   0.05% of 150 is what?

6.   0.075% of 625 is what?

7.   125% of 64 is what?

8.   210% of 12 is what?

9.   Find 10.7% of 485.

10.   Find 12.8% of 625.

11.   What is 0.25% of 3000?

12.   What is 0.06% of 250?

13.   80% of 16.25 is what?

14.   26% of 19.5 is what?

15.   What is $1\frac{1}{2}$% of 250?

16.   What is $5\frac{3}{4}$% of 65?

17. $16\frac{2}{3}$% of 120 is what?

18. $83\frac{1}{3}$% of 246 is what?

19.   What is $33\frac{1}{3}$% of 630?

20.   What is $66\frac{2}{3}$% of 891?

21.   Which is larger: 5% of 95, or 75% of 6?

22.   Which is larger: 112% of 5, or 0.45% of 800?

23.   Which is smaller: 79% of 16, or 20% of 65?

24.   Which is smaller: 15% of 80, or 95% of 15?

25.   Which is smaller: 2% of 1500, or 72% of 40?

26.   Which is larger: 22% of 120, or 84% of 32?

27.   Find 31.294% of 82,460.

28.   Find 123.94% of 275,976.

Objective B   **To solve application problems**

29.   **Health Insurance**   Approximately 30% of the 44 million people in the United States who do not have health insurance are between the ages of 18 and 24. (*Source:* U.S. Census Bureau) About how many people in the United States aged 18 to 24 do not have health insurance?

30.   **Aviation**   The Federal Aviation Administration reported that 55,422 new student pilots were flying single-engine planes last year. The number of new student pilots flying single-engine planes this year is 106% of the number flying single-engine planes last year. How many new student pilots are flying single-engine planes this year?

Copyright © Houghton Mifflin Company. All rights reserved.

Politics   The results of a survey in which 32,840 full-time college and university faculty members were asked to describe their political views is shown at the right. Use these data for Exercises 31 and 32.

Political View	Percent of Faculty Members Responding
Far Left	5.3%
Liberal	42.3%
Middle of the road	34.3%
Conservative	17.7%
Far right	0.3%

Source: Higher Education Research Institute, UCLA

31. How many more faculty members described their political views as liberal than described their views as far left?

32. How many fewer faculty members described their political views as conservative than described their views as middle of the road?

33. Taxes   A sales tax of 6% of the cost of a car was added to the purchase price of $29,500.
  a. How much was the sales tax?
  b. What is the total cost of the car, including sales tax?

34. Business   During the packaging process for oranges, spoiled oranges are discarded by an inspector. In one day an inspector found that 4.8% of the 20,000 pounds of oranges inspected were spoiled.
  a. How many pounds of oranges were spoiled?
  b. How many pounds of oranges were not spoiled?

35. **Entertainment**   A USA TODAY.com online poll asked 8878 Internet users, "Would you use software to cut out objectionable parts of movies?" 29.8% of the respondents answered yes. How many respondents did not answer yes to the question? Round to the nearest whole number.

36. Employment   Funtimes Amusement Park has 550 employees and must hire an additional 22% for the vacation season. What is the total number of employees needed for the vacation season?

APPLYING THE CONCEPTS

Sociology   The two circle graphs at the right show how surveyed employees actually spend their time and the way they would prefer to spend their time. Assuming that employees have 112 hours a week of time that is not spent sleeping, answer Exercises 37 to 39. Round to the nearest tenth of an hour.

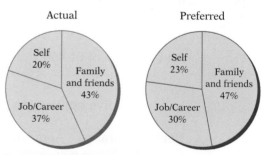

Actual / Preferred circle graphs:
Actual — Self 20%, Family and friends 43%, Job/Career 37%
Preferred — Self 23%, Family and friends 47%, Job/Career 30%

Source: WSJ Supplement, Work & Family, from *Families and Work Institute*

37. What is the actual number of hours per week that employees spend with family and friends?

38. What is the number of hours per week that employees would prefer to spend on job/career?

39. What is the difference between the number of hours an employee preferred to spend on self and the actual amount of time the employee spent on self?

Copyright © Houghton Mifflin Company. All rights reserved.

5.3 Percent Equations: Part II

Objective A   **To find the percent when the base and amount are given**

A recent promotional game at a grocery store listed the probability of winning a prize as "1 chance in 2." A percent can be used to describe the chance of winning. This requires answering the question "What percent of 2 is 1?"

The chance of winning can be found by solving the basic percent equation for *percent*.

Integrating Technology

The percent key % on a scientific calculator moves the decimal point to the left two places when pressed after a multiplication or division computation. For the example at the right, enter

1 ÷ 2 % =

The display reads 50.

What percent of   2   is   1?

$$\text{Percent } n \times \text{base } 2 = \text{amount } 1$$

$$n \times 2 = 1$$
$$n = 1 \div 2$$
$$n = 0.5$$
$$n = 50\%$$

• The solution must be written as a percent to answer the question.

There is a 50% chance of winning a prize.

Example 1 What percent of 40 is 30?

Solution
$$\text{Percent} \times \text{base} = \text{amount}$$
$$n \times 40 = 30$$
$$n = 30 \div 40$$
$$n = 0.75$$
$$n = 75\%$$

You Try It 1 What percent of 32 is 16?

Your solution

Example 2 What percent of 12 is 27?

Solution
$$\text{Percent} \times \text{base} = \text{amount}$$
$$n \times 12 = 27$$
$$n = 27 \div 12$$
$$n = 2.25$$
$$n = 225\%$$

You Try It 2 What percent of 15 is 48?

Your solution

Example 3 25 is what percent of 75?

Solution
$$\text{Percent} \times \text{base} = \text{amount}$$
$$n \times 75 = 25$$
$$n = 25 \div 75$$
$$n = \frac{1}{3} = 33\frac{1}{3}\%$$

You Try It 3 30 is what percent of 45?

Your solution

Solutions on p. S13

Copyright © Houghton Mifflin Company. All rights reserved.

Copyright © Houghton Mifflin Company. All rights reserved.

Objective B   **To solve application problems**

To solve percent problems, remember that it is necessary to identify the percent, base, and amount. Usually the base follows the phrase "percent of."

Example 4

The monthly house payment for the Kaminski family is $787.50. What percent of the Kaminskis' monthly income of $3750 is the house payment?

Strategy

To find what percent of the income the house payment is, write and solve the basic percent equation using n to represent the percent. The base is $3750 and the amount is $787.50.

Solution

$n \times 3750 = 787.50$
$n = 787.50 \div 3750$
$n = 0.21 = 21\%$

The house payment is 21% of the monthly income.

You Try It 4

Tomo Nagata had an income of $33,500 and paid $5025 in income tax. What percent of the income is the income tax?

Your strategy

Your solution

Example 5

On one Monday night, 31.39 million of the approximately 40.76 million households watching television were not watching David Letterman. What percent of these households were watching David Letterman? Round to the nearest percent.

Strategy

To find the percent of households watching David Letterman:

• Subtract to find the number of households that were watching David Letterman (40.76 million − 31.39 million).
• Write and solve the basic percent equation using n to represent the percent. The base is 40.76, and the amount is the number of households watching David Letterman.

Solution

40.76 million − 31.39 million = 9.37 million

9.37 million households were watching David Letterman.

$n \times 40.76 = 9.37$
$n = 9.37 \div 40.76$
$n \approx 0.23$

Approximately 23% of the households were watching David Letterman.

You Try It 5

According to the U.S. Department of Defense, of the 518,921 enlisted personnel in the U.S. Army in 1950, 512,370 people were men. What percent of the enlisted personnel in the U.S. Army in 1950 were women? Round to the nearest tenth of a percent.

Your strategy

Your solution

Solutions on p. S13

5.3 Exercises

Objective A   To find the percent when the base and amount are given

1. What percent of 75 is 24?

2. What percent of 80 is 20?

3. 15 is what percent of 90?

4. 24 is what percent of 60?

5. What percent of 12 is 24?

6. What percent of 6 is 9?

7. What percent of 16 is 6?

8. What percent of 24 is 18?

9. 18 is what percent of 100?

10. 54 is what percent of 100?

11. 5 is what percent of 2000?

12. 8 is what percent of 2500?

13. What percent of 6 is 1.2?

14. What percent of 2.4 is 0.6?

15. 16.4 is what percent of 4.1?

16. 5.3 is what percent of 50?

17. 1 is what percent of 40?

18. 0.3 is what percent of 20?

19. What percent of 48 is 18?

20. What percent of 11 is 88?

21. What percent of 2800 is 7?

22. What percent of 400 is 12?

23. 4.2 is what percent of 175?

24. 41.79 is what percent of 99.5?

25. What percent of 86.5 is 8.304?

26. What percent of 1282.5 is 2.565?

Objective B   To solve application problems

27. **Sociology**   Seven in ten couples disagree about financial issues. (*Source:* Yankelovich Partners for Lutheran Brotherhood) What percent of couples disagree about financial matters?

28. **Sociology**   In a survey, 1236 adults nationwide were asked, "What irks you most about the actions of other motorists?" The response "tailgaters" was given by 293 people. (*Source:* Reuters/Zogby) What percent of those surveyed were most irked by tailgaters? Round to the nearest tenth of a percent.

Copyright © Houghton Mifflin Company. All rights reserved.

29.  **Agriculture**   According to the U.S. Department of Agriculture, of the 63 billion pounds of vegetables produced in the United States in 1 year, 16 billion pounds were wasted. What percent of the vegetables produced were wasted? Round to the nearest tenth of a percent.

30. **Agriculture**   In a recent year, Wisconsin growers produced 281.72 million pounds of the 572 million pounds of cranberries grown in the United States. What percent of the total cranberry crop was produced in Wisconsin? Round to the nearest percent.

31. **Energy**   The typical American household spends $1355 a year on energy utilities. Of this amount, approximately $81.30 is spent on lighting. (*Source:* Department of Energy, Owens Corning) What percent of the total amount spent on energy utilities is spent on lighting?

32. **Education**   To receive a license to sell insurance, an insurance account executive must answer correctly 70% of the 250 questions on a test. Nicholas Mosley answered 177 questions correctly. Did he pass the test?

33. **Agriculture**   According to the U.S. Department of Agriculture, of the 356 billion pounds of food produced in the United States annually, 260 billion pounds are not wasted. What percent of the food produced in the United States is wasted? Round to the nearest percent.

34. **Construction**   In a test of the breaking strength of concrete slabs for freeway construction, 3 of the 200 slabs tested did not meet safety requirements. What percent of the slabs did meet safety requirements?

APPLYING THE CONCEPTS

Pets   The graph at the right shows several categories of average lifetime costs of dog ownership. Use this graph for Exercises 35 to 37. Round answers to the nearest tenth of a percent.

35. What percent of the total amount is spent on food?

36. What percent of the total is spent on veterinary care?

37. What percent of the total is spent on all categories except training?

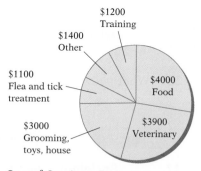

Cost of Owning a Dog

Source: Based on data from the American Kennel Club, *USA Today* research

38. **Sports**   The Fun in the Sun organization claims to have taken a survey of 350 people, asking them to give their favorite outdoor temperature for hiking. The results are given in the table at the right. Explain why these results are not possible.

Favorite Temperature	Percent
Greater than 90	5%
80–89	28%
70–79	35%
60–69	32%
Below 60	13%

Copyright © Houghton Mifflin Company. All rights reserved.

5.4 Percent Equations: Part III

Objective A   **To find the base when the percent and amount are given**

In 1780, the population of Virginia was 538,000. This was 19% of the total population of the United States at that time. To find the total population at that time, you must answer the question "19% of what number is 538,000?"

Study Tip

After completing this objective, you will have learned to solve the basic percent equation for each of the three elements: percent, base, and amount. You will need to be able to recognize these three different types of problems. To test yourself, you might do the Chapter 5 Review Exercises.

| 19% | of | what | is | 538,000? |
| ↓ | | ↓ | | ↓ |

| Percent 19% | base n | amount 538,000 |

- The population of the United States in 1780 can be found by solving the basic percent equation for the base.

$$0.19 \times n = 538{,}000$$
$$n = 538{,}000 \div 0.19$$
$$n \approx 2{,}832{,}000$$

The population of the United States in 1780 was approximately 2,832,000.

Example 1   18% of what is 900?

Solution   Percent × base = amount
$$0.18 \times n = 900$$
$$n = 900 \div 0.18$$
$$n = 5000$$

You Try It 1   86% of what is 215?

Your solution

Example 2   30 is 1.5% of what?

Solution   Percent × base = amount
$$0.015 \times n = 30$$
$$n = 30 \div 0.015$$
$$n = 2000$$

You Try It 2   15 is 2.5% of what?

Your solution

Example 3 $33\frac{1}{3}\%$ of what is 7?

Solution   Percent × base = amount
$$\frac{1}{3} \times n = 7$$
$$n = 7 \div \frac{1}{3}$$
$$n = 21$$

- Note that the percent is written as a fraction.

You Try It 3 $16\frac{2}{3}\%$ of what is 5?

Your solution

Solutions on p. S14

Objective B   **To solve application problems**

To solve percent problems, it is necessary to identify the percent, base, and amount. Usually the base follows the phrase "percent of."

Copyright © Houghton Mifflin Company. All rights reserved.

Example 4

A business office bought a used copy machine for $900, which was 75% of the original cost. What was the original cost of the copier?

Strategy

To find the original cost of the copier, write and solve the basic percent equation using n to represent the original cost (base). The percent is 75% and the amount is $900.

Solution

$75\% \times n = 900$
$0.75 \times n = 900$
$n = 900 \div 0.75$
$n = 1200$

The original cost of the copier was $1200.

You Try It 4

A used car has a value of $10,458, which is 42% of the car's original value. What was the car's original value?

Your strategy

Your solution

Example 5

A carpenter's wage this year is $26.40 per hour, which is 110% of last year's wage. What was the increase in the hourly wage over last year?

Strategy

To find the increase in the hourly wage over last year:

• Find last year's wage. Write and solve the basic percent equation using n to represent last year's wage (base). The percent is 110% and the amount is $26.40.
• Subtract last year's wage from this year's wage (26.40).

Solution

$110\% \times n = 26.40$
$1.10 \times n = 26.40$
$\quad n = 26.40 \div 1.10$
$\quad n = 24.00 \quad$ • Last year's wage

$26.40 - 24.00 = 2.40$

The increase in the hourly wage was $2.40.

You Try It 5

Chang's Sporting Goods has a tennis racket on sale for $89.60, which is 80% of the original price. What is the difference between the original price and the sale price?

Your strategy

Your solution

Solutions on p. S14

Copyright © Houghton Mifflin Company. All rights reserved.

5.4 Exercises

Objective A   **To find the base when the percent and amount are given**

For Exercises 1 to 26, solve. Round to the nearest hundredth.

1. 12% of what is 9?

2. 38% of what is 171?

3. 8 is 16% of what?

4. 54 is 90% of what?

5. 10 is 10% of what?

6. 37 is 37% of what?

7. 30% of what is 25.5?

8. 25% of what is 21.5?

9. 2.5% of what is 30?

10. 10.4% of what is 52?

11. 125% of what is 24?

12. 180% of what is 21.6?

13. 18 is 240% of what?

14. 24 is 320% of what?

15. 4.8 is 15% of what?

16. 87.5 is 50% of what?

17. 25.6 is 12.8% of what?

18. 45.014 is 63.4% of what?

19. 0.7% of what is 0.56?

20. 0.25% of what is 1?

21. 30% of what is 2.7?

22. 78% of what is 3.9?

23. 84 is $16\frac{2}{3}$% of what?

24. 120 is $33\frac{1}{3}$% of what?

25. $66\frac{2}{3}$% of what is 72?

26. $83\frac{1}{3}$% of what is 13.5?

Objective B   **To solve application problems**

27.  **Travel**   Of the travelers who, during a recent year, allowed their children to miss school to go along on a trip, approximately 1.738 million allowed their children to miss school for more than a week. This represented 11% of the travelers who allowed their children to miss school. (*Source:* Travel Industry Association) About how many travelers allowed their children to miss school to go along on a trip?

Copyright © Houghton Mifflin Company. All rights reserved.

28.   **Education**   In the United States today, 23.1% of the women and 27.5% of the men have earned a bachelor's or graduate degree. (*Source:* Census Bureau) How many women in the United States have earned a bachelor's or graduate degree?

29.   **Taxes**   A TurboTax online survey asked people how they planned to use their tax refunds. 740 people, or 22% of the respondents, said they would save the money. How many people responded to the survey?

30.   **Taxes**   The Internal Revenue Service says that the average deduction for medical expenses for taxpayers in the $40,000–$50,000 bracket is $4500. This is 26% of the medical expenses claimed by taxpayers in the over-$200,000 bracket. How much is the average deduction for medical expenses claimed by taxpayers in the over-$200,000 bracket? Round to the nearest thousand.

31.   **Manufacturing**   During a quality control test, Micronics found that 24 computer boards were defective. This amount was 0.8% of the computer boards tested.
a. How many computer boards were tested?
b. How many computer boards tested were not defective?

32.   **Directory Assistance**   Of the calls a directory assistance operator received, 441 were requests for telephone numbers listed in the current directory. This accounted for 98% of the calls for assistance that the operator received.
a. How many calls did the operator receive?
b. How many telephone numbers requested were not listed in the current directory?

APPLYING THE CONCEPTS

33.   **Demography**   At the last census, of the 281,422,000 people in the United States, 28.6% were under the age of 20. (*Source:* U.S. Census 2000) How many people in the United States were age 20 or older in 2000?

Nutrition   The table at the right contains nutrition information about a breakfast cereal. Solve Exercises 34 and 35 using information from this table.

34.   The amount of thiamin in one serving of cereal with skim milk is 0.45 milligram. Find the recommended daily allowance of thiamin for an adult.

35.   The amount of copper in one serving of cereal with skim milk is 0.08 milligram. Find the recommended daily allowance of copper for an adult.

36.   Increase a number by 10%. Now decrease the number by 10%. Is the result the original number? Explain.

NUTRITION INFORMATION

SERVING SIZE: 1.4 OZ WHEAT FLAKES WITH
0.4 OZ. RAISINS: 39.4 g. ABOUT 1/2 CUP
SERVINGS PER PACKAGE: 14

	CEREAL & RAISINS	WITH 1/2 CUP VITAMINS A & D SKIM MILK

PERCENTAGE OF U.S. RECOMMENDED DAILY ALLOWANCES (U.S. RDA)

	CEREAL & RAISINS	WITH 1/2 CUP SKIM MILK
PROTEIN	4	15
VITAMIN A	15	20
VITAMIN C	**	2
THIAMIN	25	30
RIBOFLAVIN	25	35
NIACIN	25	35
CALCIUM	**	15
IRON	100	100
VITAMIN D	10	25
VITAMIN B₆	25	25
FOLIC ACID..............	25	25
VITAMIN B₁₂	25	30
PHOSPHOROUS.........	10	15
MAGNESIUM	10	20
ZINC	25	30
COPPER..................	2	4

* 2% MILK SUPPLIES AN ADDITIONAL 20 CALORIES.
2 g FAT, AND 10 mg CHOLESTEROL.
** CONTAINS LESS THAN 2% OF THE U.S. RDA OF
THIS NUTRIENT

Copyright © Houghton Mifflin Company. All rights reserved.

5.5 Percent Problems: Proportion Method

Objective A   **To solve percent problems using proportions**

Problems that can be solved using the basic percent equation can also be solved using proportions.

The proportion method is based on writing two ratios. One ratio is the percent ratio, written as $\frac{\text{percent}}{100}$. The second ratio is the amount-to-base ratio, written as $\frac{\text{amount}}{\text{base}}$. These two ratios form the proportion

$$\frac{\text{percent}}{100} = \frac{\text{amount}}{\text{base}}$$

To use the proportion method, first identify the percent, the amount, and the base (the base usually follows the phrase "percent of").

Integrating Technology

To use a calculator to solve the proportions at the right for *n*, enter

23 × 45 ÷ 100 =

100 × 4 ÷ 25 =

100 × 12 ÷ 60 =

What is 23% of 45?

$$\frac{23}{100} = \frac{n}{45}$$
$$23 \times 45 = 100 \times n$$
$$1035 = 100 \times n$$
$$1035 \div 100 = n$$
$$10.35 = n$$

What percent of 25 is 4?

$$\frac{n}{100} = \frac{4}{25}$$
$$n \times 25 = 100 \times 4$$
$$n \times 25 = 400$$
$$n = 400 \div 25$$
$$n = 16$$
16% of 25 is 4.

12 is 60% of what number?

$$\frac{60}{100} = \frac{12}{n}$$
$$60 \times n = 100 \times 12$$
$$60 \times n = 1200$$
$$n = 1200 \div 60$$
$$n = 20$$

Example 1   15% of what is 7? Round to the nearest hundredth.

Solution
$$\frac{15}{100} = \frac{7}{n}$$
$$15 \times n = 100 \times 7$$
$$15 \times n = 700$$
$$n = 700 \div 15$$
$$n \approx 46.67$$

You Try It 1   26% of what is 22? Round to the nearest hundredth.

Your solution

Example 2   30% of 63 is what?

Solution
$$\frac{30}{100} = \frac{n}{63}$$
$$30 \times 63 = 100 \times n$$
$$1890 = 100 \times n$$
$$1890 \div 100 = n$$
$$18.90 = n$$

You Try It 2   16% of 132 is what?

Your solution

Solutions on p. S14

Copyright © Houghton Mifflin Company. All rights reserved.

Objective B   **To solve application problems**

Example 3

An antiques dealer found that 86% of the 250 items that were sold during one month sold for under $1000. How many items sold for under $1000?

Strategy

To find the number of items that sold for under $1000, write and solve a proportion using n to represent the number of items sold (amount) for less than $1000. The percent is 86% and the base is 250.

Solution

$$\frac{86}{100} = \frac{n}{250}$$
$$86 \times 250 = 100 \times n$$
$$21{,}500 = 100 \times n$$
$$21{,}500 \div 100 = n$$
$$215 = n$$

215 items sold for under $1000.

You Try It 3

Last year it snowed 64% of the 150 days of the ski season at a resort. How many days did it snow?

Your strategy

Your solution

Example 4

In a test of the strength of nylon rope, 5 pieces of the 25 pieces tested did not meet the test standards. What percent of the nylon ropes tested did meet the standards?

Strategy

To find the percent of ropes tested that met the standards:

- Find the number of ropes that met the test standards (25 − 5).
- Write and solve a proportion using n to represent the percent of ropes that met the test standards. The base is 25 and the amount is the number of ropes that met the standards.

Solution

25 − 5 = 20 ropes met test standards

$$\frac{n}{100} = \frac{20}{25}$$
$$n \times 25 = 100 \times 20$$
$$n \times 25 = 2000$$
$$n = 2000 \div 25$$
$$n = 80$$

80% of the ropes tested did meet the test standards.

You Try It 4

Five ballpoint pens in a box of 200 were found to be defective. What percent of the pens were not defective?

Your strategy

Your solution

Solutions on p. S14

Copyright © Houghton Mifflin Company. All rights reserved.

5.5 Exercises

Objective A   **To solve percent problems using proportions**

1.   26% of 250 is what?

2.   What is 18% of 150?

3.   37 is what percent of 148?

4.   What percent of 150 is 33?

5.   68% of what is 51?

6.   126 is 84% of what?

7.   What percent of 344 is 43?

8.   750 is what percent of 50?

9.   82 is 20.5% of what?

10.   2.4% of what is 21?

11.   What is 6.5% of 300?

12.   96% of 75 is what?

13.   7.4 is what percent of 50?

14.   What percent of 1500 is 693?

15.   50.5% of 124 is what?

16.   What is 87.4% of 255?

17.   120% of what is 6?

18.   14 is 175% of what?

19.   What is 250% of 18?

20.   325% of 4.4 is what?

21.   33 is 220% of what?

22.   160% of what is 40?

Objective B   **To solve application problems**

23.   **Charities**   A charitable organization spent $2940 for administrative expenses. This amount is 12% of the money it collected. What is the total amount of money that the organization collected?

24.   **Medicine**   A manufacturer of an anti-inflammatory drug claims that the drug will be effective for 6 hours. An independent testing service determined that the drug was effective for only 80% of the length of time claimed by the manufacturer. Find the length of time the drug will be effective as determined by the testing service.

25.   **Geography**   The land area of North America is approximately 9,400,000 square miles. This represents approximately 16% of the total land area of the world. What is the approximate total land area of the world?

26.   **Fire Departments**   The Rincon Fire Department received 24 false alarms out of a total of 200 alarms received. What percent of the alarms received were false alarms?

Copyright © Houghton Mifflin Company. All rights reserved.

27. **Lodging** The graph at the right shows the breakdown of the locations of the 53,500 hotels throughout the United States. How many hotels in the United States are located along highways?

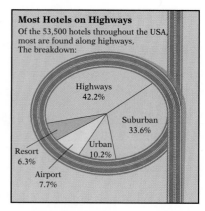

Most Hotels on Highways
Of the 53,500 hotels throughout the USA, most are found along highways, The breakdown:

Highways 42.2%
Suburban 33.6%
Resort 6.3%
Urban 10.2%
Airport 7.7%

Source: American Hotel and Lodging Association

28. **Poultry** In a recent year, North Carolina produced 1,300,000,000 pounds of turkey. This was 18.6% of the U.S. total in that year. Calculate the U.S. total turkey production for that year. Round to the nearest billion.

29. **Mining** During 1 year, approximately 2,240,000 ounces of gold went into the manufacturing of electronic equipment in the United States. This is 16% of all the gold mined in the United States that year. How many ounces of gold were mined in the United States that year?

30. **Demography** The table at the right shows the predicted increase in population from 2000 to 2040 for each of four counties in the Central Valley of California.
    a. What percent of the 2000 population of Sacramento County is the increase in population?
    b. What percent of the 2000 population of Kern County is the increase in population? Round to the nearest tenth of a percent.

County	2000 Population	Projected Increase
Sacramento	1,200,000	900,000
Kern	651,700	948,300
Fresno	794,200	705,800
San Joaquin	562,000	737,400

Source: California Department of Finance

31. **Police Officers** The graph at the right shows the causes of death for all police officers killed in the line of duty during a recent year. What percent of the deaths were due to traffic accidents? Round to the nearest tenth of a percent.

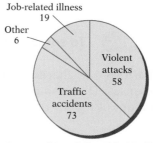

Job-related illness 19
Other 6
Violent attacks 58
Traffic accidents 73

Causes of Death for Police Officers Killed in the Line of Duty

Source: International Union of Police Associations

32. **Demography** According to a 25-city survey of the status of hunger and homelessness by the U.S. Conference of Mayors, 41% of the homeless in the United States are single men, 41% are families with children, 13% are single women, and 5% are unaccompanied minors. How many homeless people in the United States are single men?

APPLYING THE CONCEPTS

33. **The Federal Government** In the 108th Senate, there were 51 Republicans, 48 Democrats, and 1 Independent. In the 108th House of Representatives, there were 229 Republicans, 205 Democrats, and 1 Independent. Which had the larger percent of Republicans, the 108th Senate or the 108th House of Representatives?

Copyright © Houghton Mifflin Company. All rights reserved.

Focus on Problem Solving

**Using a Calculator
as a Problem-
Solving Tool**

A calculator is an important tool for problem solving. Here are a few problems to solve with a calculator. You may need to research some of the questions to find information you do not know.

1. Choose any single-digit positive number. Multiply the number by 1507 and 7373. What is the answer? Choose another positive single-digit number and again multiply by 1507 and 7373. What is the answer? What pattern do you see? Why does this work?

2. The gross domestic product in 2002 was $10,446,200,000. Is this more or less than the amount of money that would be placed on the last square of a standard checkerboard if 1 cent were placed on the first square, 2 cents were placed on the second square, 4 cents were placed on the third square, 8 cents were placed on the fourth square, and so on, until the 64th square was reached?

3. Which of the reciprocals of the first 16 natural numbers have a terminating-decimal representation and which have a repeating-decimal representation?

4. What is the largest natural number n for which $4^n > 1 \cdot 2 \cdot 3 \cdot 4 \cdot 5 \cdots \cdots n$?

5. If $1000 bills are stacked one on top of another, is the height of $1 billion less than or greater than the height of the Washington Monument?

6. What is the value of $1 + \cfrac{1}{1 + \cfrac{1}{1 + \cfrac{1}{1 + \cfrac{1}{1 + 1}}}}$?

7. Calculate 15^2, 35^2, 65^2, and 85^2. Study the results. Make a conjecture about a relationship between a number ending in 5 and its square. Use your conjecture to find 75^2 and 95^2. Does your conjecture work for 125^2?

8. Find the sum of the first 1000 natural numbers. (*Hint:* You could just start adding $1 + 2 + 3 + \cdots$, but even if you performed one operation every 3 seconds, it would take you an hour to find the sum. Instead, try pairing the numbers and then adding the pairs. Pair 1 and 1000, 2 and 999, 3 and 998, and so on. What is the sum of each pair? How many pairs are there? Use this information to answer the original question.)

9. For a borrower to qualify for a home loan, a bank requires that the monthly mortgage payment be less than 25% of a borrower's monthly take-home income. A laboratory technician has deductions for taxes, insurance, and retirement that amount to 25% of the technician's monthly gross income. What minimum monthly income must this technician earn to receive a bank loan that has a mortgage payment of $1200 per month?

Copyright © Houghton Mifflin Company. All rights reserved.

Using Estimation as a Problem-Solving Tool

You can use your knowledge of rounding, your understanding of percent, and your experience with the basic percent equation to quickly estimate the answer to a percent problem. Here is an example.

> **HOW TO** What is 11.2% of 978?
>
> Round the given numbers. $11.2\% \approx 10\%$
> $978 \approx 1000$
>
> Mentally calculate with the rounded numbers. $10\% \text{ of } 1000 = \frac{1}{10} \text{ of } 1000 = 100$
>
> 11.2% of 978 is approximately 100.

TAKE NOTE
The exact answer is $0.112 \times 978 = 109.536$. The exact answer 109.536 is close to the approximation of 100.

For Exercises 1 to 8, state which quantity is greater.

1. 49% of 51, or 201% of 15

2. 99% of 19, or 22% of 55

3. 8% of 31, or 78% of 10

4. 24% of 402, or 76% of 205

5. 10.2% of 51, or 20.9% of 41

6. 51.8% of 804, or 25.3% of 1223

7. 26% of 39.217, or 9% of 85.601

8. 66% of 31.807, or 33% of 58.203

For Exercises 9 to 12, use estimation to provide an approximate number.

9. A company survey found that 24% of its 2096 employees favored a new dental plan. How many employees favored the new dental plan?

10. A local newspaper reported that 52.3% of the 29,875 eligible voters in the town voted in the last election. How many people voted in the last election?

11. 19.8% of the 2135 first-year students at a community college have part-time jobs. How many of the first-year students at the college have part-time jobs?

12. A couple made a down payment of 33% of the $310,000 cost of the home. Find the down payment.

Projects and Group Activities

Health

The American College of Sports Medicine (ACSM) recommends that you know how to determine your target heart rate in order to get the full benefit of exercise. Your **target heart rate** is the rate at which your heart should beat during any aerobic exercise such as running, cycling, fast walking, or participating in an aerobics class. According to the ACSM, you should reach your target rate and then maintain it for 20 minutes or more to achieve cardiovascular fitness. The intensity level varies for different individuals. A sedentary person might begin at the 60% level and gradually work up to 70%, whereas athletes and very fit individuals might work at the 85% level. The ACSM suggests that you calculate both 50% and 85% of your maximum heart rate. This will give you the low and high ends of the range within which your heart rate should stay.

To calculate your target heart rate:

Copyright © Houghton Mifflin Company. All rights reserved.

	Example
Subtract your age from 220. This is your maximum heart rate.	$220 - 20 = 200$
Multiply your maximum heart rate by 50%. This is the low end of your range.	$200(0.50) = 100$
Divide the low end by 6. This is your low 10-second heart rate.	$100 \div 6 \approx 17$
Multiply your maximum heart rate by 85%. This is the high end of your range.	$200(0.85) = 170$
Divide the high end by 6. This is your high 10-second heart rate.	$170 \div 6 \approx 28$

1. Why are the low end and high end divided by 6 in order to determine the low and high 10-second heart rates?

2. Calculate your target heart rate, both the low and high end of your range.

Consumer Price Index

The consumer price index (CPI) is a percent that is written without the percent sign. For instance, a CPI of 160.1 means 160.1%. This number means that an item that cost $100 between 1982 and 1984 (the base years) would cost $160.10 today. Determining the cost is an application of the basic percent equation.

$$\text{Percent} \times \text{base} = \text{amount}$$
$$\text{CPI} \times \text{cost in base year} = \text{cost today}$$
$$1.601 \times 100 = 160.1 \qquad \bullet \ 160.1\% = 1.601$$

The table below gives the CPI for various products in July of 2003. If you have Internet access, you can obtain current data for the items below, as well as other items not on this list, by visiting the website of the Bureau of Labor Statistics.

Product	*CPI*
All items	183.9
Food and beverages	180.3
Housing	185.9
Clothes	116.2
Transportation	156.8
Medical care	297.6
Entertainment	107.7
Education	108.9

1. Of the items listed, are there any items that in 2003 cost more than twice as much as they cost during the base year? If so, which items?

2. Of the items listed, are there any items that in 2003 cost more than one-and-one-half times as much as they cost during the base years but less than twice as much as they cost during the base years? If so, which items?

3. If the cost for textbooks for one semester was $120 in the base years, how much did similar textbooks cost in 2003? Use the "Education" category.

4. If a new car cost $20,000 in 2003, what would a comparable new car have cost during the base years? Use the "Transportation" category.

Copyright © Houghton Mifflin Company. All rights reserved.

5. If a movie ticket cost $8 in 2003, what would a comparable movie ticket have cost during the base years? Use the "Entertainment" category.

6. The base year for the CPI was 1967 before the change to 1982–1984. If 1967 were still used as the base year, the CPI for all items in 2003 (not just those listed above) would be 550.9.
   a. Using the base year of 1967, explain the meaning of a CPI of 550.9.
   b. Using the base year of 1967 and a CPI of 550.9, if textbooks cost $75 for one semester in 1967, how much did similar textbooks cost in 2003?
   c. Using the base year of 1967 and a CPI of 550.9, if a family's food budget in 2003 is $800 per month, what would a comparable family budget have been in 1967?

Chapter 5 Summary

Key Words

	Examples
Percent means "parts of 100." [5.1A, p. 203]	23% means 23 of 100 equal parts.

Essential Rules and Procedures

	Examples
To write a percent as a fraction, drop the percent sign and multiply by $\frac{1}{100}$. [5.1A, p. 203]	$56\% = 56\left(\frac{1}{100}\right) = \frac{56}{100} = \frac{14}{25}$
To write a percent as a decimal, drop the percent sign and multiply by 0.01. [5.1A, p. 203]	$87\% = 87(0.01) = 0.87$
To write a fraction as a percent, multiply by 100%. [5.1B, p. 204]	$\frac{7}{20} = \frac{7}{20}(100\%) = \frac{700}{20}\% = 35\%$
To write a decimal as a percent, multiply by 100%. [5.1B, p. 204]	$0.325 = 0.325(100\%) = 32.5\%$

The Basic Percent Equation [5.2A, p. 207]
The basic percent equation is

$$\text{Percent} \times \text{base} = \text{amount}$$

Solving percent problems requires identifying the three elements of this equation. Usually the base follows the phrase "percent of."

8% of 250 is what number?
Percent × base = amount
$0.08 \times 250 = n$
$20 = n$

Proportion Method of Solving a Percent Problem [5.5A, p. 219]
The following proportion can be used to solve percent problems.

$$\frac{\text{percent}}{100} = \frac{\text{amount}}{\text{base}}$$

To use the proportion method, first identify the percent, the amount, and the base. The base usually follows the phrase "percent of."

8% of 250 is what number?
$\frac{\text{percent}}{100} = \frac{\text{amount}}{\text{base}}$
$\frac{8}{100} = \frac{n}{250}$
$8 \times 250 = 100 \times n$
$2000 = 100 \times n$
$2000 \div 100 = n$
$20 = n$

Copyright © Houghton Mifflin Company. All rights reserved.

Cumulative Review Exercises

1. Simplify: $18 \div (7 - 4)^2 + 2$.

2. Find the LCM of 16, 24, and 30.

3. Find the sum of $2\frac{1}{3}$, $3\frac{1}{2}$, and $4\frac{5}{8}$.

4. Subtract: $27\frac{5}{12} - 14\frac{9}{16}$

5. Multiply: $7\frac{1}{3} \times 1\frac{5}{7}$

6. What is $\frac{14}{27}$ divided by $1\frac{7}{9}$?

7. Simplify: $\left(\frac{3}{4}\right)^3 \cdot \left(\frac{8}{9}\right)^2$

8. Simplify: $\left(\frac{2}{3}\right)^2 - \left(\frac{3}{8} - \frac{1}{3}\right) \div \frac{1}{2}$

9. Round 3.07973 to the nearest hundredth.

10. Subtract:
$$\begin{array}{r} 3.0902 \\ -\ 1.9706 \end{array}$$

11. Divide: $0.032\overline{)1.097}$
Round to the nearest ten-thousandth.

12. Convert $3\frac{5}{8}$ to a decimal.

13. Convert 1.75 to a fraction.

14. Place the correct symbol, $<$ or $>$, between the two numbers.
$\frac{3}{8}$ 0.87

15. Solve the proportion $\frac{3}{8} = \frac{20}{n}$. Round to the nearest tenth.

16. Write "$76.80 earned in 8 hours" as a unit rate.

Copyright © Houghton Mifflin Company. All rights reserved.

17. Write $18\frac{1}{3}\%$ as a fraction.

18. Write $\frac{5}{6}$ as a percent.

19. 16.3% of 120 is what? Round to the nearest hundredth.

20. 24 is what percent of 18?

21. 12.4 is 125% of what?

22. What percent of 35 is 120? Round to the nearest tenth.

23. **Taxes**   Sergio has an income of $740 per week. One-fifth of his income is deducted for income tax payments. Find his take-home pay.

24. **Finance**   Eunice bought a used car for $8353, with a down payment of $1000. The balance was paid in 36 equal monthly payments. Find the monthly payment.

25. **Taxes**   The gasoline tax is $.19 a gallon. Find the number of gallons of gasoline used during a month in which $79.80 was paid in gasoline taxes.

26. **Taxes**   The real estate tax on a $172,000 home is $3440. At the same rate, find the real estate tax on a home valued at $250,000.

27. **Consumerism**   Ken purchased a camera for $490 and paid $29.40 in sales tax. What percent of the purchase price was the sales tax?

28. **Elections**   A survey of 300 people showed that 165 people favored a certain candidate for mayor. What percent of the people surveyed did not favor this candidate?

29. **Television**   According to the Cabletelevision Advertising Bureau, cable households watch television 36.5% of the time. On average, how many hours per week do cable households spend watching TV? Round to the nearest tenth.

30. **Health**   The Environmental Protection Agency found that 990 out of 5500 children tested had levels of lead in their blood exceeding federal guidelines. What percent of the children tested had levels of lead in the blood that exceeded federal standards?

Copyright © Houghton Mifflin Company. All rights reserved.

6 Applications for Business and Consumers

When you use a credit card to make a purchase, you are actually receiving a loan. This service allows you to defer payment on your purchase until a predetermined date in the future. In exchange for this convenient service, credit card companies will frequently charge you money for using their credit card. This added cost may be in the form of an annual fee, or it may be in the form of interest charges on balances that are not paid off after a certain deadline. These interest charges on unpaid balances are called finance charges and can be calculated using the simple interest formula. **Exercises 27 to 32 on page 257** ask you to use this formula to calculate the finance charges on unpaid balances.

Copyright © Houghton Mifflin Company. All rights reserved.

Need help? For online student resources, such as section quizzes, visit this textbook's website at **math.college.hmco.com/students.**

OBJECTIVES

Section 6.1

A To find unit cost
B To find the most economical purchase
C To find total cost

Section 6.2

A To find percent increase
B To apply percent increase to business—markup
C To find percent decrease
D To apply percent decrease to business—discount

Section 6.3

A To calculate simple interest
B To calculate finance charges on a credit card bill
C To calculate compound interest

Section 6.4

A To calculate the initial expenses of buying a home
B To calculate the ongoing expenses of owning a home

Section 6.5

A To calculate the initial expenses of buying a car
B To calculate the ongoing expenses of owning a car

Section 6.6

A To calculate commissions, total hourly wages, and salaries

Section 6.7

A To calculate checkbook balances
B To balance a checkbook

Do these exercises to prepare for Chapter 6.

For Exercises 1 to 6, add, subtract, multiply, or divide.

1. Divide: $3.75 \div 5$

2. Multiply: 3.47×15

3. Subtract: $874.50 - 369.99$

4. Multiply: $0.065 \times 150,000$

5. Multiply: $1500 \times 0.06 \times 0.5$

6. Add: $1372.47 + 36.91 + 5.00 + 2.86$

7. Divide $10 \div 3$. Round to the nearest hundredth.

8. Divide $345 \div 570$. Round to the nearest thousandth.

9. Place the correct symbol, $<$ or $>$, between the two numbers.
0.379   0.397

GO FIGURE ● ● ●

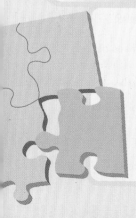

A store manager priced a pair of earrings in dollars and cents such that when 4% sales tax was added, the result was a whole number of dollars and 0 cents. Find the smallest possible number of dollars the items sold for, including the sales tax.

Copyright © Houghton Mifflin Company. All rights reserved.

6.1 Applications to Purchasing

Objective A   **To find unit cost**

Frequently, stores promote items for purchase by advertising, say, 2 Red Baron Bake to Rise Pizzas for $10.50 or 5 cans of StarKist tuna for $4.25.

The **unit cost** is the cost of *one* Red Baron Pizza or of *one* can of StarKist tuna. To find the unit cost, divide the total cost by the number of units.

2 pizzas for $10.50

$10.50 \div 2 = 5.25$

$5.25 is the cost of one pizza.

Unit cost: $5.25 per pizza

5 cans for $4.25

$4.25 \div 5 = 0.85$

$.85 is the cost of one can.

Unit cost: $.85 per can

Example 1

Find the unit cost. Round to the nearest tenth of a cent.
a. 3 gallons of mint chip ice cream for $17
b. 4 ounces of Crest toothpaste for $2.29

Strategy
To find the unit cost, divide the total cost by the number of units.

Solution
a. $17 \div 3 \approx 5.667$
   $5.667 per gallon
b. $2.29 \div 4 = 0.5725$
   $.573 per ounce

You Try It 1

Find the unit cost. Round to the nearest tenth of a cent.
a. 8 size-AA Energizer batteries for $7.67
b. 15 ounces of Revlon shampoo for $2.29

Your strategy

Your solution

Solution on p. S15

Objective B   **To find the most economical purchase**

Comparison shoppers often find the most economical buy by comparing unit costs.

One store is selling 6 twelve-ounce cans of ginger ale for $2.99, and a second store is selling 24 twelve-ounce cans of ginger ale for $11.79. To find the better buy, compare the unit costs.

$2.99 \div 6 \approx 0.498$

Unit cost: $.498 per can

$11.79 \div 24 \approx 0.491$

Unit cost: $.491 per can

Because $.491 < $.498, the better buy is 24 cans for $11.79.

Copyright © Houghton Mifflin Company. All rights reserved.

Example 2

Find the more economical purchase:
5 pounds of nails for $4.80, or 4 pounds of nails for $3.78.

Strategy
To find the more economical purchase, compare the unit costs.

Solution
$4.80 \div 5 = 0.96$
$3.78 \div 4 = 0.945$
$\$.945 < \$.96$

The more economical purchase is 4 pounds for $3.78.

You Try It 2

Find the more economical purchase: 6 cans of fruit for $5.70, or 4 cans of fruit for $3.96.

Your strategy

Your solution

Solution on p. S15

Objective C   **To find total cost**

An installer of floor tile found the unit cost of identical floor tiles at three stores.

Store 1	Store 2	Store 3
$1.22 per tile	$1.18 per tile	$1.28 per tile

By comparing the unit costs, the installer determined that store 2 would provide the most economical purchase.

The installer also uses the unit cost to find the total cost of purchasing 300 floor tiles at store 2. The **total cost** is found by multiplying the unit cost by the number of units purchased.

$$\boxed{\text{Unit cost}} \times \boxed{\text{number of units}} = \boxed{\text{total cost}}$$
$$1.18 \quad \times \quad 300 \quad = \quad 354$$

The total cost is $354.

Example 3

Clear redwood lumber costs $5.43 per foot. How much would 25 feet of clear redwood cost?

Strategy
To find the total cost, multiply the unit cost (5.43) by the number of units (25).

Solution

$$\boxed{\begin{array}{c}\text{Unit}\\\text{cost}\end{array}} \times \boxed{\begin{array}{c}\text{number}\\\text{of units}\end{array}} = \boxed{\begin{array}{c}\text{total}\\\text{cost}\end{array}}$$
$$5.43 \quad \times \quad 25 \quad = \quad 135.75$$

The total cost is $135.75.

You Try It 3

Pine saplings cost $9.96 each. How much would 7 pine saplings cost?

Your strategy

Your solution

Solution on p. S15

Copyright © Houghton Mifflin Company. All rights reserved.

6.1 Exercises

Objective A   **To find unit cost**

For Exercises 1 to 12, find the unit cost. Round to the nearest tenth of a cent.

1. Heinz B·B·Q sauce, 18 ounces for $.99

2. Birds-eye maple, 6 feet for $18.75

3. Diamond walnuts, $2.99 for 8 ounces

4. A&W root beer, 6 cans for $2.99

5. Ibuprofen, 50 tablets for $3.99

6. Visine eye drops, 0.5 ounce for $3.89

7. Adjustable wood clamps, 2 for $13.95

8. Corn, 6 ears for $1.85

9. Cheerios cereal, 15 ounces for $2.99

10. Doritos Cool Ranch chips, 14.5 ounces for $2.99

11. Sheet metal screws, 8 for $.95

12. Folgers coffee, 11.5 ounces for $4.79

Objective B   **To find the most economical purchase**

For Exercises 13 to 24, suppose your local supermarket offers the following products at the given prices. Find the more economical purchase.

13. Sutter Home pasta sauce, 25.5 ounces for $3.29, or Muir Glen Organic pasta sauce, 26 ounces for $3.79

14. Kraft mayonnaise, 40 ounces for $2.98, or Springfield mayonnaise, 32 ounces for $2.39

15. Ortega salsa, 20 ounces for $3.29 or 12 ounces for $1.99

16. L'Oréal shampoo, 13 ounces for $4.69, or Cortexx shampoo, 12 ounces for $3.99

17. Golden Sun vitamin E, 200 tablets for $7.39 or 400 tablets for $12.99

18. Ultra Mr. Clean, 20 ounces for $2.67, or Ultra Spic and Span, 14 ounces for $2.19

19. 16 ounces of Kraft cheddar cheese, $4.37, or 9 ounces of Land O' Lakes cheddar cheese, $2.29

20. Bertolli olive oil, 34 ounces for $9.49, or Pompeian olive oil, 8 ounces for $2.39

Copyright © Houghton Mifflin Company. All rights reserved.

21. Maxwell House coffee, 4 ounces for $3.99, or Sanka coffee, 2 ounces for $2.39

22. Wagner's vanilla extract, $3.29 for 1.5 ounces, or Durkee vanilla extract, 1 ounce for $2.74

23. Purina Cat Chow, $4.19 for 56 ounces, or Friskies Chef's Blend, $3.37 for 50.4 ounces

24. Kleenex tissues, $1.73 for 250 tissues, or Puffs tissues, $1.23 for 175 tissues

Objective C   **To find total cost**

25. If sliced bacon costs $4.59 per pound, find the total cost of 3 pounds.

26. Used red brick costs $.98 per brick. Find the total cost of 75 bricks.

27. Kiwi fruit cost $.23 each. Find the total cost of 8 kiwi.

28. Boneless chicken filets cost $4.69 per pound. Find the cost of 3.6 pounds. Round to the nearest cent.

29. Herbal tea costs $.98 per ounce. Find the total cost of 6.5 ounces.

30. If Stella Swiss Lorraine cheese costs $5.99 per pound, find the total cost of 0.65 pound. Round to the nearest cent.

31. Red Delicious apples cost $1.29 per pound. Find the total cost of 2.1 pounds. Round to the nearest cent.

32. Choice rib eye steak costs $8.49 per pound. Find the total cost of 2.8 pounds. Round to the nearest cent.

33. If Godiva chocolate costs $7.95 per pound, find the total cost of $\frac{3}{4}$ pound. Round to the nearest cent.

34. Color photocopying costs $.89 per page. Find the total cost for photocopying 120 pages.

APPLYING THE CONCEPTS

35. ✎ Explain in your own words the meaning of unit pricing.

36. ✎ What is the UPC (Universal Product Code) and how is it used?

Copyright © Houghton Mifflin Company. All rights reserved.

ISBN 0-395-75524-7

6.2 Percent Increase and Percent Decrease

Objective A **To find percent increase**

Percent increase is used to show how much a quantity has increased over its original value. The statements "Food prices increased by 2.3% last year" and "City council members received a 4% pay increase" are examples of percent increase.

 HOW TO   According to the Energy Information Administration, the number of alternative-fuel vehicles increased from approximately 277,000 to 352,000 in four years. Find the percent increase in alternative-fuel vehicles. Round to the nearest percent.

Point of Interest

According to the U.S. Census Bureau, the number of persons aged 65 and over in the United States will increase to about 82.0 million by 2050, a 136% increase from 2000.

$$\boxed{\text{New value}} - \boxed{\text{original value}} = \boxed{\text{amount of increase}}$$

$$352{,}000 - 277{,}000 = 75{,}000$$

Now solve the basic percent equation for percent.

$$\text{Percent} \times \text{base} = \text{amount}$$

$$\boxed{\text{Percent increase}} \times \boxed{\text{original value}} = \boxed{\text{amount of increase}}$$

$$n \times 277{,}000 = 75{,}000$$
$$n = 75{,}000 \div 277{,}000$$
$$n \approx 0.27$$

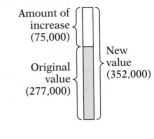

The number of alternative-fuel vehicles increased by approximately 27%.

Example 1

The average wholesale price of coffee increased from $2 per pound to $3 per pound in one year. What was the percent increase in the price of 1 pound of coffee?

Strategy
To find the percent increase:
• Find the amount of the increase.
• Solve the basic percent equation for *percent*.

Solution

$$\boxed{\text{New value}} - \boxed{\text{original value}} = \boxed{\text{amount of increase}}$$

$$3 - 2 = 1$$

$$\text{Percent} \times \text{base} = \text{amount}$$
$$n \times 2 = 1$$
$$n = 1 \div 2$$
$$n = 0.5 = 50\%$$

The percent increase was 50%.

You Try It 1

The average price of gasoline rose from $1.46 to $1.83 in 5 months. What was the percent increase in the price of gasoline? Round to the nearest percent.

Your strategy

Your solution

Solution on p. S15

Copyright © Houghton Mifflin Company. All rights reserved.

Example 2

Chris Carley was earning $13.50 an hour as a nursing assistant before receiving a 10% increase in pay. What is Chris's new hourly pay?

Strategy

To find the new hourly wage:

- Solve the basic percent equation for *amount*.
- Add the amount of the increase to the original wage.

Solution

Percent × base = amount
  0.10 × 13.50 = *n*
         1.35 = *n*

The amount of the increase was $1.35.

13.50 + 1.35 = 14.85

The new hourly wage is $14.85.

You Try It 2

Yolanda Liyama was making a wage of $12.50 an hour as a baker before receiving a 14% increase in hourly pay. What is Yolanda's new hourly wage?

Your strategy

Your solution

Solution on p. S15

Objective B To apply percent increase to business—markup

Some of the expenses involved in operating a business are salaries, rent, equipment, and utilities. To pay these expenses and earn a profit, a business must sell a product at a higher price than it paid for the product.

Cost is the price a business pays for a product, and **selling price** is the price at which a business sells a product to a customer. The difference between selling price and cost is called **markup**.

Markup
Selling price
Cost

$$\boxed{\text{Selling price}} - \boxed{\text{cost}} = \boxed{\text{markup}}$$
or
$$\boxed{\text{Cost}} + \boxed{\text{markup}} = \boxed{\text{selling price}}$$

Markup is frequently expressed as a percent of a product's cost. This percent is called the **markup rate**.

$$\boxed{\text{Markup rate}} \times \boxed{\text{cost}} = \boxed{\text{markup}}$$

Point of Interest

According to *Managing a Small Business*, from Liraz Publishing Company, goods in a store are often marked up 50% to 100% of the cost. This allows a business to make a profit of 5% to 10%.

HOW TO Suppose Bicycles Galore purchases an AMP Research B-5 bicycle for $2119.20 and sells it for $2649. What markup rate does Bicycles Galore use?

$$\boxed{\text{Selling price}} - \boxed{\text{cost}} = \boxed{\text{markup}}$$

  2649.00 − 2119.20 = 529.80   • First find the markup.

  Percent   × base = amount   • Then solve the basic percent equation for *percent*.

$$\boxed{\text{Markup rate}} \times \boxed{\text{cost}} = \boxed{\text{markup}}$$

   n   × 2119.20 = 529.80

      n = 529.80 ÷ 2119.20 = 0.25

The markup rate is 25%.

Copyright © Houghton Mifflin Company. All rights reserved.

Example 3

The manager of a sporting goods store determines that a markup rate of 36% is necessary to make a profit. What is the markup on a pair of skis that costs the store $225?

Strategy

To find the markup, solve the basic percent equation for *amount*.

Solution

Percent $\times$ base = amount

| Markup rate | $\times$ | cost | = | markup |

$$0.36 \quad \times \quad 225 \quad = \quad n$$
$$81 = n$$

The markup is $81.

You Try It 3

A bookstore manager determines that a markup rate of 20% is necessary to make a profit. What is the markup on a book that costs the bookstore $8?

Your strategy

Your solution

Example 4

A plant nursery bought a yellow twig dogwood for $4.50 and used a markup rate of 46%. What is the selling price?

Strategy

To find the selling price:

- Find the markup by solving the basic percent equation for *amount*.
- Add the markup to the cost.

Solution

Percent $\times$ base = amount

| Markup rate | $\times$ | cost | = | markup |

$$0.46 \quad \times \quad 4.50 \quad = \quad n$$
$$2.07 = n$$

| Cost | + | markup | = | selling price |

$$4.50 \quad + \quad 2.07 \quad = \quad 6.57$$

The selling price is $6.57.

You Try It 4

A clothing store bought a leather suit for $72 and used a markup rate of 55%. What is the selling price?

Your strategy

Your solution

Copyright © Houghton Mifflin Company. All rights reserved.

Solutions on p. S15

Objective C   **To find percent decrease**

Percent decrease is used to show how much a quantity has decreased from its original value. The statements "The number of family farms decreased by 2% last year" and "There has been a 50% decrease in the cost of a Pentium chip" are examples of percent decrease.

HOW TO   During a 2-year period, the value of U.S. agricultural products exported decreased from approximately $60.6 billion to $52.0 billion. Find the percent decrease in the value of U.S. agricultural exports. Round to the nearest tenth of a percent.

Original value	−	new value	=	amount of decrease

$$60.6 \quad - \quad 52.0 \quad = \quad 8.6$$

Now solve the basic percent equation for percent.

Percent × base = amount

Percent decrease	×	original value	=	amount of decrease

$$n \quad \times \quad 60.6 \quad = \quad 8.6$$
$$n = 8.6 \div 60.6$$
$$n \approx 0.142$$

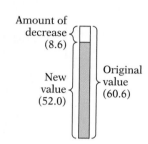

Amount of decrease (8.6)

New value (52.0)

Original value (60.6)

The value of agricultural exports decreased approximately 14.2%.

Study Tip

Note in the example below that solving a word problem includes stating a strategy and using the strategy to find a solution. If you have difficulty with a word problem, write down the known information. Be very specific. Write out a phrase or sentence that states what you are trying to find. See *AIM for Success* at the front of the book.

Example 5

During an 8-year period, the population of Baltimore, Maryland, decreased from approximately 736,000 to 646,000. Find the percent decrease in Baltimore's population. Round to the nearest tenth of a percent.

Strategy

To find the percent decrease:

• Find the amount of the decrease.
• Solve the basic percent equation for *percent*.

Solution

Original value	−	new value	=	amount of decrease

$$736,000 \quad - \quad 646,000 = \quad 90,000$$

Percent × base = amount
$$n \quad \times 736,000 = 90,000$$
$$n = 90,000 \div 736,000$$
$$n \approx 0.122$$

Baltimore's population decreased approximately 12.2%.

You Try It 5

During an 8-year period, the population of Norfolk, Virginia, decreased from approximately 261,000 to 215,000. Find the percent decrease in Norfolk's population. Round to the nearest tenth of a percent.

Your strategy

Your solution

Solution on p. S15

Copyright © Houghton Mifflin Company. All rights reserved.

Example 6

The total sales for December for a stationery store were $96,000. For January, total sales showed an 8% decrease from December's sales. What were the total sales for January?

Strategy

To find the total sales for January:

- Find the amount of decrease by solving the basic percent equation for *amount*.
- Subtract the amount of decrease from the December sales.

Solution

Percent × base = amount
0.08 × 96,000 = n
7680 = n

The decrease in sales was $7680.

96,000 − 7680 = 88,320

The total sales for January were $88,320.

You Try It 6

Fog decreased the normal 5-mile visibility at an airport by 40%. What was the visibility in the fog?

Your strategy

Your solution

Solution on p. S16

Objective D   **To apply percent decrease to business—discount**

To promote sales, a store may reduce the regular price of some of its products temporarily. The reduced price is called the **sale price.** The difference between the regular price and the sale price is called the **discount.**

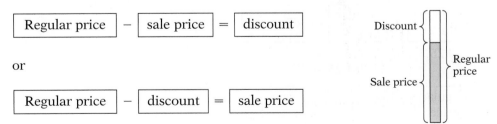

| Regular price | − | sale price | = | discount |

or

| Regular price | − | discount | = | sale price |

Discount is frequently stated as a percent of a product's regular price. This percent is called the **discount rate.**

| Discount rate | × | regular price | = | discount |

Copyright © Houghton Mifflin Company. All rights reserved.

Example 7

A GE 25-inch stereo television that regularly sells for $299 is on sale for $250. Find the discount rate. Round to the nearest tenth of a percent.

Strategy

To find the discount rate:

• Find the discount.
• Solve the basic percent equation for *percent*.

Solution

Regular price	−	sale price	=	discount
299	−	250	=	49

Percent	×	base	=	amount
Discount rate	×	regular price	=	discount
n	×	299	=	49

$$n = 49 \div 299$$
$$n \approx 0.164$$

The discount rate is 16.4%.

You Try It 7

A white azalea that regularly sells for $12.50 is on sale for $10.99. Find the discount rate. Round to the nearest tenth of a percent.

Your strategy

Your solution

Example 8

A 20-horsepower lawn mower is on sale for 25% off the regular price of $1125. Find the sale price.

Strategy

To find the sale price:

• Find the discount by solving the basic percent equation for *amount*.
• Subtract to find the sale price.

Solution

Percent	×	base	=	amount
Discount rate	×	regular price	=	discount
0.25	×	1125	=	n

$$281.25 = n$$

Regular price	−	discount	=	sale price
1125	−	281.25	=	843.75

The sale price is $843.75.

You Try It 8

A hardware store is selling a Newport security door for 15% off the regular price of $110. Find the sale price.

Your strategy

Your solution

Solutions on p. S16

Copyright © Houghton Mifflin Company. All rights reserved.

6.2 Exercises

Copyright © Houghton Mifflin Company. All rights reserved.

Objective A   **To find percent increase**

For Exercises 1 to 5, solve. Round percents to the nearest tenth of a percent.

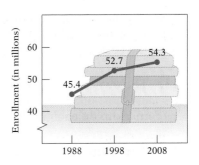

Elementary and Secondary
School Enrollments

Source: National Center for Education

1.   **Education**   The figure at the right shows the actual and projected enrollments in grades K–12 in the United States. What percent increase is expected from 1988 to 2008?

2.   **Fuel Efficiency**   An automobile manufacturer increased the average mileage on a car from 17.5 miles per gallon to 18.2 miles per gallon. Find the percent increase in mileage.

3.   **Business**   In the 1990s, the number of Target stores increased from 420 stores to 914 stores. (*Source:* Target) What was the percent increase in the number of Target stores in the 1990s?

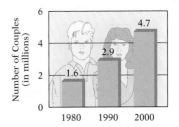

Unmarried U.S. Couples
Living Together

4.   **Demography**   The graph at the right shows the number of unmarried American couples living together. (*Source:* U.S. Census Bureau) Find the percent increase in the number of unmarried couples living together from 1980 to 2000. Round to the nearest tenth.

5.   **Sports**   In 1924, the number of events in the Winter Olympics was 14. The 2002 Winter Olympics in Salt Lake City included 78 medal events. (*Source:* David Wallenchinsky's *The Complete Book of the Winter Olympics*) Find the percent increase in the number of events in the Winter Olympics from 1924 to 2002.

Year	Number of Households Containing Millionaires
1975	350,000
1997	3,500,000
2005	5,600,000

Source: Affluent Market Institute

6.   **Wealth**   The table at the right shows the estimated number of millionaire households in the United States. Find the percent increase in the estimated number of households containing millionaires from 1975 to 2005.

7.   **Demography**   The population of Boise City, Idaho, increased 24.3% over an 8-year period. The population initially was approximately 127,000. (*Source:* U.S. Census Bureau) Find the population of Boise City 8 years later.

8.   **Television**   During 1 year, the number of people subscribing to direct broadcasting satellite systems increased 87%. If the number of subscribers at the beginning of the year was 2.3 million, how many subscribers were there at the end of the year?

9.   **Demography**   From 1970 to 2000, the average age of American mothers giving birth to their first child rose 16.4%. (*Source:* Centers for Disease Control and Prevention) If the average age in 1970 was 21.4 years, what was the average age in 2000? Round to the nearest tenth.

Objective B **To apply percent increase to business—markup**

10. A window air conditioner cost AirRite Air Conditioning Systems $285. Find the markup on the air conditioner if the markup rate is 25% of the cost.

11. The owner of Kerr Electronics purchased 300 Craig portable CD players at a cost of $85 each. If the owner uses a markup rate of 42%, what is the markup on each of the Craig CD players?

12. The manager of Brass Antiques has determined that a markup rate of 38% is necessary for a profit to be made. What is the markup on a brass doorknob that costs $45?

13. Computer Inc. uses a markup of $975 on a computer system that costs $3250. What is the markup rate on this system?

14. Saizon Pen & Office Supply uses a markup of $12 on a calculator that costs $20. What markup rate does this amount represent?

15. Giant Photo Service uses a markup rate of 48% on its Model ZA cameras, which cost the shop $162.
    a. What is the markup?
    b. What is the selling price?

16. The Circle R golf pro shop uses a markup rate of 45% on a set of Tour Pro golf clubs that costs the shop $210.
    a. What is the markup?
    b. What is the selling price?

17. Harvest Time Produce Inc. uses a 55% markup rate and pays $2 for a box of strawberries.
    a. What is the markup?
    b. What is the selling price of a box of strawberries?

18. Resner Builders' Hardware uses a markup rate of 42% for a table saw that costs $160. What is the selling price of the table saw?

19. Brad Burt's Magic Shop uses a markup rate of 48%. What is the selling price of a telescoping sword that costs $50?

Objective C **To find percent decrease**

20. **Travel**   A new bridge reduced the normal 45-minute travel time between two cities by 18 minutes. What percent decrease does this represent?

21. **Energy**   By installing energy-saving equipment, the Pala Rey Youth Camp reduced its normal $800-per-month utility bill by $320. What percent decrease does this amount represent?

22. **Depreciation**   It is estimated that the value of a new car is reduced 30% after 1 year of ownership. Using this estimate, find how much value a $18,200 new car loses after 1 year.

Copyright © Houghton Mifflin Company. All rights reserved.

23. **Employment**   A department store employs 1200 people during the holiday. At the end of the holiday season, the store reduces the number of employees by 45%. What is the decrease in the number of employees?

24. **Business**   Because of a decrease in demand for super-8 video cameras, Kit's Cameras reduced the orders for these models from 20 per month to 8 per month.
   a. What is the amount of the decrease?
   b. What percent decrease does this amount represent?

25. **Business**   A new computer system reduced the time for printing the payroll from 52 minutes to 39 minutes.
   a. What is the amount of the decrease?
   b. What percent decrease does this amount represent?

26. **Finance**   Juanita's average monthly expense for gasoline was $76. After joining a car pool, she was able to reduce the expense by 20%.
   a. What was the amount of the decrease?
   b. What is the average monthly gasoline bill now?

27. **Investments**   An oil company paid a dividend of $1.60 per share. After a reorganization, the company reduced the dividend by 37.5%.
   a. What was the amount of the decrease?
   b. What is the new dividend?

28. **Traffic Patterns**   Because of an improved traffic pattern at a sports stadium, the average amount of time a fan waits to park decreased from 3.5 minutes to 2.8 minutes. What percent decrease does this amount represent?

29. **Airplanes**   One configuration of the Boeing 777-300 has a seating capacity of 394. This is 26 more than that of the corresponding 777-200 jet. What is the percent decrease in capacity from the 777-300 to the 777-200 model? Round to nearest tenth of a percent.

30. **The Military**   In 2000, the Pentagon revised its account of the number of Americans killed in the Korean War from 54,246 to 36,940. (*Source: Time*, June 12, 2000) What is the percent decrease in the reported number of military personnel killed in the Korean War? Round to nearest tenth of a percent.

Objective D   **To apply percent decrease to business—discount**

31. The Austin College Bookstore is giving a discount of $8 on calculators that normally sell for $24. What is the discount rate?

32. A discount clothing store is selling a $72 sport jacket for $24 off the regular price. What is the discount rate?

33. A disc player that regularly sells for $340 is selling for 20% off the regular price. What is the discount?

34. Dacor Appliances is selling its $450 washing machine for 15% off the regular price. What is the discount?

35. An electric grill that regularly sells for $140 is selling for $42 off the regular price. What is the discount rate?

Copyright © Houghton Mifflin Company. All rights reserved.

36. Quick Service Gas Station has its regularly priced $45 tune-up on sale for 16% off the regular price.
   a. What is the discount?
   b. What is the sale price?

37. Tomatoes that regularly sell for $1.25 per pound are on sale for 20% off the regular price.
   a. What is the discount?
   b. What is the sale price?

38. An outdoor supply store has its regularly priced $160 sleeping bags on sale for $120.
   a. What is the discount?
   b. What is the discount rate?

39. Standard Brands paint that regularly sells for $20 per gallon is on sale for $16 per gallon.
   a. What is the discount?
   b. What is the discount rate?

40. **The Military**   The graph at the right shows the number of active-duty U.S. military personnel in 1990 and in 2000.
   a. Which branch of the military had the greatest percent decrease in personnel from 1990 to 2000?
   b. What was the percent decrease for this branch of the service?

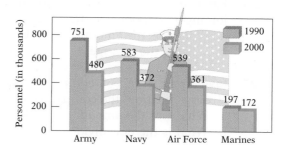

Number of Active-Duty U.S. Military Personnel

Source: Fiscal 2000 Annual Report to the President and Congress by the Secretary of Defense

APPLYING THE CONCEPTS

41. **Compensation**   A welder earning $12 per hour is given a 10% raise. To find the new wage, we can multiply $12 by 0.10 and add the product to $12. Can the new wage be found by multiplying $12 by 1.10? Try both methods and compare your answers.

42. **Business**   Grocers, florists, bakers, and other businesses must consider spoilage when deciding the markup of a product. For instance, suppose a florist purchased 200 roses at a cost of $.86 per rose. The florist wants a markup rate of 50% of the total cost of all the roses and expects 7% of the roses to wilt and therefore not be salable. Find the selling price per rose by answering each of the following questions.
   a. What is the florist's total cost for the 200 roses?
   b. Find the total selling price without spoilage.
   c. Find the number of roses the florist expects to sell. *Hint:* The number of roses the florist expects to sell is

   % of salable roses × number of roses purchased

   d. To find the selling price per rose, divide the total selling price without spoilage by the number of roses the florist expects to sell. Round to the nearest cent.

43. **Business**   A promotional sale at a department store offers 25% off the sale price. The sale price itself is 25% off the regular price. Is this the same as a sale that offers 50% off the regular price? If not, which sale gives the better price? Explain your answer.

Copyright © Houghton Mifflin Company. All rights reserved.

6.3 Interest

Objective A   **To calculate simple interest**

When you deposit money in a bank—for example, in a savings account—you are permitting the bank to use your money. The bank may use the deposited money to lend customers the money to buy cars or make renovations on their homes. The bank pays you for the privilege of using your money. The amount paid to you is called **interest.** If you are the one borrowing money from the bank, the amount you pay for the privilege of using that money is also called interest.

The original amount deposited or borrowed is called the **principal.** The amount of interest paid is usually given as a percent of the principal. The percent used to determine the amount of interest is the **interest rate.**

Interest paid on the original principal is called **simple interest.** To calculate simple interest, multiply the principal by the interest rate per period by the number of time periods. In this objective, we are working with annual interest rates, so the time periods are years. The simple interest formula for an annual interest rate is given below.

> **Simple Interest Formula for Annual Interest Rates**
>
> Principal $\times$ annual interest rate $\times$ time (in years) $=$ interest

Interest rates are generally given as percents. Before performing calculations involving an interest rate, write the interest rate as a decimal.

> **HOW TO**   Calculate the simple interest due on a 2-year loan of $1500 that has an annual interest rate of 7.5%.
>
> | Principal | $\times$ | annual interest rate | $\times$ | time (in years) | $=$ | interest |
> |---|---|---|---|---|---|---|
> | 1500 | $\times$ | 0.075 | $\times$ | 2 | $=$ | 225 |
>
> The simple interest due is $225.

When we borrow money, the total amount to be repaid to the lender is the sum of the principal and the interest. This amount is called the **maturity value of a loan.**

> **Maturity Value Formula for Simple Interest Loans**
>
> Principal $+$ interest $=$ maturity value

In the example above, the simple interest due on the loan of $1500 was $225. The maturity value of the loan is therefore $1500 + $225 = $1725.

Copyright © Houghton Mifflin Company. All rights reserved.

TAKE NOTE

If you deposit $1000 in a savings account paying 5% interest, the $1000 is the principal and 5% is the interest rate.

Copyright © Houghton Mifflin Company. All rights reserved.

HOW TO Calculate the maturity value of a simple interest, 8-month loan of $8000 if the annual interest rate is 9.75%.

First find the interest due on the loan.

Principal	×	annual interest rate	×	time (in years)	=	interest
8000	×	0.0975	×	$\frac{8}{12}$	=	520

Find the maturity value.

Principal	+	interest	=	maturity value
8000	+	520	=	8520

The maturity value of the loan is $8520.

The monthly payment on a loan can be calculated by dividing the maturity value by the length of the loan in months.

> **Monthly Payment on a Simple Interest Loan**
>
> Maturity value ÷ length of the loan in months = monthly payment

In the example above, the maturity value of the loan is $8520. To find the monthly payment on the 8-month loan, divide 8520 by 8.

Maturity value	÷	length of the loan in months	=	monthly payment
8520	÷	8	=	1065

The monthly payment on the loan is $1065.

> *TAKE NOTE*
>
> The time of the loan must be in years. Eight months is $\frac{8}{12}$ of a year.
>
> See Example 1. The time of the loan must be in years. 180 days is $\frac{180}{365}$ of a year.

Example 1

Kamal borrowed $500 from a savings and loan association for 180 days at an annual interest rate of 7%. What is the simple interest due on the loan?

Strategy

To find the simple interest due, multiply the principal (500) times the annual interest rate (7% = 0.07) times the time, in years (180 days = $\frac{180}{365}$ year).

Solution

Principal	×	annual interest rate	×	time (in years)	=	interest
500	×	0.07	×	$\frac{180}{365}$	≈	17.26

The simple interest due is $17.26.

You Try It 1

A company borrowed $15,000 from a bank for 18 months at an annual interest rate of 8%. What is the simple interest due on the loan?

Your strategy

Your solution

Solution on p. S16

Example 2

Calculate the maturity value of a simple interest, 9-month loan of $4000 if the annual interest rate is 8.75%.

Strategy

To find the maturity value:

• Use the simple interest formula to find the simple interest due.
• Find the maturity value by adding the principal and the interest.

Solution

Principal	×	annual interest rate	×	time (in years)	=	interest
4000	×	0.0875	×	$\frac{9}{12}$	=	262.5

Principal	+	interest	=	maturity value
4000	+	262.50	=	4262.50

The maturity value is $4262.50.

Example 3

The simple interest due on a 3-month loan of $1400 is $26.25. Find the monthly payment on the loan.

Strategy

To find the monthly payment:

• Find the maturity value by adding the principal and the interest.
• Divide the maturity value by the length of the loan in months (3).

Solution

Principal + interest = maturity value
   1400   + 26.25 =   1426.25

Maturity value ÷ length of the loan = payment
   1426.25   ÷   3   ≈ 475.42

The monthly payment is $475.42.

You Try It 2

Calculate the maturity value of a simple interest, 90-day loan of $3800. The annual interest rate is 6%.

Your strategy

Your solution

You Try It 3

The simple interest due on a 1-year loan of $1900 is $152. Find the monthly payment on the loan.

Your strategy

Your solution

Solutions on p. S16

Objective B   **To calculate finance charges on a credit card bill**

When a customer uses a credit card to make a purchase, the customer is actually receiving a loan. Therefore, there is frequently an added cost to the consumer who purchases on credit. This may be in the form of an annual fee and interest charges on purchases. The interest charges on purchases are called **finance charges.**

Copyright © Houghton Mifflin Company. All rights reserved.

The finance charge on a credit card bill is calculated using the simple interest formula. In the last objective, the interest rates were annual interest rates. However, credit card companies generally issue *monthly* bills and express interest rates on credit card purchases as *monthly* interest rates. Therefore, when using the simple interest formula to calculate finance charges on credit card purchases, use a monthly interest rate and express the time in months.

Note: In the simple interest formula, the time must be expressed in the same period as the rate. For an *annual* interest rate, the time must be expressed in years. For a *monthly* interest rate, the time must be in months.

Example 4

A credit card company charges a customer 1.5% per month on the unpaid balance of charges on the credit card. What is the finance charge in a month when the customer has an unpaid balance of $254?

Strategy

To find the finance charge, multiply the principal, or unpaid balance (254), times the monthly interest rate (1.5%) times the number of months (1).

Solution

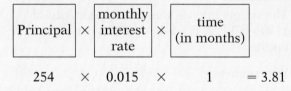

Principal	×	monthly interest rate	×	time (in months)
254	×	0.015	×	1	= 3.81

The simple interest due is $3.81.

You Try It 4

The credit card that Francesca uses charges her 1.6% per month on her unpaid balance. Find the finance charge when her unpaid balance one month is $1250.

Your strategy

Your solution

Solution on pp. S16–S17

Objective C **To calculate compound interest**

Usually, the interest paid on money deposited or borrowed is compound interest. **Compound interest** is computed not only on the original principal but also on interest already earned. Here is an illustration.

Suppose $1000 is invested for 3 years at an annual interest rate of 9% compounded annually. Because this is an *annual* interest rate, we will calculate the interest earned each year.

During the first year, the interest earned is calculated as follows:

Principal	×	annual interest rate	×	time (in years)	=	interest
1000	×	0.09	×	1	=	90

Copyright © Houghton Mifflin Company. All rights reserved.

At the end of the first year, the total amount in the account is

$$1000 + 90 = 1090$$

During the second year, the interest earned is calculated on the amount in the account at the end of the first year.

Principal	×	annual interest rate	×	time (in years)	=	interest
1090	×	0.09	×	1	=	98.10

Note that the interest earned during the second year ($98.10) is greater than the interest earned during the first year ($90). This is because the interest earned during the first year was added to the original principal, and the interest for the second year was calculated using this sum. If the account earned simple interest, the interest earned would be the same every year ($90).

At the end of the second year, the total amount in the account is the sum of the amount in the account at the end of the first year and the interest earned during the second year.

$$1090 + 98.10 = 1188.10$$

The interest earned during the third year is calculated using the amount in the account at the end of the second year ($1188.10).

Principal	×	annual interest rate	×	time (in years)	=	interest
1188.10	×	0.09	×	1	≈	106.93

The amount in the account at the end of the third year is

$$1188.10 + 106.93 = 1295.03$$

TAKE NOTE

The interest earned each year keeps increasing. This is the effect of compound interest.

To find the interest earned for the three years, subtract the original principal from the new principal.

New principal	−	original principal	=	interest earned
1295.03	−	1000	=	295.03

Note that the compound interest earned is $295.03. The simple interest earned on the investment would have been only $1000 × 0.09 × 3 = $270.

In this example, the interest was compounded annually. However, compound interest can be compounded

Compounding periods:	annually (once a year)
	semiannually (twice a year)
	quarterly (four times a year)
	monthly (12 times a year)
	daily (365 times a year)

The more frequent the compounding periods, the more interest the account earns. For example, if, in the above example, the interest had been compounded quarterly rather than annually, the interest earned would have been greater.

Copyright © Houghton Mifflin Company. All rights reserved.

Calculating compound interest can be very tedious, so there are tables that can be used to simplify these calculations. A portion of a Compound Interest Table is given in the Appendix.

> **HOW TO** What is the value after 5 years of $1000 invested at 7% annual interest compounded quarterly?
>
> To find the interest earned, multiply the original principal (1000) by the factor found in the Compound Interest Table. To find the factor, first find the table headed "Compounded Quarterly" in the Compound Interest Table in the Appendix. Then look at the number where the 7% column and the 5-year row meet.

	4%	*5%*	*6%*	*7%*	*8%*	*9%*	*10%*
				Compounded Quarterly			
1 year	1.04060	1.05094	1.06136	1.07186	1.08243	1.09308	1.10381
5 years	1.22019	1.28204	1.34686	**1.41478**	1.48595	1.56051	1.63862
10 years	1.48886	1.64362	1.81402	2.00160	2.20804	2.43519	2.68506
15 years	1.81670	2.10718	2.44322	2.83182	3.28103	3.80013	4.39979
20 years	2.21672	2.70148	3.29066	4.00639	4.87544	5.93015	7.20957

The factor is 1.41478.

$$1000 \times 1.41478 = 1414.78$$

The value of the investment after 5 years is $1414.78.

Example 5

An investment of $650 pays 8% annual interest compounded semiannually. What is the interest earned in 5 years?

Strategy

To find the interest earned:

• Find the new principal by multiplying the original principal (650) by the factor found in the Compound Interest Table (1.48024).
• Subtract the original principal from the new principal.

Solution

$650 \times 1.48024 \approx 962.16$

The new principal is $962.16.

$962.16 - 650 = 312.16$

The interest earned is $312.16.

You Try It 5

An investment of $1000 pays 6% annual interest compounded quarterly. What is the interest earned in 20 years?

Your strategy

Your solution

Solution on p. S17

Copyright © Houghton Mifflin Company. All rights reserved.

6.3 Exercises

Copyright © Houghton Mifflin Company. All rights reserved.

Objective A   **To calculate simple interest**

1. A 2-year student loan of $10,000 is made at an annual simple interest rate of 4.25%. The simple interest on the loan is $850. Identify **a.** the principal, **b.** the interest, **c.** the interest rate, and **d.** the time period of the loan.

2. A contractor obtained a 9-month loan for $80,000 at an annual simple interest rate of 9.75%. The simple interest on the loan is $5850. Identify **a.** the principal, **b.** the interest, **c.** the interest rate, and **d.** the time period of the loan.

3. Find the simple interest Jacob Zucker owes on a 2-year student loan of $8000 at an annual interest rate of 6%.

4. Find the simple interest Kara Tanamachi owes on a $1\frac{1}{2}$-year loan of $1500 at an annual interest rate of 7.5%.

5. To finance the purchase of 15 new cars, the Tropical Car Rental Agency borrowed $100,000 for 9 months at an annual interest rate of 4.5%. What is the simple interest due on the loan?

6. A home builder obtained a preconstruction loan of $50,000 for 8 months at an annual interest rate of 9.5%. What is the simple interest due on the loan?

7. A bank lent Gloria Masters $20,000 at an annual interest rate of 8.8%. The period of the loan was 9 months. Find the simple interest due on the loan.

8. Eugene Madison obtained an 8-month loan of $4500 at an annual interest rate of 6.2%. Find the simple interest Eugene owes on the loan.

9. Shannon O'Hara borrowed $5000 for 90 days at an annual simple interest rate of 7.5%. Find the simple interest due on the loan.

10. Jon McCloud borrowed $8500 for 180 days at an annual simple interest rate of 9.25%. Find the simple interest due on the loan.

11. Jorge Elizondo took out a 75-day loan of $7500 at an annual interest rate of 5.5%. Find the simple interest due on the loan.

12. Kristi Yang borrowed $15,000. The term of the loan was 90 days, and the annual simple interest rate was 7.4%. Find the simple interest due on the loan.

13. The simple interest due on a 4-month loan of $4800 is $320. What is the maturity value of the loan?

14. The simple interest due on a 60-day loan of $6500 is $80.14. Find the maturity value of the loan.

15. An auto parts dealer borrowed $150,000 at a 9.5% annual simple interest rate for 1 year. Find the maturity value of the loan.

16. A corporate executive took out a $25,000 loan at an 8.2% annual simple interest rate for 1 year. Find the maturity value of the loan.

17. William Carey borrowed $12,500 for 8 months at an annual simple interest rate of 4.5%. Find the total amount due on the loan.

18. You arrange for a 9-month bank loan of $9000 at an annual simple interest rate of 8.5%. Find the total amount you must repay to the bank.

19. Capital City Bank approves a home-improvement loan application for $14,000 at an annual simple interest rate of 5.25% for 270 days. What is the maturity value of the loan?

20. A credit union lends a member $5000 for college tuition. The loan is made for 18 months at an annual simple interest rate of 6.9%. What is the maturity value of this loan?

21. Action Machining Company purchased a robot-controlled lathe for $225,000 and financed the full amount at 8% annual simple interest for 4 years. The simple interest on the loan is $72,000. Find the monthly payment.

22. For the purchase of an entertainment center, a $1900 loan is obtained for 2 years at an annual simple interest rate of 9.4%. The simple interest due on the loan is $357.20. What is the monthly payment on the loan?

23. To attract new customers, Heller Ford is offering car loans at an annual simple interest rate of 4.5%.
a. Find the interest charged to a customer who finances a car loan of $12,000 for 2 years.
b. Find the monthly payment.

24. Cimarron Homes Inc. purchased a snow plow for $57,000 and financed the full amount for 5 years at an annual simple interest rate of 9%.
a. Find the interest due on the loan.
b. Find the monthly payment.

Copyright © Houghton Mifflin Company. All rights reserved.

25. Dennis Pappas decided to build onto his present home instead of buying a new, larger house. He borrowed $42,000 for $3\frac{1}{2}$ years at an annual simple interest rate of 9.5%. Find the monthly payment.

26. Rosalinda Johnson took out a 6-month, $12,000 loan. The annual simple interest rate on the loan was 8.5%. Find the monthly payment.

Objective B **To calculate finance charges on a credit card bill**

27. A credit card company charges a customer 1.25% per month on the unpaid balance of charges on the credit card. What is the finance charge in a month when the customer has an unpaid balance of $118.72?

28. The credit card that Dee Brown uses charges her 1.75% per month on her unpaid balance. Find the finance charge when her unpaid balance one month is $391.64.

29. What is the finance charge on an unpaid balance of $12,368.92 on a credit card that charges 1.5% per month on any unpaid balance?

30. Suppose you have an unpaid balance of $995.04 on a credit card that charges 1.2% per month on any unpaid balance. What finance charge do you owe the company?

31. A credit card customer has an unpaid balance of $1438.20. What is the difference between monthly finance charges of 1.15% per month on the unpaid balance and monthly finance charges of 1.85% per month?

32. One credit card company charges 1.25% per month on any unpaid balance, and a second company charges 1.75%. What is the difference between the finance charges that these two companies assess on an unpaid balance of $687.45?

Objective C **To calculate compound interest**

33. North Island Federal Credit Union pays 4% annual interest, compounded daily, on time savings deposits. Find the value of $750 deposited in this account after 1 year.

34. Tanya invested $2500 in a tax-sheltered annuity that pays 8% annual interest compounded daily. Find the value of her investment after 20 years.

35. Sal Travato invested $3000 in a corporate retirement account that pays 6% annual interest compounded semiannually. Find the value of his investment after 15 years.

Copyright © Houghton Mifflin Company. All rights reserved.

36. To replace equipment, a farmer invested $20,000 in an account that pays 7% annual interest compounded monthly. What is the value of the investment after 5 years?

37. Green River Lodge invests $75,000 in a trust account that pays 8% interest compounded quarterly.
  a. What will the value of the investment be in 5 years?
  b. How much interest will be earned in the 5 years?

38. To save for retirement, a couple deposited $3000 in an account that pays 7% annual interest compounded daily.
  a. What will the value of the investment be in 10 years?
  b. How much interest will be earned in the 10 years?

39. To save for a child's education, the Petersens deposited $2500 into an account that pays 6% annual interest compounded daily. Find the amount of interest earned on this account over a 20-year period.

40. How much interest is earned in 2 years on $4000 deposited in an account that pays 6% interest, compounded quarterly?

APPLYING THE CONCEPTS

41. **Banking**   The Mission Valley Credit Union charges its customers an interest rate of 2% per month on money that is transferred into an account that is overdrawn. Find the interest owed to the credit union for 1 month when $800 is transferred into an overdrawn account.

42. **Banking**   Suppose you have a savings account that earns interest at the rate of 6% per year compounded monthly. On January 1, you open this account with a deposit of $100.
  a. On February 1, you deposit an additional $100 into the account. What is the value of the account after the deposit?
  b. On March 1, you deposit an additional $100 into the account. What is the value of the account after the deposit?
  Note: This type of savings plan, wherein equal amounts ($100) are saved at equal time intervals (every month), is called an annuity.

43.   **Banking**   At 4 P.M. on July 31, you open a savings account that pays 5% annual interest and you deposit $500 in the account. Your deposit is credited as of August 1. At the beginning of September, you receive a statement from the bank that shows that during the month of August, you received $2.12 in interest. The interest has been added to your account, bringing the total on deposit to $502.12. At the beginning of October, you receive a statement from the bank that shows that during the month of September, you received $2.06 in interest on the $502.12 on deposit. Explain why you received less interest during the second month when there was more money on deposit.

Copyright © Houghton Mifflin Company. All rights reserved.

6.4 Real Estate Expenses

Objective A   **To calculate the initial expenses of buying a home**

One of the largest investments most people ever make is the purchase of a home. The major initial expense in the purchase is the **down payment,** which is normally a percent of the purchase price. This percent varies among banks, but it usually ranges from 5% to 25%.

The **mortgage** is the amount that is borrowed to buy real estate. The mortgage amount is the difference between the purchase price and the down payment.

> **HOW TO**   A home is purchased for $140,000, and a down payment of $21,000 is made. Find the mortgage.
>
> | Purchase price | − | down payment | = | mortgage |
> |:---:|:---:|:---:|:---:|:---:|
> | 140,000 | − | 21,000 | = | 119,000 |
>
> The mortgage is $119,000.

Another initial expense in buying a home is the **loan origination fee,** which is a fee that the bank charges for processing the mortgage papers. The loan origination fee is usually a percent of the mortgage and is expressed in **points,** which is the term banks use to mean percent. For example, "5 points" means "5 percent."

Points	×	mortgage	=	loan origination fee

TAKE NOTE

Because *points* means percent, a loan origination fee of

$2\frac{1}{2}$ points $= 2\frac{1}{2}\% =$

2.5% = 0.025.

Example 1

A house is purchased for $125,000, and a down payment, which is 20% of the purchase price, is made. Find the mortgage.

Strategy

To find the mortgage:

- Find the down payment by solving the basic percent equation for *amount*.
- Subtract the down payment from the purchase price.

Solution

Percent	×	base	=	amount
Percent	×	purchase price	=	down payment
0.20	×	125,000	=	n
		25,000	=	n

Purchase price	−	down payment	=	mortgage
125,000	−	25,000	=	100,000

The mortgage is $100,000.

You Try It 1

An office building is purchased for $216,000, and a down payment, which is 25% of the purchase price, is made. Find the mortgage.

Your strategy

Your solution

Solution on p. S17

Copyright © Houghton Mifflin Company. All rights reserved.

Copyright © Houghton Mifflin Company. All rights reserved.

Example 2

A home is purchased with a mortgage of $65,000. The buyer pays a loan origination fee of $3\frac{1}{2}$ points. How much is the loan origination fee?

Strategy

To find the loan origination fee, solve the basic percent equation for *amount*.

Solution

Percent	×	base	= amount
Points	×	mortgage	= fee
0.035	×	65,000	= n
		2275	= n

The loan origination fee is $2275.

You Try It 2

The mortgage on a real estate investment is $80,000. The buyer paid a loan origination fee of $4\frac{1}{2}$ points. How much was the loan origination fee?

Your strategy

Your solution

Solution on p. S17

Objective B   **To calculate the ongoing expenses of owning a home**

Point of Interest

The number-one response of adults asked what they would spend money on first if they suddenly became wealthy (for example, by winning the lottery) was a house; 31% gave this response. (*Source:* Yankelovich Partners for Lutheran Brotherhood)

Besides the initial expenses of buying a house, there are continuing monthly expenses involved in owning a home. The **monthly mortgage payment** (one of 12 payments due each year to the lender of money to buy real estate), utilities, insurance, and **property tax** (a tax based on the value of real estate) are some of these ongoing expenses. Of these expenses, the largest one is normally the monthly mortgage payment.

For a **fixed-rate mortgage,** the monthly mortgage payment remains the same throughout the life of the loan. The calculation of the monthly mortgage payment is based on the amount of the loan, the interest rate on the loan, and the number of years required to pay back the loan. Calculating the monthly mortgage payment is fairly difficult, so tables such as the one in the Appendix are used to simplify these calculations.

Integrating Technology

In general, when a problem requests a monetary payment, the answer is rounded to the nearest cent. For the example at the right, enter

60000 × 0.0080462 =

The display reads 482.772. Round this number to the nearest hundredth: 482.77. The answer is $482.77.

HOW TO   Find the monthly mortgage payment on a 30-year $60,000 mortgage at an interest rate of 9%. Use the Monthly Payment Table in the Appendix.

$$60,000 \times \underset{\substack{\uparrow \\ \text{From the} \\ \text{table}}}{0.0080462} \approx 482.77$$

The monthly mortgage payment is $482.77.

The monthly mortgage payment includes the payment of both principal and interest on the mortgage. The interest charged during any one month is charged on the unpaid balance of the loan. Therefore, during the early years of the mortgage, when the unpaid balance is high, most of the monthly mortgage payment is interest charged on the loan. During the last few years of a mortgage, when the unpaid balance is low, most of the monthly mortgage payment goes toward paying off the loan.

Point of Interest

Home buyers rated the following characteristics "extremely important" in their purchase decision.

Natural, open space: 77%
Walking and biking paths: 74%
Gardens with native plants: 56%
Clustered retail stores: 55%
Wilderness area: 52%
Outdoor pool: 52%
Community recreation center: 52%
Interesting little parks: 50%

(*Sources:* American Lives, Inc; Intercommunications, Inc.)

HOW TO Find the interest paid on a mortgage during a month when the monthly mortgage payment is $186.26 and $58.08 of that amount goes toward paying off the principal.

Monthly mortgage payment	−	principal	=	interest
186.26	−	58.08	=	128.18

The interest paid on the mortgage is $128.18.

Property tax is another ongoing expense of owning a house. Property tax is normally an annual expense that may be paid on a monthly basis. The monthly property tax, which is determined by dividing the annual property tax by 12, is usually added to the monthly mortgage payment.

HOW TO A homeowner must pay $534 in property tax annually. Find the property tax that must be added each month to the homeowner's monthly mortgage payment.

$534 ÷ 12 = 44.5$

Each month, $44.50 must be added to the monthly mortgage payment for property tax.

Example 3

Serge purchased some land for $120,000 and made a down payment of $25,000. The savings and loan association charges an annual interest rate of 8% on Serge's 25-year mortgage. Find the monthly mortgage payment.

Strategy

To find the monthly mortgage payment:

- Subtract the down payment from the purchase price to find the mortgage.
- Multiply the mortgage by the factor found in the Monthly Payment Table in the Appendix.

Solution

Purchase price	−	down payment	=	mortgage
120,000	−	25,000	=	95,000

$95,000 × 0.0077182 ≈ 733.23$
↑
From the table

The monthly mortgage payment is $733.23.

You Try It 3

A new condominium project is selling townhouses for $75,000. A down payment of $15,000 is required, and a 20-year mortgage at an annual interest rate of 9% is available. Find the monthly mortgage payment.

Your strategy

Your solution

Solution on p. S17

Copyright © Houghton Mifflin Company. All rights reserved.

Example 4

A home has a mortgage of $134,000 for 25 years at an annual interest rate of 7%. During a month when $375.88 of the monthly mortgage payment is principal, how much of the payment is interest?

Strategy

To find the interest:

- Multiply the mortgage by the factor found in the Monthly Payment Table in the Appendix to find the monthly mortgage payment.
- Subtract the principal from the monthly mortgage payment.

Solution

$$134,000 \times 0.0070678 \approx 947.09$$

↑ From the table

↑ Monthly mortgage payment

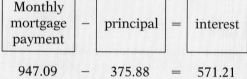

| Monthly mortgage payment | − | principal | = | interest |

| 947.09 | − | 375.88 | = | 571.21 |

$571.21 of the payment is interest on the mortgage.

You Try It 4

An office building has a mortgage of $125,000 for 25 years at an annual interest rate of 9%. During a month when $492.65 of the monthly mortgage payment is principal, how much of the payment is interest?

Your strategy

Your solution

Example 5

The monthly mortgage payment for a home is $598.75. The annual property tax is $900. Find the total monthly payment for the mortgage and property tax.

Strategy

To find the monthly payment:

- Divide the annual property tax by 12 to find the monthly property tax.
- Add the monthly property tax to the monthly mortgage payment.

Solution

$900 \div 12 = 75$ • **Monthly property tax**

$598.75 + 75 = 673.75$

The total monthly payment is $673.75.

You Try It 5

The monthly mortgage payment for a home is $415.20. The annual property tax is $744. Find the total monthly payment for the mortgage and property tax.

Your strategy

Your solution

Solutions on pp. S17–S18

Copyright © Houghton Mifflin Company. All rights reserved.

6.4 Exercises

Objective A   **To calculate the initial expenses of buying a home**

1. A condominium at Mt. Baldy Ski Resort was purchased for $97,000, and a down payment of $14,550 was made. Find the mortgage.

2. An insurance business was purchased for $173,000, and a down payment of $34,600 was made. Find the mortgage.

3. A building lot was purchased for $25,000. The lender requires a down payment of 30% of the purchase price. Find the down payment.

4. Ian Goldman purchased a new home for $188,500. The lender requires a down payment of 20% of the purchase price. Find the down payment.

5. Brian Stedman made a down payment of 25% of the $850,000 purchase price of an apartment building. How much was the down payment?

6. A clothing store was purchased for $125,000, and a down payment that was 25% of the purchase price was made. How much was the down payment?

7. A loan of $150,000 is obtained to purchase a home. The loan origination fee is $2\frac{1}{2}$ points. Find the amount of the loan origination fee.

8. Security Savings & Loan requires a borrower to pay $3\frac{1}{2}$ points for a loan. Find the amount of the loan origination fee for a loan of $90,000.

9. Baja Construction Inc. is selling homes for $150,000. A down payment of 5% is required.
    a. Find the down payment.
    b. Find the mortgage.

10. A cattle rancher purchased some land for $240,000. The bank requires a down payment of 15% of the purchase price.
    a. Find the down payment.
    b. Find the mortgage.

11. Vivian Tom purchased a home for $210,000. Find the mortgage if the down payment Vivian made is 10% of the purchase price.

12. A mortgage lender requires a down payment of 5% of the $80,000 purchase price of a condominium. How much is the mortgage?

Objective B   **To calculate the ongoing expenses of owning a home**

For Exercises 13 to 24, solve. Use the Monthly Payment Table in the Appendix. Round to the nearest cent.

13. An investor obtained a loan of $150,000 to buy a car wash business. The monthly mortgage payment was based on 25 years at 8%. Find the monthly mortgage payment.

14. A beautician obtained a 20-year mortgage of $90,000 to expand the business. The credit union charges an annual interest rate of 6%. Find the monthly mortgage payment.

Copyright © Houghton Mifflin Company. All rights reserved.

15. A couple interested in buying a home determines that they can afford a monthly mortgage payment of $800. Can they afford to buy a home with a 30-year $110,000 mortgage at 8% interest?

16. A lawyer is considering purchasing a new office building with a 15-year $400,000 mortgage at 6% interest. The lawyer can afford a monthly mortgage payment of $3500. Can the lawyer afford the monthly mortgage payment on the new office building?

17. The county tax assessor has determined that the annual property tax on a $125,000 house is $1348.20. Find the monthly property tax.

18. The annual property tax on a $155,000 home is $1992. Find the monthly property tax.

19. Abacus Imports Inc. has a warehouse with a 25-year mortgage of $200,000 at an annual interest rate of 9%.
   a. Find the monthly mortgage payment.
   b. During a month when $941.72 of the monthly mortgage payment is principal, how much of the payment is interest?

20. A vacation home has a mortgage of $135,000 for 30 years at an annual interest rate of 7%.
   a. Find the monthly mortgage payment.
   b. During a month when $392.47 of the monthly mortgage payment is principal, how much of the payment is interest?

21. The annual mortgage payment on a duplex is $10,844.40. The owner must pay an annual property tax of $948. Find the total monthly payment for the mortgage and property tax.

22. The monthly mortgage payment on a home is $716.40, and the homeowner pays an annual property tax of $792. Find the total monthly payment for the mortgage and property tax.

23. Maria Hernandez purchased a home for $210,000 and made a down payment of $15,000. The balance was financed for 15 years at an annual interest rate of 6%. Find the monthly mortgage payment.

24. A customer of a savings and loan purchased a $185,000 home and made a down payment of $20,000. The savings and loan charges its customers an annual interest rate of 7% for 30 years for a home mortgage. Find the monthly mortgage payment.

APPLYING THE CONCEPTS

25. **Mortgages**   A couple considering a mortgage of $100,000 have a choice of loans. One loan is an 8% loan for 20 years, and the other loan is at 8% for 30 years. Find the amount of interest that the couple can save by choosing the 20-year loan.

26. **Mortgages**   Find out what an adjustable-rate mortgage is. What is the difference between this type of loan and a fixed-rate mortgage? List some of the advantages and disadvantages of each.

Copyright © Houghton Mifflin Company. All rights reserved.

6.5 Car Expenses

Objective A   **To calculate the initial expenses of buying a car**

The initial expenses in the purchase of a car usually include the down payment, the **license fees** (fees charged for authorization to operate a vehicle), and the **sales tax** (a tax levied by a state or municipality on purchases). The down payment may be very small or as much as 25% or 30% of the purchase price of the car, depending on the lending institution. License fees and sales tax are regulated by each state, so these expenses vary from state to state.

Example 1

A car is purchased for $18,500, and the lender requires a down payment of 15% of the purchase price. Find the amount financed.

Strategy

To find the amount financed:

- Find the down payment by solving the basic percent equation for *amount*.
- Subtract the down payment from the purchase price.

Solution

Percent × base = amount

| Percent | × | purchase price | = | down payment |

$$0.15 \times 18{,}500 = n$$
$$2775 = n$$

$$18{,}500 - 2775 = 15{,}725$$

The amount financed is $15,725.

You Try It 1

A down payment of 20% of the $19,200 purchase price of a new car is made. Find the amount financed.

Your strategy

Your solution

Example 2

A sales clerk purchases a car for $16,500 and pays a sales tax that is 5% of the purchase price. How much is the sales tax?

Strategy

To find the sales tax, solve the basic percent equation for *amount*.

Solution

Percent × base = amount

| Percent | × | purchase price | = | sales tax |

$$0.05 \times 16{,}500 = n$$
$$825 = n$$

The sales tax is $825.

You Try It 2

A car is purchased for $17,350. The car license fee is 1.5% of the purchase price. How much is the license fee?

Your strategy

Your solution

Solutions on p. S18

Copyright © Houghton Mifflin Company. All rights reserved.

Copyright © Houghton Mifflin Company. All rights reserved.

Objective B   **To calculate the ongoing expenses of owning a car**

> TAKE NOTE
> The same formula that is used to calculate a monthly mortgage payment is used to calculate a monthly car payment.

Besides the initial expenses of buying a car, there are continuing expenses involved in owning a car. These ongoing expenses include car insurance, gas and oil, general maintenance, and the monthly car payment. The monthly car payment is calculated in the same manner as the monthly mortgage payment on a home loan. A monthly payment table, such as the one in the Appendix, is used to simplify the calculation of monthly car payments.

Example 3

At a cost of $.33 per mile, how much does it cost to operate a car during a year in which the car is driven 15,000 miles?

Strategy
To find the cost, multiply the cost per mile (0.33) by the number of miles driven (15,000).

Solution
15,000 × 0.33 = 4950

The cost is $4950.

You Try It 3

At a cost of $.31 per mile, how much does it cost to operate a car during a year in which the car is driven 23,000 miles?

Your strategy

Your solution

Example 4

During one month, your total gasoline bill was $84 and the car was driven 1200 miles. What was the cost per mile for gasoline?

Strategy
To find the cost per mile, divide the cost for gasoline (84) by the number of miles driven (1200).

Solution
84 ÷ 1200 = 0.07

The cost per mile was $.07.

You Try It 4

In a year in which your total car insurance bill was $360 and the car was driven 15,000 miles, what was the cost per mile for car insurance?

Your strategy

Your solution

Example 5

A car is purchased for $18,500 with a down payment of $3700. The balance is financed for 3 years at an annual interest rate of 6%. Find the monthly car payment.

Strategy
To find the monthly payment:
• Subtract the down payment from the purchase price to find the amount financed.
• Multiply the amount financed by the factor found in the Monthly Payment Table in the Appendix.

Solution
18,500 − 3700 = 14,800
14,800 × 0.0304219 ≈ 450.24

The monthly payment is $450.24.

You Try It 5

A truck is purchased for $25,900 with a down payment of $6475. The balance is financed for 4 years at an annual interest rate of 8%. Find the monthly car payment.

Your strategy

Your solution

Solutions on p. S18

6.5 Exercises

Objective A   **To calculate the initial expenses of buying a car**

1. Amanda has saved $780 to make a down payment on a used minivan that costs $7100. The car dealer requires a down payment of 12% of the purchase price. Has she saved enough money to make the down payment?

2. A sedan was purchased for $23,500. A down payment of 15% of the purchase price was required. How much was the down payment?

3. A drapery installer bought a van to carry drapery samples. The purchase price of the van was $26,500, and a 4.5% sales tax was paid. How much was the sales tax?

4. A & L Lumber Company purchased a delivery truck for $28,500. A sales tax of 4% of the purchase price was paid. Find the sales tax.

5. A license fee of 2% of the purchase price is paid on a pickup truck costing $22,500. Find the license fee for the truck.

6. Your state charges a license fee of 1.5% on the purchase price of a car. How much is the license fee for a car that costs $16,998?

7. An electrician bought a $32,000 flatbed truck. A state license fee of $275 and a sales tax of 3.5% of the purchase price are required.
    a. Find the sales tax.
    b. Find the total cost of the sales tax and the license fee.

8. A physical therapist bought a used car for $9375 and made a down payment of $1875. The sales tax is 5% of the purchase price.
    a. Find the sales tax.
    b. Find the total cost of the sales tax and the down payment.

9. Martin bought a motorcycle for $16,200 and made a down payment of 25% of the purchase price.
    a. Find the down payment.
    b. Find the amount financed.

10. A carpenter bought a utility van for $24,900 and made a down payment of 15% of the purchase price.
    a. Find the down payment.
    b. Find the amount financed.

11. An author bought a sports car for $35,000 and made a down payment of 20% of the purchase price. Find the amount financed.

12. Tania purchased a used car for $13,500 and made a down payment of 25% of the cost. Find the amount financed.

Objective B   **To calculate the ongoing expenses of owning a car**

For Exercises 13 to 24, solve. Use the Monthly Payment Table in the Appendix. Round to the nearest cent.

13. A rancher financed $24,000 for the purchase of a truck through a credit union at 5% interest for 4 years. Find the monthly truck payment.

Copyright © Houghton Mifflin Company. All rights reserved.

14. A car loan of $18,000 is financed for 3 years at an annual interest rate of 4%. Find the monthly car payment.

15. An estimate of the cost of owning a compact car is $.32 per mile. Using this estimate, find how much it costs to operate a car during a year in which the car is driven 16,000 miles.

16. An estimate of the cost of care and maintenance of automobile tires is $.015 per mile. Using this estimate, find how much it costs for care and maintenance of tires during a year in which the car is driven 14,000 miles.

17. A family spent $1600 on gas, oil, and car insurance during a period in which the car was driven 14,000 miles. Find the cost per mile for gas, oil, and car insurance.

18. Last year you spent $2100 for gasoline for your car. The car was driven 15,000 miles. What was your cost per mile for gasoline?

19. Elena's monthly car payment is $143.50. During a month in which $68.75 of the monthly payment is principal, how much of the payment is interest?

20. The cost for a pizza delivery truck for the year included $2868 in truck payments, $2400 for gasoline, and $675 for insurance. Find the total cost for truck payments, gasoline, and insurance for the year.

21. The city of Colton purchased a fire truck for $164,000 and made a down payment of $10,800. The balance is financed for 5 years at an annual interest rate of 6%.
    a. Find the amount financed.
    b. Find the monthly truck payment.

22. A used car is purchased for $14,999, and a down payment of $2999 is made. The balance is financed for 3 years at an annual interest rate of 5%.
    a. Find the amount financed.
    b. Find the monthly car payment.

23. An artist purchased a new car costing $27,500 and made a down payment of $5500. The balance is financed for 3 years at an annual interest rate of 4%. Find the monthly car payment.

24. A camper is purchased for $39,500, and a down payment of $5000 is made. The balance is financed for 4 years at an annual interest rate of 6%. Find the monthly payment.

APPLYING THE CONCEPTS

25. **Car Loans**   One bank offers a 4-year car loan at an annual interest rate of 7% plus a loan application fee of $45. A second bank offers 4-year car loans at an annual interest rate of 8% but charges no loan application fee. If you need to borrow $5800 to purchase a car, which of the two bank loans has the lesser loan costs? Assume you keep the car for 4 years.

26. **Car Loans**   How much interest is paid on a 5-year car loan of $9000 if the interest rate is 9%?

Copyright © Houghton Mifflin Company. All rights reserved.

6.6 Wages

Objective A   **To calculate commissions, total hourly wages, and salaries**

Commissions, hourly wage, and salary are three ways to receive payment for doing work.

Commissions are usually paid to salespersons and are calculated as a percent of total sales.

HOW TO As a real estate broker, Emma Smith receives a commission of 4.5% of the selling price of a house. Find the commission she earned for selling a home for $175,000.

To find the commission Emma earned, solve the basic percent equation for *amount*.

Percent	×	base	=	amount
Commission rate	×	total sales	=	commission
0.045	×	175,000	=	7875

The commission is $7875.

An employee who receives an **hourly wage** is paid a certain amount for each hour worked.

HOW TO A plumber receives an hourly wage of $18.25. Find the plumber's total wages for working 37 hours.

To find the plumber's total wages, multiply the hourly wage by the number of hours worked.

Hourly wage	×	number of hours worked	=	total wages
18.25	×	37	=	675.25

The plumber's total wages for working 37 hours are $675.25.

An employee who is paid a **salary** receives payment based on a weekly, biweekly (every other week), monthly, or annual time schedule. Unlike the employee who receives an hourly wage, the salaried worker does not receive additional pay for working more than the regularly scheduled workday.

HOW TO Ravi Basar is a computer operator who receives a weekly salary of $695. Find his salary for 1 month (4 weeks).

To find Ravi's salary for 1 month, multiply the salary per pay period by the number of pay periods.

Salary per pay period	×	number of pay periods	=	total salary
695	×	4	=	2780

Ravi's total salary for 1 month is $2780.

Copyright © Houghton Mifflin Company. All rights reserved.

Example 1

A pharmacist's hourly wage is $28. On Saturday, the pharmacist earns time and a half (1.5 times the regular hourly wage). How much does the pharmacist earn for working 6 hours on Saturday?

Strategy

To find the pharmacist's earnings:

• Find the hourly wage for working on Saturday by multiplying the hourly wage by 1.5.
• Multiply the hourly wage by the number of hours worked.

Solution

$28 \times 1.5 = 42$ $42 \times 6 = 252$

The pharmacist earns $252.

You Try It 1

A construction worker, whose hourly wage is $18.50, earns double time (2 times the regular hourly wage) for working overtime. Find the worker's wages for working 8 hours of overtime.

Your strategy

Your solution

Example 2

An efficiency expert received a contract for $3000. The expert spent 75 hours on the project. Find the consultant's hourly wage.

Strategy

To find the hourly wage, divide the total earnings by the number of hours worked.

Solution

$3000 \div 75 = 40$

The hourly wage was $40.

You Try It 2

A contractor for a bridge project receives an annual salary of $48,228. What is the contractor's salary per month?

Your strategy

Your solution

Example 3

Dani Greene earns $28,500 per year plus a 5.5% commission on sales over $100,000. During one year, Dani sold $150,000 worth of computers. Find Dani's total earnings for the year.

Strategy

To find the total earnings:

• Find the sales over $100,000.
• Multiply the commission rate by sales over $100,000.
• Add the commission to the annual pay.

Solution

$150,000 - 100,000 = 50,000$
$50,000 \times 0.055 = 2750$ • Commission
$28,500 + 2750 = 31,250$

Dani earned $31,250.

You Try It 3

An insurance agent earns $27,000 per year plus a 9.5% commission on sales over $50,000. During one year, the agent's sales totaled $175,000. Find the agent's total earnings for the year.

Your strategy

Your solution

Solutions on pp. S18–S19

Copyright © Houghton Mifflin Company. All rights reserved.

6.6 Exercises

Objective A   **To calculate commissions, total hourly wages, and salaries**

1. Lewis works in a clothing store and earns $9.50 per hour. How much does he earn in a 40-hour week?

2. Sasha pays a gardener an hourly wage of $11. How much does she pay the gardener for working 25 hours?

3. A real estate agent receives a 3% commission for selling a house. Find the commission that the agent earned for selling a house for $131,000.

4. Ron Caruso works as an insurance agent and receives a commission of 40% of the first year's premium. Find Ron's commission for selling a life insurance policy with a first-year premium of $1050.

5. A stockbroker receives a commission of 1.5% of the price of stock that is bought or sold. Find the commission on 100 shares of stock that were bought for $5600.

6. The owner of the Carousel Art Gallery receives a commission of 20% on paintings that are sold on consignment. Find the commission on a painting that sold for $22,500.

7. Keisha Brown receives an annual salary of $38,928 as a teacher of Italian. How much does Keisha receive each month?

8. An apprentice plumber receives an annual salary of $27,900. How much does the plumber receive per month?

9. An electrician's hourly wage is $25.80. For working overtime, the electrician earns double time. What is the electrician's hourly wage for working overtime?

10. Carlos receives a commission of 12% of his weekly sales as a sales representative for a medical supply company. Find the commission he earned during a week in which sales were $4500.

11. A golf pro receives a commission of 25% for selling a golf set. Find the commission the pro earned for selling a golf set costing $450.

12. Steven receives $3.75 per square yard to install carpet. How much does he receive for installing 160 square yards of carpet?

13. A typist charges $2.75 per page for typing technical material. How much does the typist earn for typing a 225-page book?

Copyright © Houghton Mifflin Company. All rights reserved.

14. A nuclear chemist received $15,000 in consulting fees while working on a nuclear power plant. The chemist worked 120 hours on the project. Find the chemist's hourly wage.

15. Maxine received $3400 for working on a project as a computer consultant for 40 hours. Find her hourly wage.

16. Gil Stratton's hourly wage is $10.78. For working overtime, he receives double time.
 a. What is Gil's hourly wage for working overtime?
 b. How much does he earn for working 16 hours of overtime?

17. Mark is a lathe operator and receives an hourly wage of $15.90. When working on Saturday, he receives time and a half.
 a. What is Mark's hourly wage on Saturday?
 b. How much does he earn for working 8 hours on Saturday?

18. A stock clerk at a supermarket earns $8.20 an hour. For working the night shift, the clerk's wage increases by 15%.
 a. What is the increase in hourly pay for working the night shift?
 b. What is the clerk's hourly wage for working the night shift?

19. A nurse earns $21.50 an hour. For working the night shift, the nurse receives a 10% increase in pay.
 a. What is the increase in hourly pay for working the night shift?
 b. What is the hourly pay for working the night shift?

20. Tony's hourly wage as a service station attendant is $9.40. For working the night shift, his wage is increased 25%. What is Tony's hourly wage for working the night shift?

21. Nicole Tobin, a salesperson, receives a salary of $250 per week plus a commission of 15% on all sales over $1500. Find her earnings during a week in which sales totaled $3000.

APPLYING THE CONCEPTS

Compensation   The table at the right shows the average starting salaries for recent college graduates. Use this table for Exercises 22 to 25. Round to the nearest dollar.

22. What was the starting salary in the previous year for an accountant?

23. How much did the starting salary for a chemical engineer increase over that of the previous year?

24. What was the starting salary in the previous year for a computer science major?

25. How much did the starting salary for a political science major decrease from that of the previous year?

Average Starting Salaries		
Bachelor's Degree	Average Starting Salary	Change from Previous Year
Chemical Engineering	$52,169	1.8% increase
Electrical Engineering	$50,566	0.4% increase
Computer Science	$46,536	7.6% decrease
Accounting	$41,360	2.6% increase
Business	$36,515	3.7% increase
Biology	$29,554	1.0% decrease
Political Science	$28,546	12.6% decrease
Psychology	$26,738	10.7% decrease

Source: National Association of Colleges

Copyright © Houghton Mifflin Company. All rights reserved.

6.7 Bank Statements

Copyright © Houghton Mifflin Company. All rights reserved.

Objective A

To calculate checkbook balances

TAKE NOTE

A **checking account** is a bank account that enables you to withdraw money or make payments to other people using checks. A **check** is a printed form that, when filled out and signed, instructs a bank to pay a specified sum of money to the person named on it.

A **deposit slip** is a form for depositing money in a checking account.

A checking account can be opened at most banks and savings and loan associations by depositing an amount of money in the bank. A checkbook contains checks and deposit slips and a checkbook register in which to record checks written and amounts deposited in the checking account. A sample check is shown below.

Payee | Date Check is Written

Check Number

East Phoenix Rental Equipment
3011 N.W. Ventura Street
Phoenix, Arizona 85280

NO. 2023
$\frac{68 - 461}{1052}$

Date _October 11, 2005_

PAY TO THE ORDER OF _Tellas Manufacturing Co._ $ _827 $\frac{00}{100}$_

Amount of Check

Eight Hundred Twenty-Seven and $\frac{00}{100}$ ——— DOLLARS

MEYERS' NATIONAL BANK
11 N.W. Nova Street
Phoenix, Arizona 85215

Memo _____

I: 1052 III 0461 I: 5008 2023 III

Eugene L. Madison

Amount of Check in Words | Depositor's Signature

Each time a check is written, the amount of the check is subtracted from the amount in the account. When a deposit is made, the amount deposited is added to the amount in the account.

A portion of a checkbook register is shown below. The account holder had a balance of $587.93 before writing two checks, one for $286.87 and the other for $202.38, and making one deposit of $345.00.

Point of Interest

There are a number of computer programs that serve as "electronic" checkbooks. With these programs, you can pay your bills by using a computer to write the check and then transmit the check over telephone lines using a modem.

		RECORD ALL CHARGES OR CREDITS THAT AFFECT YOUR ACCOUNT					BALANCE	
NUMBER	DATE	DESCRIPTION OF TRANSACTION	PAYMENT/DEBIT (−)	√ T	FEE (IF ANY) (−)	DEPOSIT/CREDIT (+)	$ 587	93
108	8/4	Plumber	$286 87	$	$		301	06
109	8/10	Car Payment	202 38				98	68
	8/14	Deposit				345 00	443	68

To find the current checking account balance, subtract the amount of each check from the previous balance. Then add the amount of the deposit.

The current checking account balance is $443.68.

Example 1

A mail carrier had a checking account balance of $485.93 before writing two checks, one for $18.98 and another for $35.72, and making a deposit of $250. Find the current checking account balance.

You Try It 1

A cement mason had a checking account balance of $302.46 before writing a check for $20.59 and making two deposits, one in the amount of $176.86 and another in the amount of $94.73. Find the current checking account balance.

Strategy

To find the current balance:

• Subtract the amount of each check from the old balance.
• Add the amount of the deposit.

Your strategy

Solution

$$
\begin{array}{rl}
485.93 & \\
-\ \ 18.98 & \text{first check} \\
\hline
466.95 & \\
-\ \ 35.72 & \text{second check} \\
\hline
431.23 & \\
+\ 250.00 & \text{deposit} \\
\hline
681.23 &
\end{array}
$$

The current checking account balance is $681.23.

Your solution

Solution on p. S19

Objective B   **To balance a checkbook**

Each month a bank statement is sent to the account holder. A **bank statement** is a document showing all the transactions in a bank account during the month. It shows the checks that the bank has paid, the deposits received, and the current bank balance.

A bank statement and checkbook register are shown on the next page.

Balancing a checkbook, or determining whether the checking account balance is accurate, requires a number of steps.

1. In the checkbook register, put a check mark (✓) by each check paid by the bank and by each deposit recorded by the bank.

Copyright © Houghton Mifflin Company. All rights reserved.

		RECORD ALL CHARGES OR CREDITS THAT AFFECT YOUR ACCOUNT						BALANCE	
NUMBER	DATE	DESCRIPTION OF TRANSACTION	PAYMENT/DEBIT (−)	√ T	FEE (IF ANY) (−)	DEPOSIT/CREDIT (+)	$	840	27
263	5/20	Dentist	$ 75 00	√	$	$		765	27
264	5/22	Post Office	33 61	√				731	66
265	5/22	Gas Company	67 14					664	52
	5/29	Deposit		√		192 00		856	52
266	5/29	Pharmacy	38 95	√				817	57
267	5/30	Telephone	63 85					753	72
268	6/2	Groceries	73 19	√				680	53
	6/3	Deposit		√		215 00		895	53
269	6/7	Insurance	103 00	√				792	53
	6/10	Deposit				225 00		1017	53
270	6/15	Photo Shop	16 63	√				1000	90
271	6/18	Newspaper	27 00					973	90

CHECKING ACCOUNT Monthly Statement   Account Number: 924-297-8

Date	Transaction	Amount	Balance
5/20	OPENING BALANCE		840.27
5/21	CHECK	75.00	765.27
5/23	CHECK	33.61	731.66
5/29	DEPOSIT	192.00	923.66
6/1	CHECK	38.95	884.71
6/1	INTEREST	4.47	889.18
6/3	CHECK	73.19	815.99
6/3	DEPOSIT	215.00	1030.99
6/9	CHECK	103.00	927.99
6/16	CHECK	16.63	911.36
6/20	SERVICE CHARGE	3.00	908.36
6/20	CLOSING BALANCE		908.36

2. Add to the current checkbook balance all checks that have been written but have not yet been paid by the bank and any interest paid on the account.

Current checkbook balance:	973.90
Checks: 265	67.14
267	63.85
271	27.00
Interest:	+ 4.47
	1136.36

3. Subtract any service charges and any deposits not yet recorded by the bank. This is the checkbook balance.

Service charge:	− 3.00
	1133.36
Deposit:	− 225.00
Checkbook balance:	908.36

4. Compare the balance with the bank balance listed on the bank statement. If the two numbers are equal, the bank statement and the checkbook "balance."

Closing bank balance from bank statement = Checkbook balance

$908.36 = $908.36

The bank statement and checkbook balance.

Copyright © Houghton Mifflin Company. All rights reserved.

TAKE NOTE

A **service charge** is an amount of money charged by a bank for handling a transaction.

Copyright © Houghton Mifflin Company. All rights reserved.

HOW TO

		RECORD ALL CHARGES OR CREDITS THAT AFFECT YOUR ACCOUNT							
NUMBER	DATE	DESCRIPTION OF TRANSACTION	PAYMENT/DEBIT (−)	√T	FEE (IF ANY) (−)	DEPOSIT/CREDIT (+)		BALANCE $ 1620	42
413	3/2	Car Payment	$232 15	√	$	$		1388	27
414	3/2	Utilities	67 14	√				1321	13
415	3/5	Restaurant	78 14					1242	99
	3/8	Deposit		√		1842 66		3085	65
416	3/10	House Payment	672 14	√				2413	51
417	3/14	Insurance	177 10					2236	41

CHECKING ACCOUNT Monthly Statement   Account Number: 924-297-8

Date	Transaction	Amount	Balance
3/1	OPENING BALANCE		1620.42
3/4	CHECK	232.15	1388.27
3/5	CHECK	67.14	1321.13
3/8	DEPOSIT	1842.66	3163.79
3/10	INTEREST	6.77	3170.56
3/12	CHECK	672.14	2498.42
3/25	SERVICE CHARGE	2.00	2496.42
3/30	CLOSING BALANCE		2496.42

Balance the bank statement shown above.

1. In the checkbook register, put a check mark (√) by each check paid by the bank and by each deposit recorded by the bank.

2. Add to the current checkbook balance all checks that have been written but have not yet been paid by the bank and any interest paid on the account.

3. Subtract any service charges and any deposits not yet recorded by the bank. This is the checkbook balance.

4. Compare the balance with the bank balance listed on the bank statement. If the two numbers are equal, the bank statement and the checkbook "balance."

Current checkbook balance: 2236.41
Checks: 415   78.14
        417   177.10
Interest: +   6.77
                                           2498.42
Service charge: −   2.00
Checkbook balance: 2496.42

Closing bank balance   Checkbook
from bank statement   balance
    $2496.42   = $2496.42

The bank statement and checkbook balance.

Example 2 Balance the bank statement shown below.

		RECORD ALL CHARGES OR CREDITS THAT AFFECT YOUR ACCOUNT							BALANCE	
NUMBER	DATE	DESCRIPTION OF TRANSACTION	PAYMENT/DEBIT (−)		√ T	FEE (IF ANY) (−)	DEPOSIT/CREDIT (+)		$ 412	64
345	1/14	Phone Bill	$ 54	75	√	$	$		357	89
346	1/19	News Shop	18	98	√				338	91
347	1/23	Theater Tickets	95	00					243	91
	1/31	Deposit			√		947	00	1190	91
348	2/5	Cash	250	00	√				940	91
349	2/12	Rent	840	00					100	91

CHECKING ACCOUNT Monthly Statement			Account Number: 924-297-8
Date	Transaction	Amount	Balance
1/10	OPENING BALANCE		412.64
1/18	CHECK	54.75	357.89
1/23	CHECK	18.98	338.91
1/31	DEPOSIT	947.00	1285.91
2/1	INTEREST	4.52	1290.43
2/10	CHECK	250.00	1040.43
2/10	CLOSING BALANCE		1040.43

Solution

Current checkbook
balance: 100.91

Checks: 347   95.00

         349   840.00

Interest: +   4.52

                            1040.43

Service charge: −   0.00

                            1040.43

Deposit: −   0.00

Checkbook balance: 1040.43

Closing bank balance from
bank statement: $1040.43

Checkbook balance:
$1040.43

The bank statement and
the checkbook balance.

Copyright © Houghton Mifflin Company. All rights reserved.

You Try It 2

Balance the bank statement shown below.

NUMBER	DATE	DESCRIPTION OF TRANSACTION	PAYMENT/DEBIT (–)		√ T	FEE (IF ANY) (–)	DEPOSIT/CREDIT (+)		BALANCE	
		RECORD ALL CHARGES OR CREDITS THAT AFFECT YOUR ACCOUNT							$ 903	17
	2/15	Deposit	$			$	$ 523	84	1427	01
234	2/20	Mortgage	773	21					653	80
235	2/27	Cash	200	00					453	80
	3/1	Deposit					523	84	977	64
236	3/12	Insurance	275	50					702	14
237	3/12	Telephone	78	73					623	41

CHECKING ACCOUNT Monthly Statement   Account Number: 314-271-4

Date	Transaction	Amount	Balance
2/14	OPENING BALANCE		903.17
2/15	DEPOSIT	523.84	1427.01
2/21	CHECK	773.21	653.80
2/28	CHECK	200.00	453.80
3/1	INTEREST	2.11	455.91
3/14	CHECK	275.50	180.41
3/14	CLOSING BALANCE		180.41

Your solution

Solution on p. S19

Copyright © Houghton Mifflin Company. All rights reserved.

6.7 Exercises

Objective A   **To calculate checkbook balances**

1. You had a checking account balance of $342.51 before making a deposit of $143.81. What is your new checking account balance?

2. Carmen had a checking account balance of $493.26 before writing a check for $48.39. What is the current checking account balance?

3. A real estate firm had a balance of $2431.76 in its rental-property checking account. What is the balance in this account after a check for $1209.29 has been written?

4. The business checking account for R and R Tires showed a balance of $1536.97. What is the balance in this account after a deposit of $439.21 has been made?

5. A nutritionist had a checking account balance of $1204.63 before writing one check for $119.27 and another check for $260.09. Find the current checkbook balance.

6. Sam had a checking account balance of $3046.93 before writing a check for $1027.33 and making a deposit of $150.00. Find the current checkbook balance.

7. The business checking account for Rachael's Dry Cleaning had a balance of $3476.85 before a deposit of $1048.53 was made. The store manager then wrote two checks, one for $848.37 and another for $676.19. Find the current checkbook balance.

8. Joel had a checking account balance of $427.38 before a deposit of $127.29 was made. Joel then wrote two checks, one for $43.52 and one for $249.78. Find the current checkbook balance.

9. A carpenter had a checkbook balance of $404.96 before making a deposit of $350 and writing a check for $71.29. Is there enough money in the account to purchase a refrigerator for $675?

10. A taxi driver had a checkbook balance of $149.85 before making a deposit of $245 and writing a check for $387.68. Is there enough money in the account for the bank to pay the check?

11. A sporting goods store has the opportunity to buy downhill skis and cross-country skis at a manufacturer's closeout sale. The downhill skis will cost $3500, and the cross-country skis will cost $2050. There is currently $5625.42 in the sporting goods store's checking account. Is there enough money in the account to make both purchases by check?

12. A lathe operator's current checkbook balance is $1143.42. The operator wants to purchase a utility trailer for $525 and a used piano for $650. Is there enough money in the account to make the two purchases?

Copyright © Houghton Mifflin Company. All rights reserved.

To balance a checkbook

13. Balance the checkbook.

		RECORD ALL CHARGES OR CREDITS THAT AFFECT YOUR ACCOUNT								
NUMBER	DATE	DESCRIPTION OF TRANSACTION	PAYMENT/DEBIT (−)		√T	FEE (IF ANY) (−)	DEPOSIT/CREDIT (+)		BALANCE $ 2466	79
223	3/2	Groceries	$ 167	32		$	$		2299	47
	3/5	Deposit					960	70	3260	17
224	3/5	Rent	860	00					2400	17
225	3/7	Gas & Electric	142	35					2257	82
226	3/7	Cash	300	00					1957	82
227	3/7	Insurance	218	44					1739	38
228	3/7	Credit Card	419	32					1320	06
229	3/12	Dentist	92	00					1228	06
230	3/13	Drug Store	47	03					1181	03
	3/19	Deposit					960	70	2141	73
231	3/22	Car Payment	241	35					1900	38
232	3/25	Cash	300	00					1600	38
233	3/25	Oil Company	166	40					1433	98
234	3/28	Plumber	155	73					1278	25
235	3/29	Department Store	288	39					989	86

CHECKING ACCOUNT Monthly Statement   Account Number: 122-345-1

Date	Transaction	Amount	Balance
3/1	OPENING BALANCE		2466.79
3/5	DEPOSIT	960.70	3427.49
3/7	CHECK	167.32	3260.17
3/8	CHECK	860.00	2400.17
3/8	CHECK	300.00	2100.17
3/9	CHECK	142.35	1957.82
3/12	CHECK	218.44	1739.38
3/14	CHECK	92.00	1647.38
3/18	CHECK	47.03	1600.35
3/19	DEPOSIT	960.70	2561.05
3/25	CHECK	241.35	2319.70
3/27	CHECK	300.00	2019.70
3/29	CHECK	155.73	1863.97
3/30	INTEREST	13.22	1877.19
4/1	CLOSING BALANCE		1877.19

Copyright © Houghton Mifflin Company. All rights reserved.

14. Balance the checkbook.

RECORD ALL CHARGES OR CREDITS THAT AFFECT YOUR ACCOUNT

NUMBER	DATE	DESCRIPTION OF TRANSACTION	PAYMENT/DEBIT (–)		√ T	FEE (IF ANY) (–)	DEPOSIT/CREDIT (+)		BALANCE $ 1219	43
	5/1	Deposit	$			$	$ 619	14	1838	57
515	5/2	Electric Bill	42	35					1796	22
516	5/2	Groceries	95	14					1701	08
517	5/4	Insurance	122	17					1578	91
518	5/5	Theatre Tickets	84	50					1494	41
	5/8	Deposit					619	14	2113	55
519	5/10	Telephone	37	39					2076	16
520	5/12	Newspaper	22	50					2053	66
	5/15	Deposit					619	14	2672	80
521	5/20	Computer Store	172	90					2499	90
522	5/21	Credit Card	313	44					2186	46
523	5/22	Eye Exam	82	00					2104	46
524	5/24	Groceries	107	14					1997	32
525	5/24	Deposit					619	14	2616	46
526	5/25	Oil Company	144	16					2472	30
527	5/30	Car Payment	288	62					2183	68
528	5/30	Mortgage Payment	877	42					1306	26

CHECKING ACCOUNT Monthly Statement   Account Number: 122-345-1

Date	Transaction	Amount	Balance
5/1	OPENING BALANCE		1219.43
5/1	DEPOSIT	619.14	1838.57
5/3	CHECK	95.14	1743.43
5/4	CHECK	42.35	1701.08
5/6	CHECK	84.50	1616.58
5/8	CHECK	122.17	1494.41
5/8	DEPOSIT	619.14	2113.55
5/15	INTEREST	7.82	2121.37
5/15	CHECK	37.39	2083.98
5/15	DEPOSIT	619.14	2703.12
5/23	CHECK	82.00	2621.12
5/23	CHECK	172.90	2448.22
5/24	CHECK	107.14	2341.08
5/24	DEPOSIT	619.14	2960.22
5/30	CHECK	288.62	2671.60
6/1	CLOSING BALANCE		2671.60

Copyright © Houghton Mifflin Company. All rights reserved.

15. Balance the checkbook.

NUMBER	DATE	DESCRIPTION OF TRANSACTION	PAYMENT/DEBIT (−)		√ T	FEE (IF ANY) (−)	DEPOSIT/CREDIT (+)		BALANCE $	
									2035	18
218	7/2	Mortgage	$ 984	60		$	$		1050	58
219	7/4	Telephone	63	36					987	22
220	7/7	Cash	200	00					787	22
	7/12	Deposit					792	60	1579	82
221	7/15	Insurance	292	30					1287	52
222	7/18	Investment	500	00					787	52
223	7/20	Credit Card	414	83					372	69
	7/26	Deposit					792	60	1165	29
224	7/27	Department Store	113	37					1051	92

CHECKING ACCOUNT Monthly Statement   Account Number: 122-345-1

Date	Transaction	Amount	Balance
7/1	OPENING BALANCE		2035.18
7/1	INTEREST	5.15	2040.33
7/4	CHECK	984.60	1055.73
7/6	CHECK	63.36	992.37
7/12	DEPOSIT	792.60	1784.97
7/20	CHECK	292.30	1492.67
7/24	CHECK	500.00	992.67
7/26	DEPOSIT	792.60	1785.27
7/28	CHECK	200.00	1585.27
7/30	CLOSING BALANCE		1585.27

APPLYING THE CONCEPTS

16. When a check is written, the amount is _____ from the balance.

17. When a deposit is made, the amount is _____ to the balance.

18. In checking the bank statement, _____ to the checkbook balance all checks that have been written but not processed.

19. In checking the bank balance, _____ any service charge and any deposits not yet recorded.

20. Define the words *credit* and *debit* as they apply to checkbooks.

Copyright © Houghton Mifflin Company. All rights reserved.

Focus on Problem Solving

Counterexamples   An example that is given to show that a statement is not true is called a **counter-example.** For instance, suppose someone makes the statement "All colors are red." A counterexample to that statement would be to show someone the color blue or some other color.

If a statement is *always* true, there are no counterexamples. The statement "All even numbers are divisible by 2" is always true. It is not possible to give an example of an even number that is not divisible by 2.

In mathematics, statements that are always true are called *theorems*, and mathematicians are always searching for theorems. Sometimes a conjecture by a mathematician appears to be a theorem. That is, the statement appears to be always true, but later on someone finds a counterexample.

TAKE NOTE
Recall that a prime number is a natural number greater than 1 that can be divided by only itself and 1. For instance, 17 is a prime number. 12 is not a prime number because 12 is divisible by numbers other than 1 and 12—for example, 4.

One example of this occurred when the French mathematician Pierre de Fermat (1601–1665) conjectured that $2^{(2^n)} + 1$ is always a prime number for any natural number n. For instance, when $n = 3$, we have $2^{(2^3)} + 1 = 2^8 + 1 = 257$, and 257 is a prime number. However, in 1732 Leonhard Euler (1707–1783) showed that when $n = 5$, $2^{(2^5)} + 1 = 4{,}294{,}967{,}297$ and that $4{,}294{,}967{,}297 = 641 \cdot 6{,}700{,}417$—without a calculator! Because $4{,}294{,}967{,}297$ is the product of two numbers (other than itself and 1), it is not a prime number. This counterexample showed that Fermat's conjecture is not a theorem.

For Exercises 1 and 5, find at least one counterexample.

1. All composite numbers are divisible by 2.

2. All prime numbers are odd numbers.

3. The square of any number is always bigger than the number.

4. The reciprocal of a number is always less than 1.

5. A number ending in 9 is always larger than a number ending in 3.

When a problem is posed, it may not be known whether the problem statement is true or false. For instance, Christian Goldbach (1690–1764) stated that every even number greater than 2 can be written as the sum of two prime numbers. For example,

$$12 = 5 + 7 \qquad 32 = 3 + 29$$

Although this problem is approximately 250 years old, mathematicians have not been able to prove it is a theorem, nor have they been able to find a counterexample.

For Exercises 6 to 9, answer true if the statement is always true. If there is an instance in which the statement is false, give a counterexample.

6. The sum of two positive numbers is always larger than either of the two numbers.

7. The product of two positive numbers is always larger than either of the two numbers.

8. Percents always represent a number less than or equal to 1.

9. It is never possible to divide by zero.

Copyright © Houghton Mifflin Company. All rights reserved.

Projects and Group Activities

Buying a Car Suppose a student has an after-school job to earn money to buy and maintain a car. We will make assumptions about the monthly costs in several categories in order to determine how many hours per week the student must work to support the car. Assume the student earns $8.50 per hour.

1. Monthly payment

   Assume that the car cost $8500 with a down payment of $1020. The remainder is financed for 3 years at an annual simple interest rate of 9%.

   Monthly payment = _____

2. Insurance

   Assume that insurance costs $1500 per year.

   Monthly insurance payment = _____

3. Gasoline

   Assume that the student travels 750 miles per month, that the car travels 25 miles per gallon of gasoline, and that gasoline costs $1.50 per gallon.

   Number of gallons of gasoline purchased per month = _____

   Monthly cost for gasoline = _____

4. Miscellaneous

   Assume $.33 per mile for upkeep.

   Monthly expense for upkeep = _____

5. Total monthly expenses for the monthly payment, insurance, gasoline, and miscellaneous = _____

6. To find the number of hours per month that the student must work to finance the car, divide the total monthly expenses by the hourly rate.

   Number of hours per month = _____

7. To find the number of hours per week that the student must work, divide the number of hours per month by 4.

   Number of hours per week = _____

   The student has to work _____ hours per week to pay the monthly car expenses.

If you own a car, make out your own expense record. If you do not own a car, make assumptions about the kind of car that you would like to purchase, and calculate the total monthly expenses that you would have. An insurance company will give you rates on different kinds of insurance. An automobile club can give you approximations of miscellaneous expenses.

Copyright © Houghton Mifflin Company. All rights reserved.

Chapter 6 Summary

Key Words	Examples
The *unit cost* is the cost of one item. [6.1A, p. 235]	Three paperback books cost $36. The unit cost is the cost of one paperback book, $12.
Percent increase is used to show how much a quantity has increased over its original value. [6.2A, p. 239]	The city's population increased 5%, from 10,000 people to 10,500 people.
Cost is the price a business pays for a product. *Selling price* is the price at which a business sells a product to a customer. *Markup* is the difference between selling price and cost. *Markup rate* is the markup expressed as a percent of a product's cost. [6.2B, p. 240]	A business pays $90 for a pair of cross trainers; the cost is $90. The business sells the cross trainers for $135; the selling price is $135. The markup is $135 − $90 = $45.
Percent decrease is used to show how much a quantity has decreased from its original value. [6.2C, p. 242]	Sales decreased 10%, from 10,000 units in the third quarter to 9000 units in the fourth quarter.
Sale price is the price after a reduction from the regular price. *Discount* is the difference between the regular price and the sale price. *Discount rate* is the discount as a percent of a product's regular price. [6.2D, p. 243]	A movie video that regularly sells for $50 is on sale for $40. The regular price is $50. The sale price is $40. The discount is $50 − $40 = $10.
Interest is the amount paid for the privilege of using someone else's money. *Principal* is the amount of money originally deposited or borrowed. The percent used to determine the amount of interest is the *interest rate*. Interest computed on the original amount is called *simple interest*. The principal plus the interest owed on a loan is called the *maturity value*. [6.3A, p. 249]	Consider a 1-year loan of $5000 at an annual simple interest rate of 8%. The principal is $5000. The interest rate is 8%. The interest paid on the loan is $5000 × 0.08 = $400. The maturity value is $5000 + $400 = $5400.
The interest charged on purchases made with a credit card are called *finance charges*. [6.3B, p. 251]	A credit card company charges 1.5% per month on any unpaid balance. The finance charge on an unpaid balance of $1000 is $1000 × 0.015 × 1 = $15.
Compound interest is computed not only on the original principal but also on the interest already earned. [6.3C, p. 252]	$10,000 is invested at 5% annual interest compounded monthly. The value of the investment after 5 years can be found by multiplying 10,000 by the factor found in the Compound Interest Table in the Appendix. $10,000 × 1.283359 = $12,833.59
A *mortgage* is an amount that is borrowed to buy real estate. The *loan origination fee* is usually a percent of the mortgage and is expressed as *points*. [6.4A, p. 259]	The loan origination fee of 3 points paid on a mortgage of $200,000 is 0.03 × $200,000 = $6000.
A *commission* is usually paid to a salesperson and is calculated as a percent of sales. [6.6A, p. 269]	A commission of 5% on sales of $50,000 is 0.05 × $50,000 = $2500.

Copyright © Houghton Mifflin Company. All rights reserved.

An employee who receives an *hourly wage* is paid a certain amount for each hour worked. [6.6A, p. 269]	An employee is paid an hourly wage of $15. The employee's wages for working 10 hours are $15 × 10 = $150.
An employee who is paid a *salary* receives payment based on a weekly, biweekly, monthly, or annual time schedule. [6.6A, p. 269]	An employee paid an annual salary of $60,000 is paid $60,000 ÷ 12 = $5000 per month.
Balancing a checkbook is determining whether the checkbook balance is accurate. [6.7B, pp. 274–275]	To balance a checkbook: (1) Put a checkmark in the checkbook register by each check paid by the bank and by each deposit recorded by the bank. (2) Add to the current checkbook balance all checks that have been written but have not yet been paid by the bank and any interest paid on the account. (3) Subtract any charges and any deposits not yet recorded by the bank. This is the checkbook balance. (4) Compare the balance with the bank balance listed on the bank statement. If the two numbers are equal, the bank statement and the checkbook "balance."

Essential Rules and Procedures | Examples

Essential Rules and Procedures	Examples
To find unit cost, divide the total cost by the number of units. [6.1A, p. 235]	Three paperback books cost $36. The unit cost is $36 ÷ 3 = $12 per book.
To find total cost, multiply the unit cost by the number of units purchased. [6.1C, p. 236]	One melon costs $3. The total cost for 5 melons is $3 × 5 = $15.
Basic Markup Equations [6.2B, p. 240] Selling price − cost = markup Cost + markup = selling price Markup rate × cost = markup	A pair of cross trainers that cost a business $90 has a 50% markup rate. The markup is 0.50 × $90 = $45. The selling price is $90 + $45 = $135.
Basic Discount Equations [6.2D, p. 243] Regular price − sale price = discount Regular price − discount = sale price Discount rate × regular price = discount	A movie video is on sale for 20% off the regular price of $50. The discount is 0.20 × $50 = $10. The sale price is $50 − $10 = $40.
Simple Interest Formula for Annual Interest Rates [6.3A, p. 249]	Principal × annual interest rate × time (in years) = interest The simple interest due on a 2-year loan of $5000 that has an annual interest rate of 5% is $5000 × 0.05 × 2 = $500.
Maturity Value Formula for Simple Interest Loan [6.3A, p. 249]	Principal + interest = maturity value The interest to be paid on a 2-year loan of $5000 is $500. The maturity value of the loan is $5000 + $500 = $5500.
Monthly Payment on a Simple Interest Loan [6.3A, p. 250] Maturity value ÷ length of the loan in months = monthly payment	The maturity value of a simple interest 8-month loan is $8000. The monthly payment is $8000 ÷ 8 = $1000.

Copyright © Houghton Mifflin Company. All rights reserved.

Chapter 6 Review Exercises

1. **Consumerism**   A 20-ounce box of cereal costs $3.90. Find the unit cost.

2. **Car Expenses**   An account executive had car expenses of $1025.58 for insurance, $605.82 for gas, $37.92 for oil, and $188.27 for maintenance during a year in which 11,320 miles were driven. Find the cost per mile for these four items taken as a group. Round to the nearest tenth of a cent.

3. **Investments**   An oil stock was bought for $42.375 per share. Six months later, the stock was selling for $55.25 per share. Find the percent increase in the price of the stock for the 6 months. Round to the nearest tenth of a percent.

4. **Markup**   A sporting goods store uses a markup rate of 40%. What is the markup on a ski suit that costs the store $180?

5. **Simple Interest**   A contractor borrowed $100,000 from a credit union for 9 months at an annual interest rate of 4%. What is the simple interest due on the loan?

6. **Compound Interest**   A computer programmer invested $25,000 in a retirement account that pays 6% interest, compounded daily. What is the value of the investment in 10 years? Use the Compound Interest Table in the Appendix. Round to the nearest cent.

7. **Investments**   Last year an oil company had earnings of $4.12 per share. This year the earnings are $4.73 per share. What is the percent increase in earnings per share? Round to the nearest percent.

8. **Real Estate**   The monthly mortgage payment for a condominium is $523.67. The owner must pay an annual property tax of $658.32. Find the total monthly payment for the mortgage and property tax.

9. **Car Expenses**   A used pickup truck is purchased for $24,450. A down payment of 8% is made, and the remaining cost is financed for 4 years at an annual interest rate of 5%. Find the monthly payment. Use the Monthly Payment Table in the Appendix. Round to the nearest cent.

10. **Compound Interest**   A fast-food restaurant invested $50,000 in an account that pays 7% annual interest compounded quarterly. What is the value of the investment in 1 year? Use the Compound Interest Table in the Appendix.

11. **Real Estate**   Paula Mason purchased a home for $125,000. The lender requires a down payment of 15%. Find the amount of the down payment.

12. **Car Expenses**   A plumber bought a truck for $18,500. A state license fee of $315 and a sales tax of 6.25% of the purchase price are required. Find the total cost of the sales tax and the license fee.

Copyright © Houghton Mifflin Company. All rights reserved.

13. **Markup** Techno-Center uses a markup rate of 35% on all computer systems. Find the selling price of a computer system that costs the store $1540.

14. **Car Expenses** Mien pays a monthly car payment of $122.78. During a month in which $25.45 is principal, how much of the payment is interest?

15. **Compensation** The manager of the retail store at a ski resort receives a commission of 3% on all sales at the alpine shop. Find the total commission received during a month in which the shop had $108,000 in sales.

16. **Discount** A suit that regularly costs $235 is on sale for 40% off the regular price. Find the sale price.

17. **Banking** Luke had a checking account balance of $1568.45 before writing checks for $123.76, $756.45, and $88.77. He then deposited a check for $344.21. Find Luke's current checkbook balance.

18. **Simple Interest** Pros' Sporting Goods borrowed $30,000 at an annual interest rate of 8% for 6 months. Find the maturity value of the loan.

19. **Real Estate** A credit union requires a borrower to pay $2\frac{1}{2}$ points for a loan. Find the origination fee for a $75,000 loan.

20. **Consumerism** Sixteen ounces of mouthwash cost $3.49. A 33-ounce container of the same brand of mouthwash costs $6.99. Which is the better buy?

21. **Real Estate** The Sweeneys bought a home for $156,000. The family made a 10% down payment and financed the remainder with a 30-year loan at an annual interest rate of 7%. Find the monthly mortgage payment. Use the Monthly Payment Table in the Appendix. Round to the nearest cent.

22. **Compensation** Richard Valdez receives $12.60 per hour for working 40 hours a week and time and a half for working over 40 hours. Find his total income during a week in which he worked 48 hours.

23. **Banking** The business checking account of a donut shop showed a balance of $9567.44 before checks of $1023.55, $345.44, and $23.67 were written and checks of $555.89 and $135.91 were deposited. Find the current checkbook balance.

24. **Simple Interest** The simple interest due on a 4-month loan of $55,000 is $1375. Find the monthly payment on the loan.

25. **Simple Interest** A credit card company charges a customer 1.25% per month on the unpaid balance of charges on the card. What is the finance charge in a month in which the customer has an unpaid balance of $576?

Copyright © Houghton Mifflin Company. All rights reserved.

Chapter 6 Test

1. **Consumerism**   Twenty feet of lumber cost $138.40. What is the cost per foot?

2. **Consumerism**   Which is the more economical purchase: 3 pounds of tomatoes for $7.49 or 5 pounds of tomatoes for $12.59?

3. **Consumerism**   Red snapper costs $4.15 per pound. Find the cost of $3\frac{1}{2}$ pounds. Round to the nearest cent.

4. **Business**   An exercise bicycle increased in price from $415 to $498. Find the percent increase in the cost of the exercise bicycle.

5. **Markup**   A department store uses a 40% markup rate. Find the selling price of a compact disc player that the store purchased for $215.

6. **Investments**   The price of gold dropped from $390 per ounce to $360 per ounce. What percent decrease does this amount represent? Round to the nearest tenth of a percent.

7. **Consumerism**   The price of a video camera dropped from $1120 to $896. What percent decrease does this price drop represent?

8. **Discount**   A corner hutch with a regular price of $299 is on sale for 30% off the regular price. Find the sale price.

9. **Discount**   A box of stationery that regularly sells for $9.50 is on sale for $5.70. Find the discount rate.

10. **Simple Interest**   A construction company borrowed $75,000 for 4 months at an annual interest rate of 8%. Find the simple interest due on the loan.

11. **Simple Interest**   Craig Allen borrowed $25,000 at an annual interest rate of 9.2% for 9 months. Find the maturity value of the loan.

12. **Simple Interest**   A credit card company charges a customer 1.2% per month on the unpaid balance of charges on the card. What is the finance charge in a month in which the customer has an unpaid balance of $374.95?

13. **Compound Interest**   Jorge, who is self-employed, placed $30,000 in an account that pays 6% annual interest compounded quarterly. How much interest was earned in 10 years? Use the Compound Interest Table in the Appendix.

14. **Real Estate**   A savings and loan institution is offering mortgage loans that have a loan origination fee of $2\frac{1}{2}$ points. Find the loan origination fee when a home is purchased with a loan of $134,000.

Copyright © Houghton Mifflin Company. All rights reserved.

15. **Real Estate**   A new housing development offers homes with a mortgage of $222,000 for 25 years at an annual interest rate of 8%. Find the monthly mortgage payment. Use the Monthly Payment Table in the Appendix.

16. **Car Expenses**   A Chevrolet was purchased for $23,750, and a 20% down payment was made. Find the amount financed.

17. **Car Expenses**   A rancher purchased an SUV for $23,714 and made a down payment of 15% of the cost. The balance was financed for 4 years at an annual interest rate of 7%. Find the monthly truck payment. Use the Monthly Payment Table in the Appendix.

18. **Compensation**   Shaney receives an hourly wage of $20.40 an hour as an emergency room nurse. When called in at night, she receives time and a half. How much does Shaney earn in a week when she works 30 hours at normal rates and 15 hours during the night?

19. **Banking**   The business checking account for a pottery store had a balance of $7349.44 before checks for $1349.67 and $344.12 were written. The store manager then made a deposit of $956.60. Find the current checkbook balance.

20. **Banking**   Balance the checkbook shown.

| | | RECORD ALL CHARGES OR CREDITS THAT AFFECT YOUR ACCOUNT | | | | | BALANCE | |
NUMBER	DATE	DESCRIPTION OF TRANSACTION	PAYMENT/DEBIT (−)	√ T	FEE (IF ANY) (−)	DEPOSIT/CREDIT (+)	$ 1422	13
843	8/1	House Payment	$ 713 72		$	$	708	41
	8/4	Deposit				852 60	1561	01
844	8/5	Loan Payment	162 40				1398	61
845	8/6	Groceries	166 44				1232	17
846	8/10	Car Payment	322 37				909	80
	8/15	Deposit				852 60	1762	40
847	8/16	Credit Card	413 45				1348	95
848	8/18	Pharmacy	92 14				1256	81
849	8/22	Utilities	72 30				1184	51
850	8/28	Telephone	78 20				1106	31

CHECKING ACCOUNT Monthly Statement		Account Number: 122-345-1	
Date	Transaction	Amount	Balance
8/1	OPENING BALANCE		1422.13
8/3	CHECK	713.72	708.41
8/4	DEPOSIT	852.60	1561.01
8/8	CHECK	166.44	1394.57
8/8	CHECK	162.40	1232.17
8/15	DEPOSIT	852.60	2084.77
8/23	CHECK	72.30	2012.47
8/24	CHECK	92.14	1920.33
9/1	CLOSING BALANCE		1920.33

Copyright © Houghton Mifflin Company. All rights reserved.

Cumulative Review Exercises

1. Simplify: $12 - (10 - 8)^2 \div 2 + 3$

2. Add: $3\frac{1}{3} + 4\frac{1}{8} + 1\frac{1}{12}$

3. Find the difference between $12\frac{3}{16}$ and $9\frac{5}{12}$.

4. Find the product of $5\frac{5}{8}$ and $1\frac{9}{15}$.

5. Divide: $3\frac{1}{2} \div 1\frac{3}{4}$

6. Simplify $\left(\frac{3}{4}\right)^2 \div \left(\frac{3}{8} - \frac{1}{4}\right) + \frac{1}{2}$.

7. Divide: $0.059\overline{)3.0792}$
   Round to the nearest tenth.

8. Convert $\frac{17}{12}$ to a decimal. Round to the nearest thousandth.

9. Write "$410 in 8 hours" as a unit rate.

10. Solve the proportion $\frac{5}{n} = \frac{16}{35}$.
    Round to the nearest hundredth.

11. Write $\frac{5}{8}$ as a percent.

12. Find 6.5% of 420.

13. Write 18.2% as a decimal.

14. What percent of 20 is 8.4?

15. 30 is 12% of what?

16. 65 is 42% of what? Round to the nearest hundredth.

Copyright © Houghton Mifflin Company. All rights reserved.

17. **Meteorology**   A series of late-summer storms produced rainfall of $3\frac{3}{4}$, $8\frac{1}{2}$, and $1\frac{2}{3}$ inches during a 3-week period. Find the total rainfall during the 3 weeks.

18. **Taxes**   The Homer family pays $\frac{1}{5}$ of its total monthly income for taxes. The family has a total monthly income of $4850. Find the amount of their monthly income that the Homers pay in taxes.

19. **Consumerism**   In 5 years, the cost of a scientific calculator went from $75 to $30. What is the ratio of the decrease in price to the original price?

20. **Fuel Efficiencies**   A compact car was driven 417.5 miles on 12.5 gallons of gasoline. Find the number of miles driven per gallon of gasoline.

21. **Consumerism**   A 14-pound turkey costs $12.96. Find the unit cost. Round to the nearest cent.

22. **Investments**   Eighty shares of a stock paid a dividend of $112. At the same rate, find the dividend on 200 shares of the stock.

23. **Discount**   A video camera that regularly sells for $900 is on sale for 20% off the regular price. What is the sale price?

24. **Markup**   A department store bought a portable disc player for $85 and used a markup rate of 40%. Find the selling price of the disc player.

25. **Compensation**   Sook Kim, an elementary school teacher, received an increase in salary from $2800 per month to $3024 per month. Find the percent increase in her salary.

26. **Simple Interest**   A contractor borrowed $120,000 for 6 months at an annual interest rate of 4.5%. How much simple interest is due on the loan?

27. **Car Expenses**   A red Ford Mustang was purchased for $26,900, and a down payment of $2000 was made. The balance is financed for 3 years at an annual interest rate of 9%. Find the monthly payment. Use the Monthly Payment Table in the Appendix. Round to the nearest cent.

28. **Banking**   A family had a checking account balance of $1846.78. A check of $568.30 was deposited into the account, and checks of $123.98 and $47.33 were written. Find the new checking account balance.

29. **Car Expenses**   During 1 year, Anna Gonzalez spent $840 on gasoline and oil, $520 on insurance, $185 on tires, and $432 on repairs. Find the cost per mile to drive the car 10,000 miles during the year. Round to the nearest cent.

30. **Real Estate**   A house has a mortgage of $172,000 for 20 years at an annual interest rate of 6%. Find the monthly mortgage payment. Use the Monthly Payment Table in the Appendix. Round to the nearest cent.

Copyright © Houghton Mifflin Company. All rights reserved.

Appendix

Equations and Properties

Equations

Basic Percent Equation

Percent × base = amount

Proportion Method of Solving Percent Equations

$$\frac{\text{Percent}}{100} = \frac{\text{amount}}{\text{base}}$$

Basic Markup Equations

Selling price = cost + markup
Markup = markup rate × cost

Basic Discount Equations

Sale price = regular price − discount
Discount = discount rate × regular price

Annual Simple Interest Formula

Principal × annual simple interest rate × time in years = interest

Maturity Value Formula for Simple Interest

Principal + interest = maturity value

Monthly Payment on a Simple Interest Loan

Maturity value ÷ length of the loan in months = monthly payment

Properties

Addition Property of Zero

Zero added to a number does not change the number.

Commutative Property of Addition

Two numbers can be added in either order; the sum will be the same.

Associative Property of Addition

Grouping addition in any order gives the same result.

Multiplication Property of Zero

The product of a number and zero is zero.

Multiplication Property of One

The product of a number and one is the number.

Commutative Property of Multiplication

Two numbers can be multiplied in either order. The product will be the same.

Associative Property of Multiplication

Grouping the numbers to be multiplied in any order gives the same result.

Properties of Zero and One in Division

Any number divided by 1 is the number.
Any number other than zero divided by itself is 1.
Division by zero is not allowed.
Zero divided by any number other than zero is zero.

Copyright © Houghton Mifflin Company. All rights reserved.

Compound Interest Table

Compounded Annually

	4%	5%	6%	7%	8%	9%	10%
1 year	1.04000	1.05000	1.06000	1.07000	1.08000	1.09000	1.10000
5 years	1.21665	1.27628	1.33823	1.40255	1.46933	1.53862	1.61051
10 years	1.48024	1.62890	1.79085	1.96715	2.15893	2.36736	2.59374
15 years	1.80094	2.07893	2.39656	2.75903	3.17217	3.64248	4.17725
20 years	2.19112	2.65330	3.20714	3.86968	4.66095	5.60441	6.72750

Compounded Semiannually

	4%	5%	6%	7%	8%	9%	10%
1 year	1.04040	1.05062	1.06090	1.07123	1.08160	1.09203	1.10250
5 years	1.21899	1.28008	1.34392	1.41060	1.48024	1.55297	1.62890
10 years	1.48595	1.63862	1.80611	1.98979	2.19112	2.41171	2.65330
15 years	1.81136	2.09757	2.42726	2.80679	3.24340	3.74531	4.32194
20 years	2.20804	2.68506	3.26204	3.95926	4.80102	5.81634	7.03999

Compounded Quarterly

	4%	5%	6%	7%	8%	9%	10%
1 year	1.04060	1.05094	1.06136	1.07186	1.08243	1.09308	1.10381
5 years	1.22019	1.28204	1.34686	1.41478	1.48595	1.56051	1.63862
10 years	1.48886	1.64362	1.81402	2.00160	2.20804	2.43519	2.68506
15 years	1.81670	2.10718	2.44322	2.83182	3.28103	3.80013	4.39979
20 years	2.21672	2.70148	3.29066	4.00639	4.87544	5.93015	7.20957

Compounded Monthly

	4%	5%	6%	7%	8%	9%	10%
1 year	1.04074	1.051162	1.061678	1.072290	1.083000	1.093807	1.104713
5 years	1.220997	1.283359	1.348850	1.417625	1.489846	1.565681	1.645309
10 years	1.490833	1.647009	1.819397	2.009661	2.219640	2.451357	2.707041
15 years	1.820302	2.113704	2.454094	2.848947	3.306921	3.838043	4.453920
20 years	2.222582	2.712640	3.310204	4.038739	4.926803	6.009152	7.328074

Compounded Daily

	4%	5%	6%	7%	8%	9%	10%
1 year	1.04080	1.05127	1.06183	1.07250	1.08328	1.09416	1.10516
5 years	1.22139	1.28400	1.34983	1.41902	1.49176	1.56823	1.64861
10 years	1.49179	1.64866	1.82203	2.01362	2.22535	2.45933	2.71791
15 years	1.82206	2.11689	2.45942	2.85736	3.31968	3.85678	4.48077
20 years	2.22544	2.71810	3.31979	4.05466	4.95217	6.04830	7.38703

To use this table:
1. Locate the section that gives the desired compounding period.
2. Locate the interest rate in the top row of that section.
3. Locate the number of years in the left-hand column of that section.
4. Locate the number where the interest-rate column and the number-of-years row meet. This is the compound interest factor.

Example An investment yields an annual interest rate of 10% compounded quarterly for 5 years.
The compounding period is "compounded quarterly."
The interest rate is 10%.
The number of years is 5.
The number where the row and column meet is 1.63862. This is the compound interest factor.

Copyright © Houghton Mifflin Company. All rights reserved.

Compound Interest Table

Compounded Annually

	11%	12%	13%	14%	15%	16%	17%
1 year	1.11000	1.12000	1.13000	1.14000	1.15000	1.16000	1.17000
5 years	1.68506	1.76234	1.84244	1.92542	2.01136	2.10034	2.19245
10 years	2.83942	3.10585	3.39457	3.70722	4.04556	4.41144	4.80683
15 years	4.78459	5.47357	6.25427	7.13794	8.13706	9.26552	10.53872
20 years	8.06239	9.64629	11.52309	13.74349	16.36654	19.46076	23.10560

Compounded Semiannually

	11%	12%	13%	14%	15%	16%	17%
1 year	1.11303	1.12360	1.13423	1.14490	1.15563	1.16640	1.17723
5 years	1.70814	1.79085	1.87714	1.96715	2.06103	2.15893	2.26098
10 years	2.91776	3.20714	3.52365	3.86968	4.24785	4.66096	5.11205
15 years	4.98395	5.74349	6.61437	7.61226	8.75496	10.06266	11.55825
20 years	8.51331	10.28572	12.41607	14.97446	18.04424	21.72452	26.13302

Compounded Quarterly

	11%	12%	13%	14%	15%	16%	17%
1 year	1.11462	1.12551	1.13648	1.14752	1.15865	1.16986	1.18115
5 years	1.72043	1.80611	1.89584	1.98979	2.08815	2.19112	2.29891
10 years	2.95987	3.26204	3.59420	3.95926	4.36038	4.80102	5.28497
15 years	5.09225	5.89160	6.81402	7.87809	9.10513	10.51963	12.14965
20 years	8.76085	10.64089	12.91828	15.67574	19.01290	23.04980	27.93091

Compounded Monthly

	11%	12%	13%	14%	15%	16%	17%
1 year	1.115719	1.126825	1.138032	1.149342	1.160755	1.172271	1.183892
5 years	1.728916	1.816697	1.908857	2.005610	2.107181	2.213807	2.325733
10 years	2.989150	3.300387	3.643733	4.022471	4.440213	4.900941	5.409036
15 years	5.167988	5.995802	6.955364	8.067507	9.356334	10.849737	12.579975
20 years	8.935015	10.892554	13.276792	16.180270	19.715494	24.019222	29.257669

Compounded Daily

	11%	12%	13%	14%	15%	16%	17%
1 year	1.11626	1.12747	1.13880	1.15024	1.16180	1.17347	1.18526
5 years	1.73311	1.82194	1.91532	2.01348	2.11667	2.22515	2.33918
10 years	3.00367	3.31946	3.66845	4.05411	4.48031	4.95130	5.47178
15 years	5.20569	6.04786	7.02625	8.16288	9.48335	11.01738	12.79950
20 years	9.02203	11.01883	13.45751	16.43582	20.07316	24.51534	29.94039

Copyright © Houghton Mifflin Company. All rights reserved.

Monthly Payment Table

	4%	5%	6%	7%	8%	9%
1 year	0.0851499	0.0856075	0.0860664	0.0865267	0.0869884	0.0874515
2 years	0.0434249	0.0438714	0.0443206	0.0447726	0.0452273	0.0456847
3 years	0.0295240	0.0299709	0.0304219	0.0308771	0.0313364	0.0317997
4 years	0.0225791	0.0230293	0.0234850	0.0239462	0.0244129	0.0248850
5 years	0.0184165	0.0188712	0.0193328	0.0198012	0.0202764	0.0207584
15 years	0.0073969	0.0079079	0.0084386	0.0089883	0.0095565	0.0101427
20 years	0.0060598	0.0065996	0.0071643	0.0077530	0.0083644	0.0089973
25 years	0.0052784	0.0058459	0.0064430	0.0070678	0.0077182	0.0083920
30 years	0.0047742	0.0053682	0.0059955	0.0066530	0.0073376	0.0080462

	10%	11%	12%	13%
1 year	0.0879159	0.0883817	0.0888488	0.0893173
2 years	0.0461449	0.0466078	0.0470735	0.0475418
3 years	0.0322672	0.0327387	0.0332143	0.0336940
4 years	0.0253626	0.0258455	0.0263338	0.0268275
5 years	0.0212470	0.0217424	0.0222445	0.0227531
15 years	0.0107461	0.0113660	0.0120017	0.0126524
20 years	0.0096502	0.0103219	0.0110109	0.0117158
25 years	0.0090870	0.0098011	0.0105322	0.0112784
30 years	0.0087757	0.0095232	0.0102861	0.0110620

To use this table:
1. Locate the desired interest rate in the top row.
2. Locate the number of years in the left-hand column.
3. Locate the number where the interest-rate column and the number-of-years row meet. This is the monthly payment factor.

Example   A home has a 30-year mortgage at an annual interest rate of 12%.
The interest rate is 12%.
The number of years is 30.
The number where the row and column meet is 0.0102861. This is the monthly payment factor.

Copyright © Houghton Mifflin Company. All rights reserved.

Solutions to Chapter 1 "You Try It"

SECTION 1.1

You Try It 1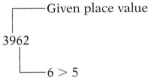

You Try It 2   **a.** $45 > 29$ **b.** $27 > 0$

You Try It 3   Thirty-six million four hundred sixty-two thousand seventy-five

You Try It 4   452,007

You Try It 5 $60,000 + 8000 + 200 + 80 + 1$

You Try It 6 $100,000 + 9000 + 200 + 7$

You Try It 7

```
        ┌──── Given place value
        │
   368,492
        │
        └──── 8 > 5
```

368,492 rounded to the nearest ten-thousand is 370,000.

You Try It 8

```
        ┌──── Given place value
        │
   3962
        │
        └──── 6 > 5
```

3962 rounded to the nearest hundred is 4000.

SECTION 1.2

You Try It 1

```
      1 1
     347       • 7 + 3 = 10
 + 12,453        Write the 0 in the ones
   12,800        column. Carry the 1 to the
                 tens column.
                 1 + 4 + 5 = 10
                 Write the 0 in the tens
                 column. Carry the 1 to the
                 hundreds column.
                 1 + 3 + 4 = 8
```

347 increased by 12,453 is 12,800.

You Try It 2

```
    2
   95       • 5 + 8 + 7 = 20
   88         Write the 0 in the ones column.
 + 67         Carry the 2 to the tens column.
  250
```

You Try It 3

```
    1 1 2 1
     392
   4,079
  89,035
 + 4,992
  98,498
```

You Try It 4

Strategy   To find the total number of Wal-Mart stores outside of the United States, read the table to find the number of stores outside the United States in each of the four categories listed. Then add the four numbers.

Solution

```
    942
    238
     71
 +   37
   1288
```

Wal-Mart has 1288 stores outside the United States.

SECTION 1.3

You Try It 1

```
   8925      Check:   6413
 − 6413             + 2512
   2512               8925
```

You Try It 2

```
   17,504    Check:    9,302
 −  9,302            +  8,202
    8,202             17,504
```

You Try It 3

```
   2  14 7 11
   ß   Å 8 ʆ    Check:    865
 −    8 6 5             + 2616
   2  6 1 6               3481
```

You Try It 4

```
           15
        4  5 12
   5 4,ß 6 2    Check:   14,485
 − 1 4,4 8 5           + 40,077
   4 0,0 7 7             54,562
```

You Try It 5

```
      3 10
   6 Å,Ø 0 3    • There are 2 zeros in the
 − 5 4,9 3 6      minuend.
                  Borrow 1 thousand from
                  the thousands column
                  and write 10 in the
                  hundreds column.
                  (Continued)
```

Copyright © Houghton Mifflin Company. All rights reserved.

(Continued)

$$\begin{array}{r}9\\3\ \ 10\ \ 10\\6\ \cancel{4},\cancel{0}\ \cancel{0}\ 3\\-5\ 4,9\ 3\ 6\end{array}$$

- Borrow 1 hundred from the hundreds column and write 10 in the tens column.

$$\begin{array}{r}13\ \ 9\ \ 9\\5\ \ 3\ \ 10\ \ 10\ \ 13\\\cancel{6}\ \cancel{4},\cancel{0}\ \cancel{0}\ \cancel{3}\\-5\ 4,9\ 3\ 6\\\hline 9,0\ 6\ 7\end{array}$$

- Borrow 1 ten from the tens column and add 10 to the 3 in the ones column.

Check:
$$\begin{array}{r}54,936\\+\ \ \ 9,067\\\hline 64,003\end{array}$$

You Try It 6

Strategy To find the difference, subtract the number of personnel on active duty in the Air Force in 1945 (2,282,259) from the number of personnel on active duty in the Navy in 1945 (3,380,817).

Solution
$$\begin{array}{r}3,380,817\\-\ 2,282,259\\\hline 1,098,558\end{array}$$

The difference was 1,098,558 personnel.

You Try It 7

Strategy To find your take-home pay:
- Add to find the total of the deductions (127 + 18 + 35).
- Subtract the total of the deductions from your total salary (638).

Solution
$$\begin{array}{r}127\\18\\+\ 35\\\hline 180\end{array}\qquad\begin{array}{r}638\\-\ 180\\\hline 458\end{array}$$

180 in deductions

Your take-home pay is $458.

SECTION 1.4

You Try It 1

$$\begin{array}{r}3\ 5\\648\\\times\ \ \ 7\\\hline 4536\end{array}$$

- $7 \times 8 = 56$
  Write the 6 in the ones column.
  Carry the 5 to the tens column.
  $7 \times 4 = 28,\ 28 + 5 = 33$
  $7 \times 6 = 42,\ 42 + 3 = 45$

You Try It 2

$$\begin{array}{r}756\\\times\ 305\\\hline 3780\\22680\\\hline 230,580\end{array}$$

- $5 \times 756 = 3780$
- Write a zero in the tens column for 0×756.
  $3 \times 756 = 2268$

You Try It 3

Strategy To find the number of cars the dealer will receive in 12 months, multiply the number of months (12) by the number of cars received each month (37).

Solution
$$\begin{array}{r}37\\\times\ \ 12\\\hline 74\\37\\\hline 444\end{array}$$

The dealer will receive 444 cars in 12 months.

You Try It 4

Strategy To find the total cost of the order:
- Find the cost of the sports jackets by multiplying the number of jackets (25) by the cost for each jacket (23).
- Add the product to the cost for the suits (4800).

Solution
$$\begin{array}{r}23\\\times\ 25\\\hline 115\\46\\\hline 575\end{array}\quad\text{cost for jackets}\qquad\begin{array}{r}4800\\+\ \ \ 575\\\hline 5375\end{array}$$

The total cost of the order is $5375.

SECTION 1.5

You Try It 1

$$9\overline{)63}\quad\dfrac{7}{}$$

Check: $7 \times 9 = 63$

You Try It 2

$$\begin{array}{r}453\\9\overline{)\ 4077}\\-36\ \ \ \ \\\hline 47\ \ \\-45\ \ \\\hline 27\\-27\\\hline 0\end{array}$$

Check: $453 \times 9 = 4077$

Copyright © Houghton Mifflin Company. All rights reserved.

You Try It 3

```
       705
  9) 6345
    -63
      04
    -  0
      45
     -45
       0
```

- Think 9)4. Place 0 in quotient.
- Subtract 0 × 9.
- Bring down the 5.

Check: 705 × 9 = 6345

You Try It 4

```
       870 r5
  6) 5225
    -48
      42
     -42
      05
     - 0
       5
```

- Think 6)5. Place 0 in quotient.
- Subtract 0 × 6.

Check: (870 × 6) + 5 =
        5220 + 5 = 5225

You Try It 5

```
       3,058 r3
  7) 21,409
    -21
      0 4
    -  0
      40
     -35
      59
     -56
       3
```

- Think 7)4. Place 0 in quotient.
- Subtract 0 × 7.

Check: (3058 × 7) + 3 =
        21,406 + 3 = 21,409

You Try It 6

```
        109
  42) 4578
     -42
       37
     -  0
      378
     -378
        0
```

- Think 42)37. Place 0 in quotient.
- Subtract 0 × 42.

Check: 109 × 42 = 4578

You Try It 7

```
        470 r29
  39) 18,359
     -15 6
       2 75
      -2 73
         29
       -  0
         29
```

- Think 3)18. 6 × 39 is too large. Try 5. 5 × 39 is too large. Try 4.

Check: (470 × 39) + 29 =
        18,330 + 29 = 18,359

You Try It 8

```
         62 r111
  534) 33,219
      -32 04
        1 179
       -1 068
          111
```

Check: (62 × 534) + 111 =
        33,108 + 111 = 33,219

You Try It 9

```
          421 r33
  515) 216,848
      -206 0
        10 84
       -10 30
          548
         -515
           33
```

Check: (421 × 515) + 33 =
        216,815 + 33 = 216,848

You Try It 10

Strategy   To find the number of tires that can be stored on each shelf, divide the number of tires (270) by the number of shelves (15).

Solution
```
        18
  15) 270
     -15
      120
     -120
        0
```

Each shelf can store 18 tires.

You Try It 11

Strategy   To find the number of cases produced in 8 hours:
- Find the number of cases produced in 1 hour by dividing the number of cans produced (12,600) by the number of cans to a case (24).
- Multiply the number of cases produced in 1 hour by 8.

(Continued)

Copyright © Houghton Mifflin Company. All rights reserved.

(Continued)

Solution

$$24\overline{)12{,}600} \quad \begin{array}{l}\text{cases produced}\\ \text{in 1 hour}\end{array}$$

$$
\begin{array}{r}
525 \\
24\overline{)12{,}600} \\
-12\ 0 \\
\hline
60 \\
-48 \\
\hline
120 \\
-120 \\
\hline
0
\end{array}
$$

$$
\begin{array}{r}
525 \\
\times\quad 8 \\
\hline
4200
\end{array}
$$

In 8 hours, 4200 cases are produced.

SECTION 1.6

You Try It 1 $2 \cdot 2 \cdot 2 \cdot 2 \cdot 3 \cdot 3 \cdot 3 = 2^4 \cdot 3^3$

You Try It 2 $10 \cdot 10 \cdot 10 \cdot 10 \cdot 10 \cdot 10 \cdot 10 = 10^7$

You Try It 3 $2^3 \cdot 5^2 = (2 \cdot 2 \cdot 2) \cdot (5 \cdot 5) = 8 \cdot 25$
$$= 200$$

You Try It 4 $5 \cdot (8 - 4)^2 \div 4 - 2$

$\begin{aligned} &= 5 \cdot 4^2 \div 4 - 2 &&\bullet \text{ Parentheses}\\ &= 5 \cdot 16 \div 4 - 2 &&\bullet \text{ Exponents}\\ &= 80 \div 4 - 2 &&\bullet \text{ Multiplication}\\ & &&\text{ and division}\\ &= 20 - 2\\ &= 18 &&\bullet \text{ Subtraction} \end{aligned}$

SECTION 1.7

You Try It 1

$40 \div 1 = 40$
$40 \div 2 = 20$
$40 \div 3 \qquad$ Will not divide evenly
$40 \div 4 = 10$
$40 \div 5 = 8$
$40 \div 6 \qquad$ Will not divide evenly
$40 \div 7 \qquad$ Will not divide evenly
$40 \div 8 = 5$

1, 2, 4, 5, 8, 10, 20, and 40 are factors of 40.

You Try It 2

$$
\begin{array}{r|l}
\multicolumn{2}{c}{44} \\
2 & 22 \\
2 & 11 \\
11 & 1
\end{array}
$$

• $44 \div 2 = 22$
• $22 \div 2 = 11$
• $11 \div 11 = 1$

$44 = 2 \cdot 2 \cdot 11$

You Try It 3

$$
\begin{array}{r|l}
\multicolumn{2}{c}{177} \\
3 & 59 \\
59 & 1
\end{array}
$$

• Try only 2, 3, 4, 7, and 11 because $11^2 > 59$.

$177 = 3 \cdot 59$

Solutions to Chapter 2 "You Try It"

SECTION 2.1

You Try It 1

	2	3	5
12 =	(2 · 2)	3	
27 =		(3 · 3 · 3)	
50 =	2		(5 · 5)

The LCM $= 2 \cdot 2 \cdot 3 \cdot 3 \cdot 3 \cdot 5 \cdot 5$
$\qquad\quad = 2700$

You Try It 2

	2	3	5
36 =	(2 · 2)	3 · 3	
60 =	2 · 2	(3)	5
72 =	2 · 2 · 2	3 · 3	

The GCF $= 2 \cdot 2 \cdot 3 = 12$.

You Try It 3

	2	3	5	11
11 =				11
24 =	2 · 2 · 2	3		
30 =	2	3	5	

Because no numbers are circled, the GCF $= 1$.

SECTION 2.2

You Try It 1 $4\dfrac{1}{4}$

You Try It 2 $\dfrac{17}{4}$

You Try It 3

$$
\begin{array}{r}
4 \\
5\overline{)22} \\
-20 \\
\hline
2
\end{array}
$$

$\dfrac{22}{5} = 4\dfrac{2}{5}$

You Try It 4

$$
\begin{array}{r}
4 \\
7\overline{)28} \\
-28 \\
\hline
0
\end{array}
$$

$\dfrac{28}{7} = 4$

Copyright © Houghton Mifflin Company. All rights reserved.

You Try It 5 $14\dfrac{+5}{\times 8} = \dfrac{112 + 5}{8} = \dfrac{117}{8}$

SECTION 2.3

You Try It 1 $45 \div 5 = 9$ $\dfrac{3}{5} = \dfrac{3 \cdot 9}{5 \cdot 9} = \dfrac{27}{45}$

$\dfrac{27}{45}$ is equivalent to $\dfrac{3}{5}$.

You Try It 2   Write 6 as $\dfrac{6}{1}$.

$18 \div 1 = 18$ $6 = \dfrac{6 \cdot 18}{1 \cdot 18} = \dfrac{108}{18}$

$\dfrac{108}{18}$ is equivalent to 6.

You Try It 3 $\dfrac{16}{24} = \dfrac{\overset{1}{\cancel{2}} \cdot \overset{1}{\cancel{2}} \cdot \overset{1}{\cancel{2}} \cdot 2}{\underset{1}{\cancel{2}} \cdot \underset{1}{\cancel{2}} \cdot \underset{1}{\cancel{2}} \cdot 3} = \dfrac{2}{3}$

You Try It 4 $\dfrac{8}{56} = \dfrac{\overset{1}{\cancel{2}} \cdot \overset{1}{\cancel{2}} \cdot \overset{1}{\cancel{2}}}{\underset{1}{\cancel{2}} \cdot \underset{1}{\cancel{2}} \cdot \underset{1}{\cancel{2}} \cdot 7} = \dfrac{1}{7}$

You Try It 5 $\dfrac{15}{32} = \dfrac{3 \cdot 5}{2 \cdot 2 \cdot 2 \cdot 2 \cdot 2} = \dfrac{15}{32}$

You Try It 6 $\dfrac{48}{36} = \dfrac{\overset{1}{\cancel{2}} \cdot \overset{1}{\cancel{2}} \cdot 2 \cdot 2 \cdot \overset{1}{\cancel{3}}}{\underset{1}{\cancel{2}} \cdot \underset{1}{\cancel{2}} \cdot \underset{1}{\cancel{3}} \cdot 3} = \dfrac{4}{3} = 1\dfrac{1}{3}$

SECTION 2.4

You Try It 1

$\begin{aligned} &\dfrac{3}{8} \\ +\;&\dfrac{7}{8} \\ \hline &\dfrac{10}{8} = \dfrac{5}{4} = 1\dfrac{1}{4} \end{aligned}$

• The denominators are the same. Add the numerators. Place the sum over the common denominator.

You Try It 2

$\begin{aligned} &\dfrac{5}{12} = \dfrac{20}{48} \\ +\;&\dfrac{9}{16} = \dfrac{27}{48} \\ \hline &\qquad\;\; \dfrac{47}{48} \end{aligned}$

• The LCM of 12 and 16 is 48.

You Try It 3

$\begin{aligned} &\dfrac{7}{8} = \dfrac{105}{120} \\ +\;&\dfrac{11}{15} = \dfrac{88}{120} \\ \hline &\qquad\;\; \dfrac{193}{120} = 1\dfrac{73}{120} \end{aligned}$

• The LCM of 8 and 15 is 120.

You Try It 4

$\begin{aligned} &\dfrac{3}{4} = \dfrac{30}{40} \\ &\dfrac{4}{5} = \dfrac{32}{40} \\ +\;&\dfrac{5}{8} = \dfrac{25}{40} \\ \hline &\qquad\;\; \dfrac{87}{40} = 2\dfrac{7}{40} \end{aligned}$

• The LCM of 4, 5, and 8 is 40.

You Try It 5 $7 + \dfrac{6}{11} = 7\dfrac{6}{11}$

You Try It 6

$\begin{aligned} &\;\;29 \\ +\;&17\dfrac{5}{12} \\ \hline &\;\;46\dfrac{5}{12} \end{aligned}$

You Try It 7

$\begin{aligned} &\;\;7\dfrac{4}{5} \;=\; 7\dfrac{24}{30} \\ &\;\;6\dfrac{7}{10} \;=\; 6\dfrac{21}{30} \\ +\;&13\dfrac{11}{15} = 13\dfrac{22}{30} \\ \hline &\qquad 26\dfrac{67}{30} = 28\dfrac{7}{30} \end{aligned}$

• LCM = 30

You Try It 8

$\begin{aligned} &\;\;9\dfrac{3}{8} \;=\; 9\dfrac{45}{120} \\ &17\dfrac{7}{12} = 17\dfrac{70}{120} \\ +\;&10\dfrac{14}{15} = 10\dfrac{112}{120} \\ \hline &\qquad 36\dfrac{227}{120} = 37\dfrac{107}{120} \end{aligned}$

• LCM = 120

You Try It 9

Strategy   To find the total time spent on the activities, add the three times $\left(4\dfrac{1}{2}, \; 3\dfrac{3}{4}, \; 1\dfrac{1}{3}\right)$.

Solution

$\begin{aligned} &\;\;4\dfrac{1}{2} = 4\dfrac{6}{12} \\ &\;\;3\dfrac{3}{4} = 3\dfrac{9}{12} \\ +\;&\;\;1\dfrac{1}{3} = 1\dfrac{4}{12} \\ \hline &\qquad 8\dfrac{19}{12} = 9\dfrac{7}{12} \end{aligned}$

The total time spent on the three activities was $9\dfrac{7}{12}$ hours.

Copyright © Houghton Mifflin Company. All rights reserved.

You Try It 10

Strategy To find the overtime pay:
- Find the total number of overtime hours $\left(1\frac{2}{3} + 3\frac{1}{3} + 2\right)$.
- Multiply the total number of hours by the overtime hourly wage (36).

Solution

$$1\frac{2}{3}$$
$$3\frac{1}{3}$$
$$+ 2$$
$$\overline{6\frac{3}{3}} = 7 \text{ hours}$$

$$\begin{array}{r} 36 \\ \times\ 7 \\ \hline 252 \end{array}$$

Jeff earned $252 in overtime pay.

SECTION 2.5

You Try It 1

$$\begin{array}{r} \frac{16}{27} \\ - \frac{7}{27} \\ \hline \frac{9}{27} = \frac{1}{3} \end{array}$$

- The denominators are the same. Subtract the numerators. Place the difference over the common denominator.

You Try It 2

$$\begin{array}{r} \frac{13}{18} = \frac{52}{72} \\ - \frac{7}{24} = \frac{21}{72} \\ \hline \frac{31}{72} \end{array}$$

- LCM = 72

You Try It 3

$$\begin{array}{r} 17\frac{5}{9} = 17\frac{20}{36} \\ - 11\frac{5}{12} = 11\frac{15}{36} \\ \hline 6\frac{5}{36} \end{array}$$

- LCM = 36

You Try It 4

$$\begin{array}{r} 8\ \ = 7\frac{13}{13} \\ - 2\frac{4}{13} = 2\frac{4}{13} \\ \hline 5\frac{9}{13} \end{array}$$

- LCM = 13

You Try It 5

$$\begin{array}{r} 21\frac{7}{9} = 21\frac{28}{36} = 20\frac{64}{36} \\ - 7\frac{11}{12} = 7\frac{33}{36} = 7\frac{33}{36} \\ \hline 13\frac{31}{36} \end{array}$$

- LCM = 36

You Try It 6

Strategy To find the time remaining before the plane lands, subtract the number of hours already in the air $\left(2\frac{3}{4}\right)$ from the total time of the trip $\left(5\frac{1}{2}\right)$.

Solution

$$\begin{array}{r} 5\frac{1}{2} = 5\frac{2}{4} = 4\frac{6}{4} \\ - 2\frac{3}{4} = 2\frac{3}{4} = 2\frac{3}{4} \\ \hline 2\frac{3}{4} \text{ hours} \end{array}$$

The plane will land in $2\frac{3}{4}$ hours.

You Try It 7

Strategy To find the amount of weight to be lost during the third month:
- Find the total weight loss during the first two months $\left(7\frac{1}{2} + 5\frac{3}{4}\right)$.
- Subtract the total weight loss from the goal (24 pounds).

Solution

$$\begin{array}{r} 7\frac{1}{2} = \ \ 7\frac{2}{4} \\ + 5\frac{3}{4} = \ \ 5\frac{3}{4} \\ \hline 12\frac{5}{4} = 13\frac{1}{4} \text{ pounds lost} \end{array}$$

$$\begin{array}{r} 24\ \ = 23\frac{4}{4} \\ - 13\frac{1}{4} = 13\frac{1}{4} \\ \hline 10\frac{3}{4} \text{ pounds} \end{array}$$

The patient must lose $10\frac{3}{4}$ pounds to achieve the goal.

SECTION 2.6

You Try It 1

$$\frac{4}{21} \times \frac{7}{44} = \frac{4 \cdot 7}{21 \cdot 44}$$

$$= \frac{\overset{1}{2} \cdot \overset{1}{2} \cdot \overset{1}{7}}{3 \cdot 7 \cdot 2 \cdot 2 \cdot 11}_{1\ \ 1\ \ 1} = \frac{1}{33}$$

You Try It 2

$$\frac{2}{21} \times \frac{10}{33} = \frac{2 \cdot 10}{21 \cdot 33}$$

$$= \frac{2 \cdot 2 \cdot 5}{3 \cdot 7 \cdot 3 \cdot 11} = \frac{20}{693}$$

Copyright © Houghton Mifflin Company. All rights reserved.

You Try It 3
$$\frac{16}{5} \times \frac{15}{24} = \frac{16 \cdot 15}{5 \cdot 24}$$
$$= \frac{\overset{1}{\cancel{2}} \cdot \overset{1}{\cancel{2}} \cdot \overset{1}{\cancel{2}} \cdot 2 \cdot 3 \cdot \overset{1}{\cancel{3}} \cdot \overset{1}{\cancel{5}}}{\underset{1}{\cancel{5}} \cdot \underset{1}{\cancel{2}} \cdot \underset{1}{\cancel{2}} \cdot \underset{1}{\cancel{2}} \cdot \underset{1}{\cancel{3}}} = \frac{2}{1} = 2$$

You Try It 4
$$5\frac{2}{5} \times \frac{5}{9} = \frac{27}{5} \times \frac{5}{9} = \frac{27 \cdot 5}{5 \cdot 9}$$
$$= \frac{\overset{1}{\cancel{3}} \cdot \overset{1}{\cancel{3}} \cdot 3 \cdot \overset{1}{\cancel{5}}}{\underset{1}{\cancel{5}} \cdot \underset{1}{\cancel{3}} \cdot \underset{1}{\cancel{3}}} = \frac{3}{1} = 3$$

You Try It 5
$$3\frac{2}{5} \times 6\frac{1}{4} = \frac{17}{5} \times \frac{25}{4} = \frac{17 \cdot 25}{5 \cdot 4}$$
$$= \frac{17 \cdot \overset{1}{\cancel{5}} \cdot 5}{\underset{1}{\cancel{5}} \cdot 2 \cdot 2} = \frac{85}{4} = 21\frac{1}{4}$$

You Try It 6
$$3\frac{2}{7} \times 6 = \frac{23}{7} \times \frac{6}{1} = \frac{23 \cdot 6}{7 \cdot 1}$$
$$= \frac{23 \cdot 2 \cdot 3}{7 \cdot 1} = \frac{138}{7} = 19\frac{5}{7}$$

You Try It 7

Strategy To find the value of the house today, multiply the old value of the house (170,000) by $2\frac{1}{2}$.

Solution
$$170,000 \times 2\frac{1}{2} = \frac{170,000}{1} \times \frac{5}{2}$$
$$= \frac{170,000 \cdot 5}{1 \cdot 2}$$
$$= 425,000$$

The value of the house today is $425,000.

You Try It 8

Strategy To find the cost of the air compressor:
- Multiply to find the value of the drying chamber $\left(\frac{4}{5} \times 160,000\right)$.
- Subtract the value of the drying chamber from the total value of the two items (160,000).

Solution
$$\frac{4}{5} \times \frac{160,000}{1} = \frac{640,000}{5}$$
$$= 128,000 \quad \bullet \text{ Value of the}$$
$$\text{drying chamber}$$
$$160,000 - 128,000 = 32,000$$

The cost of the air compressor was $32,000.

SECTION 2.7

You Try It 1
$$\frac{3}{7} \div \frac{2}{3} = \frac{3}{7} \times \frac{3}{2} = \frac{3 \cdot 3}{7 \cdot 2} = \frac{9}{14}$$

You Try It 2
$$\frac{3}{4} \div \frac{9}{10} = \frac{3}{4} \times \frac{10}{9}$$
$$= \frac{3 \cdot 10}{4 \cdot 9} = \frac{\overset{1}{\cancel{3}} \cdot \overset{1}{\cancel{2}} \cdot 5}{2 \cdot 2 \cdot \underset{1}{\cancel{3}} \cdot 3} = \frac{5}{6}$$

You Try It 3
$$\frac{5}{7} \div 6 = \frac{5}{7} \div \frac{6}{1} \qquad \bullet \; 6 = \frac{6}{1}. \text{ The}$$
$$= \frac{5}{7} \times \frac{1}{6} = \frac{5 \cdot 1}{7 \cdot 6} \qquad \text{reciprocal of } \frac{6}{1}$$
$$= \frac{5}{7 \cdot 2 \cdot 3} = \frac{5}{42} \qquad \text{is } \frac{1}{6}.$$

You Try It 4
$$12\frac{3}{5} \div 7 = \frac{63}{5} \div \frac{7}{1} = \frac{63}{5} \times \frac{1}{7}$$
$$= \frac{63 \cdot 1}{5 \cdot 7} = \frac{3 \cdot 3 \cdot \overset{1}{\cancel{7}}}{5 \cdot \underset{1}{\cancel{7}}} = \frac{9}{5} = 1\frac{4}{5}$$

You Try It 5
$$3\frac{2}{3} \div 2\frac{2}{5} = \frac{11}{3} \div \frac{12}{5}$$
$$= \frac{11}{3} \times \frac{5}{12} = \frac{11 \cdot 5}{3 \cdot 12}$$
$$= \frac{11 \cdot 5}{3 \cdot 2 \cdot 2 \cdot 3} = \frac{55}{36} = 1\frac{19}{36}$$

You Try It 6
$$2\frac{5}{6} \div 8\frac{1}{2} = \frac{17}{6} \div \frac{17}{2}$$
$$= \frac{17}{6} \times \frac{2}{17} = \frac{17 \cdot 2}{6 \cdot 17}$$
$$= \frac{\overset{1}{\cancel{17}} \cdot \overset{1}{\cancel{2}}}{2 \cdot 3 \cdot \underset{1}{\cancel{17}}} = \frac{1}{3}$$

You Try It 7
$$6\frac{2}{5} \div 4 = \frac{32}{5} \div \frac{4}{1}$$
$$= \frac{32}{5} \times \frac{1}{4} = \frac{32 \cdot 1}{5 \cdot 4}$$
$$= \frac{2 \cdot 2 \cdot 2 \cdot \overset{1}{\cancel{2}} \cdot \overset{1}{\cancel{2}}}{5 \cdot \underset{1}{\cancel{2}} \cdot \underset{1}{\cancel{2}}} = \frac{8}{5} = 1\frac{3}{5}$$

You Try It 8

Strategy To find the number of products, divide the number of minutes in 1 hour (60) by the time to assemble one product $\left(7\frac{1}{2}\right)$.

Solution
$$60 \div 7\frac{1}{2} = \frac{60}{1} \div \frac{15}{2} = \frac{60}{1} \cdot \frac{2}{15}$$
$$= \frac{60 \cdot 2}{1 \cdot 15} = 8$$

The factory worker can assemble eight products in 1 hour.

Copyright © Houghton Mifflin Company. All rights reserved.

You Try It 9

Strategy To find the length of the remaining piece:
- Divide the total length of the board (16) by the length of each shelf $\left(3\frac{1}{3}\right)$. This will give you the number of shelves cut, with a certain fraction of a shelf left over.
- Multiply the fractional part of the result in step 1 by the length of one shelf to determine the length of the remaining piece.

Solution
$$16 \div 3\frac{1}{3} = 16 \div \frac{10}{3}$$
$$= \frac{16}{1} \times \frac{3}{10} = \frac{16 \cdot 3}{1 \cdot 10}$$
$$= \frac{\overset{1}{2} \cdot 2 \cdot 2 \cdot 2 \cdot 3}{\underset{1}{2} \cdot 5} = \frac{24}{5}$$
$$= 4\frac{4}{5}$$

There are 4 pieces $3\frac{1}{3}$ feet long. There is 1 piece that is $\frac{4}{5}$ of $3\frac{1}{3}$ feet long.

$$\frac{4}{5} \times 3\frac{1}{3} = \frac{4}{5} \times \frac{10}{3}$$
$$= \frac{4 \cdot 10}{5 \cdot 3} = \frac{2 \cdot 2 \cdot 2 \cdot \overset{1}{5}}{\underset{1}{5} \cdot 3}$$
$$= \frac{8}{3} = 2\frac{2}{3}$$

The length of the piece remaining is $2\frac{2}{3}$ feet.

SECTION 2.8

You Try It 1 $\dfrac{9}{14} = \dfrac{27}{42}$ $\dfrac{13}{21} = \dfrac{26}{42}$ $\dfrac{9}{14} > \dfrac{13}{21}$

You Try It 2 $\left(\dfrac{7}{11}\right)^2 \cdot \left(\dfrac{2}{7}\right) = \left(\dfrac{7}{11} \cdot \dfrac{7}{11}\right) \cdot \left(\dfrac{2}{7}\right)$
$$= \frac{\overset{1}{7} \cdot 7 \cdot 2}{11 \cdot 11 \cdot \underset{1}{7}} = \frac{14}{121}$$

You Try It 3 $\left(\dfrac{1}{13}\right)^2 \cdot \left(\dfrac{1}{4} + \dfrac{1}{6}\right) \div \dfrac{5}{13} =$
$$\left(\frac{1}{13}\right)^2 \cdot \left(\frac{5}{12}\right) \div \frac{5}{13} =$$
$$\left(\frac{1}{169}\right) \cdot \left(\frac{5}{12}\right) \div \frac{5}{13} =$$
$$\left(\frac{1 \cdot 5}{13 \cdot 13 \cdot 12}\right) \div \frac{5}{13} =$$
$$\left(\frac{1 \cdot 5}{13 \cdot 13 \cdot 12}\right) \times \frac{13}{5} =$$
$$\frac{1 \cdot \overset{1}{5} \cdot \overset{1}{13}}{\underset{1}{13} \cdot 13 \cdot 12 \cdot \underset{1}{5}} = \frac{1}{156}$$

Solutions to Chapter 3 "You Try It"

SECTION 3.1

You Try It 1 The digit 4 is in the thousandths place.

You Try It 2 $\dfrac{501}{1000} = 0.501$
(five hundred one thousandths)

You Try It 3 $0.67 = \dfrac{67}{100}$ (sixty-seven hundredths)

You Try It 4 fifty-five and six thousand eighty-three ten-thousandths

You Try It 5 806.00491   • 1 is in the hundred-thousandths place.

You Try It 6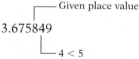
3.675849 rounded to the nearest ten-thousandth is 3.6758.

You Try It 7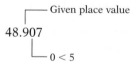
48.907 rounded to the nearest tenth is 48.9.

Copyright © Houghton Mifflin Company. All rights reserved.

You Try It 8

┌─────── Given place value

31.8652

└─────── 8 > 5

31.8652 rounded to the nearest whole number is 32.

You Try It 9   2.65 rounded to the nearest whole number is 3.

To the nearest inch, the average annual precipitation in Yuma is 3 inches.

SECTION 3.2

You Try It 1

```
  1 2
  4.62
 27.9
+ 0.62054
─────────
 33.14054
```
• **Place the decimal points on a vertical line.**

You Try It 2

```
     1
  6.05
 12.
+ 0.374
───────
 18.424
```

You Try It 3

Strategy   To determine the number, add the numbers of hearing-impaired Americans of ages 45 to 54, 55 to 64, and 65 to 74, and 75 and over.

Solution
```
  4.48
  4.31
  5.41
 +3.80
──────
 18.00
```

18 million Americans ages 45 and older are hearing-impaired.

You Try It 4

Strategy   To find the total income, add the four commissions (985.80, 791.46, 829.75, and 635.42) to the salary (875).

Solution $875 + 985.80 + 791.46 + 829.75$
$+ 635.42 = 4117.43$

Anita's total income was $4117.43.

SECTION 3.3

You Try It 1

```
    11 9
   6 1 10 13
   7 2.0 3 9
 −    8.4 7
 ──────────
   6 3.5 6 9
```
Check:
```
  1 1 1
   8.47
 + 63.569
 ────────
  72.039
```

You Try It 2

```
   14 9
  2 4 10 10
  3 5.0 0
 −   9.6 7
 ─────────
  2 5.3 3
```
Check:
```
   1 1 1
   9.67
 + 25.33
 ───────
  35.00
```

You Try It 3

```
     16
  2 6 9 9 10
  3.7 0 0 0
 − 1.9 7 1 5
 ───────────
  1.7 2 8 5
```
Check:
```
  1 1 11
  1.9715
 + 1.7285
 ────────
  3.7000
```

You Try It 4

Strategy   To find the amount of change, subtract the amount paid (6.85) from 10.00.

Solution
```
  10.00
 − 6.85
 ──────
  3.15
```

Your change was $3.15.

You Try It 5

Strategy   To find the new balance:
• Add to find the total of the three checks (1025.60 + 79.85 + 162.47).
• Subtract the total from the previous balance (2472.69).

Solution
```
  1025.60              2472.69
    79.85            − 1267.92
 + 162.47            ─────────
 ────────            1204.77
  1267.92
```

The new balance is $1204.77.

SECTION 3.4

You Try It 1

```
    870
 ×  4.6
 ──────
  522 0
 3480
 ──────
 4002.0
```
• **1 decimal place**

• **1 decimal place**

You Try It 2

```
  0.000086
 ×   0.057
 ─────────
      602
      430
 ─────────
 0.000004902
```
• **6 decimal places**
• **3 decimal places**

• **9 decimal places**

You Try It 3

```
    4.68
 × 6.03
 ──────
   1404
 28 080
 ──────
 28.2204
```
• **2 decimal places**
• **2 decimal places**

• **4 decimal places**

Copyright © Houghton Mifflin Company. All rights reserved.

You Try It 4 $6.9 \times 1000 = 6900$

You Try It 5 $4.0273 \times 10^2 = 402.73$

You Try It 6

Strategy   To find the total bill:
- Find the number of gallons of water used by multiplying the number of gallons used per day (5000) by the number of days (62).
- Find the cost of water by multiplying the cost per 1000 gallons (1.39) by the number of 1000-gallon units used.
- Add the cost of the water to the meter fee (133.70).

Solution

Number of gallons = 5000(62) = 310,000

Cost of water = $\dfrac{310,000}{1000} \times 1.39 = 430.90$

Total cost = 430.90 + 133.70 = 564.60

The total bill is $564.60.

You Try It 7

Strategy   To find the cost of running the freezer for 210 hours, multiply the hourly cost (0.035) by the number of hours the freezer has run (210).

Solution

$$
\begin{array}{r}
0.035 \\
\times\ \ 210 \\
\hline
350 \\
7\ 0 \\
\hline
7.350
\end{array}
$$

The cost of running the freezer for 210 hours is $7.35.

You Try It 8

Strategy   To find the total cost of the stereo:
- Multiply the monthly payment (37.18) by the number of months (18).
- Add that product to the down payment (175.00).

Solution

$$
\begin{array}{r}
37.18 \\
\times\ \ 18 \\
\hline
297\ 44 \\
371\ 8 \\
\hline
669.24
\end{array}
\qquad
\begin{array}{r}
175.00 \\
+\ 669.24 \\
\hline
844.24
\end{array}
$$

The total cost of the stereo is $844.24.

SECTION 3.5

You Try It 1

$$
\begin{array}{r}
2.7 \\
0.052.\overline{)0.140.4} \\
-104 \\
\hline
36\ 4 \\
-36\ 4 \\
\hline
0
\end{array}
$$

- Move the decimal point 3 places to the right in the divisor and the dividend. Write the decimal point in the quotient directly over the decimal point in the dividend.

You Try It 2

$$
\begin{array}{r}
0.4873 \approx 0.487 \\
76\overline{)37.0420} \\
-30\ 4 \\
\hline
6\ 64 \\
-6\ 08 \\
\hline
562 \\
-532 \\
\hline
300 \\
-228
\end{array}
$$

- Write the decimal point in the quotient directly over the decimal point in the dividend.

You Try It 3

$$
\begin{array}{r}
72.73 \approx 72.7 \\
5.09.\overline{)370.20.00} \\
-356\ 3 \\
\hline
13\ 90 \\
10\ 18 \\
\hline
3\ 720 \\
-3\ 563 \\
\hline
1570 \\
-1527
\end{array}
$$

You Try It 4 $309.21 \div 10,000 = 0.030921$

You Try It 5 $42.93 \div 10^4 = 0.004293$

You Try It 6

Strategy   To find how many times greater the average hourly earnings were, divide the 1990 average hourly earnings (10.02) by the 1970 average hourly earnings (3.23).

Solution $10.02 \div 3.23 \approx 3.1$

The average hourly earnings in 1990 were about 3.1 times greater than in 1970.

You Try It 7

Strategy   To find the average number of people watching TV:
- Add the number of people watching each day of the week.
- Divide the total number of people watching by 7.

Copyright © Houghton Mifflin Company. All rights reserved.

Solution $91.9 + 89.8 + 90.6 + 93.9 + 78.0$
$+ 77.1 + 87.7 = 609$

$$\frac{609}{7} = 87$$

An average of 87 million people watch television per day.

SECTION 3.6

You Try It 1

$$\begin{array}{r} 0.56 \approx 0.6 \\ 16\overline{)9.00} \end{array}$$

You Try It 2 $4\dfrac{1}{6} = \dfrac{25}{6}$

$$\begin{array}{r} 4.166 \approx 4.17 \\ 6\overline{)25.000} \end{array}$$

You Try It 3 $0.56 = \dfrac{56}{100} = \dfrac{14}{25}$

$5.35 = 5\dfrac{35}{100} = 5\dfrac{7}{20}$

You Try It 4 $0.12\dfrac{7}{8} = \dfrac{12\dfrac{7}{8}}{100} = 12\dfrac{7}{8} \div 100$

$= \dfrac{103}{8} \times \dfrac{1}{100} = \dfrac{103}{800}$

You Try It 5 $\dfrac{5}{8} = 0.625$ • Convert the fraction $\dfrac{5}{8}$ to a decimal.

$0.630 > 0.625$ • Compare the two decimals.

$0.63 > \dfrac{5}{8}$ • Convert 0.625 back to a fraction.

Solutions to Chapter 4 "You Try It"

SECTION 4.1

You Try It 1 $\dfrac{20 \text{ pounds}}{24 \text{ pounds}} = \dfrac{20}{24} = \dfrac{5}{6}$

20 pounds : 24 pounds = 20 : 24 = 5 : 6

20 pounds to 24 pounds = 20 to 24
$= 5 \text{ to } 6$

You Try It 2 $\dfrac{64 \text{ miles}}{8 \text{ miles}} = \dfrac{64}{8} = \dfrac{8}{1}$

64 miles : 8 miles = 64 : 8 = 8 : 1

64 miles to 8 miles = 64 to 8 = 8 to 1

You Try It 3

Strategy   To find the ratio, write the ratio of board feet of cedar (12,000) to board feet of ash (18,000) in simplest form.

Solution $\dfrac{12,000}{18,000} = \dfrac{2}{3}$

The ratio is $\dfrac{2}{3}$.

You Try It 4

Strategy   To find the ratio, write the ratio of the amount spent on radio advertising (45,000) to the amount spent on radio and television advertising (45,000 + 60,000) in simplest form.

Solution $\dfrac{\$45,000}{\$45,000 + \$60,000} = \dfrac{45,000}{105,000} = \dfrac{3}{7}$

The ratio is $\dfrac{3}{7}$.

SECTION 4.2

You Try It 1 $\dfrac{15 \text{ pounds}}{12 \text{ trees}} = \dfrac{5 \text{ pounds}}{4 \text{ trees}}$

You Try It 2 $\dfrac{260 \text{ miles}}{8 \text{ hours}}$

$$\begin{array}{r} 32.5 \\ 8\overline{)260.0} \end{array}$$

32.5 miles/hour

You Try It 3

Strategy   To find Erik's profit per ounce:
• Find the total profit by subtracting the cost ($1625) from the selling price ($1720).
• Divide the total profit by the number of ounces (5).

Solution $1720 - 1625 = 95$

$95 \div 5 = 19$

The profit was $19/ounce.

SECTION 4.3

You Try It 1 $\dfrac{6}{10} \diagdown\!\!\!\diagup \dfrac{9}{15}$ $\begin{array}{l} 10 \times 9 = 90 \\ 6 \times 15 = 90 \end{array}$

The cross products are equal. The proportion is true.

You Try It 2 $\dfrac{32}{6} \diagdown\!\!\!\diagup \dfrac{90}{8}$ $\begin{array}{l} 6 \times 90 = 540 \\ 32 \times 8 = 256 \end{array}$

The cross products are not equal. The proportion is not true.

Copyright © Houghton Mifflin Company. All rights reserved.

You Try It 3 $\dfrac{n}{14} = \dfrac{3}{7}$ • Find the cross products. Then solve for n.

$$n \times 7 = 14 \times 3$$
$$n \times 7 = 42$$
$$n = 42 \div 7$$
$$n = 6$$

Check: $\dfrac{6}{14} \underset{}{\overset{}{\bowtie}} \dfrac{3}{7}$ ➞ $14 \times 3 = 42$
  ➞ $6 \times 7 = 42$

You Try It 4 $\dfrac{5}{7} = \dfrac{n}{20}$ • Find the cross products. Then solve for n.

$$5 \times 20 = 7 \times n$$
$$100 = 7 \times n$$
$$100 \div 7 = n$$
$$14.3 \approx n$$

You Try It 5 $\dfrac{15}{20} = \dfrac{12}{n}$ • Find the cross products. Then solve for n.

$$15 \times n = 20 \times 12$$
$$15 \times n = 240$$
$$n = 240 \div 15$$
$$n = 16$$

Check: $\dfrac{15}{20} \underset{}{\overset{}{\bowtie}} \dfrac{12}{16}$ ➞ $20 \times 12 = 240$
  ➞ $15 \times 16 = 240$

You Try It 6 $\dfrac{12}{n} = \dfrac{7}{4}$

$$12 \times 4 = n \times 7$$
$$48 = n \times 7$$
$$48 \div 7 = n$$
$$6.86 \approx n$$

You Try It 7 $\dfrac{n}{12} = \dfrac{4}{1}$

$$n \times 1 = 12 \times 4$$
$$n \times 1 = 48$$
$$n = 48 \div 1$$
$$n = 48$$

Check: $\dfrac{48}{12} \underset{}{\overset{}{\bowtie}} \dfrac{4}{1}$ ➞ $12 \times 4 = 48$
  ➞ $48 \times 1 = 48$

You Try It 8

Strategy   To find the number of tablespoons of fertilizer needed, write and solve a proportion using n to represent the number of tablespoons of fertilizer.

Solution
$$\dfrac{3 \text{ tablespoons}}{4 \text{ gallons}} = \dfrac{n \text{ tablespoons}}{10 \text{ gallons}}$$ • The unit "tablespoons" is in the numerator. The unit "gallons" is in the denominator.

$$3 \times 10 = 4 \times n$$
$$30 = 4 \times n$$
$$30 \div 4 = n$$
$$7.5 = n$$

For 10 gallons of water, 7.5 tablespoons of fertilizer are required.

You Try It 9

Strategy   To find the number of jars that can be packed in 15 boxes, write and solve a proportion using n to represent the number of jars.

Solution $\dfrac{24 \text{ jars}}{6 \text{ boxes}} = \dfrac{n \text{ jars}}{15 \text{ boxes}}$

$$24 \times 15 = 6 \times n$$
$$360 = 6 \times n$$
$$360 \div 6 = n$$
$$60 = n$$

60 jars can be packed in 15 boxes.

Solutions to Chapter 5 "You Try It"

SECTION 5.1

You Try It 1   **a.** $125\% = 125 \times \dfrac{1}{100} = \dfrac{125}{100} = 1\dfrac{1}{4}$

b. $125\% = 125 \times 0.01 = 1.25$

You Try It 2 $33\dfrac{1}{3}\% = 33\dfrac{1}{3} \times \dfrac{1}{100}$

$$= \dfrac{100}{3} \times \dfrac{1}{100}$$
$$= \dfrac{100}{300} = \dfrac{1}{3}$$

Copyright © Houghton Mifflin Company. All rights reserved.

You Try It 3 $0.25\% = 0.25 \times 0.01 = 0.0025$

You Try It 4 $0.048 = 0.048 \times 100\% = 4.8\%$

You Try It 5 $3.67 = 3.67 \times 100\% = 367\%$

You Try It 6 $0.62\frac{1}{2} = 0.62\frac{1}{2} \times 100\%$

$$= 62\frac{1}{2}\%$$

You Try It 7 $\frac{5}{6} = \frac{5}{6} \times 100\% = \frac{500}{6}\% = 83\frac{1}{3}\%$

You Try It 8 $1\frac{4}{9} = \frac{13}{9} = \frac{13}{9} \times 100\%$

$$= \frac{1300}{9}\% \approx 144.4\%$$

SECTION 5.2

You Try It 1   Percent $\times$ base = amount
$$0.063 \times 150 = n$$
$$9.45 = n$$

You Try It 2   Percent $\times$ base = amount
$$\frac{1}{6} \times 66 = n \qquad \bullet\; 16\frac{2}{3}\% = \frac{1}{6}$$
$$11 = n$$

You Try It 3

Strategy   To determine the amount that came from corporations, write and solve the basic percent equation using n to represent the amount. The percent is 4%. The base is $212 billion.

Solution   Percent $\times$ base = amount
$$4\% \times 212 = n$$
$$0.04 \times 212 = n$$
$$8.48 = n$$

Corporations gave $8.48 billion to charities.

You Try It 4

Strategy   To find the new hourly wage:
- Find the amount of the raise. Write and solve the basic percent equation using n to represent the amount of the raise (amount). The percent is 8%. The base is $33.50.
- Add the amount of the raise to the old wage (33.50).

Solution $8\% \times 33.50 = n$ 33.50
$0.08 \times 33.50 = n$ + 2.68
$2.68 = n$ 36.18

The new hourly wage is $36.18.

SECTION 5.3

You Try It 1   Percent $\times$ base = amount
$$n \times 32 = 16$$
$$n = 16 \div 32$$
$$n = 0.50$$
$$n = 50\%$$

You Try It 2   Percent $\times$ base = amount
$$n \times 15 = 48$$
$$n = 48 \div 15$$
$$n = 3.2$$
$$n = 320\%$$

You Try It 3   Percent $\times$ base = amount
$$n \times 45 = 30$$
$$n = 30 \div 45$$
$$n = \frac{2}{3} = 66\frac{2}{3}\%$$

You Try It 4

Strategy   To find what percent of the income the income tax is, write and solve the basic percent equation using n to represent the percent. The base is $33,500 and the amount is $5025.

Solution $n \times 33,500 = 5025$
$$n = 5025 \div 33,500$$
$$n = 0.15 = 15\%$$

The income tax is 15% of the income.

You Try It 5

Strategy   To find the percent who were women:
- Subtract to find the number of enlisted personnel who were women ($518,921 - 512,370$).
- Write and solve the basic percent equation using n to represent the percent. The base is 518,921, and the amount is the number of enlisted personnel who were women.

Solution $518,921 - 512,370 = 6551$

$$n \times 518,921 = 6551$$
$$n = 6551 \div 518,921$$
$$n \approx 0.013$$

In 1950, 1.3% of the enlisted personnel in the U.S. Army were women.

Copyright © Houghton Mifflin Company. All rights reserved.

SECTION 5.4

You Try It 1   Percent × base = amount
$$0.86 \times n = 215$$
$$n = 215 \div 0.86$$
$$n = 250$$

You Try It 2   Percent × base = amount
$$0.025 \times n = 15$$
$$n = 15 \div 0.025$$
$$n = 600$$

You Try It 3   Percent × base = amount
$$\frac{1}{6} \times n = 5 \qquad \bullet \ 16\frac{2}{3}\% = \frac{1}{6}$$
$$n = 5 \div \frac{1}{6}$$
$$n = 30$$

You Try It 4

Strategy   To find the original value of the car, write and solve the basic percent equation using n to represent the original value (base). The percent is 42% and the amount is $10,458.

Solution $42\% \times n = 10,458$
$0.42 \times n = 10,458$
$\qquad\qquad n = 10,458 \div 0.42$
$\qquad\qquad n = 24,900$

The original value of the car was $24,900.

You Try It 5

Strategy   To find the difference between the original price and the sale price:
• Find the original price. Write and solve the basic percent equation using n to represent the original price (base). The percent is 80% and the amount is $89.60.
• Subtract the sale price (89.60) from the original price.

Solution $80\% \times n = 89.60$
$0.80 \times n = 89.60$
$\qquad\qquad n = 89.60 \div 0.80$
$\qquad\qquad n = 112.00 \qquad$ (original price)

$112.00 - 89.60 = 22.40$

The difference between the original price and the sale price is $22.40.

SECTION 5.5

You Try It 1 $$\frac{26}{100} = \frac{22}{n}$$
$$26 \times n = 100 \times 22$$
$$26 \times n = 2200$$
$$n = 2200 \div 26$$
$$n \approx 84.62$$

You Try It 2 $$\frac{16}{100} = \frac{n}{132}$$
$$16 \times 132 = 100 \times n$$
$$2112 = 100 \times n$$
$$2112 \div 100 = n$$
$$21.12 = n$$

You Try It 3

Strategy   To find the number of days it snowed, write and solve a proportion using n to represent the number of days it snowed (amount). The percent is 64% and the base is 150.

Solution $$\frac{64}{100} = \frac{n}{150}$$
$$64 \times 150 = 100 \times n$$
$$9600 = 100 \times n$$
$$9600 \div 100 = n$$
$$96 = n$$

It snowed 96 days.

You Try It 4

Strategy   To find the percent of pens that were not defective:
• Subtract to find the number of pens that were not defective $(200 - 5)$.
• Write and solve a proportion using n to represent the percent of pens that were not defective. The base is 200, and the amount is the number of pens not defective.

Solution $200 - 5 = 195$ (number of pens not defective)

$$\frac{n}{100} = \frac{195}{200}$$
$$n \times 200 = 100 \times 195$$
$$n \times 200 = 19,500$$
$$n = 19,500 \div 200$$
$$n = 97.5$$

97.5% of the pens were not defective.

Copyright © Houghton Mifflin Company. All rights reserved.

Solutions to Chapter 6 "You Try It"

SECTION 6.1

You Try It 1

Strategy   To find the unit cost, divide the total cost by the number of units.

Solution   **a.** $7.67 \div 8 = 0.95875$
$.959 per battery
b. $2.29 \div 15 \approx 0.153$
$.153 per ounce

You Try It 2

Strategy   To find the more economical purchase, compare the unit costs.

Solution $5.70 \div 6 = 0.95$
$3.96 \div 4 = 0.99$
$.95 < $.99

The more economical purchase is 6 cans for $5.70.

You Try It 3

Strategy   To find the total cost, multiply the unit cost (9.96) by the number of units (7).

Solution

Unit cost	$\times$	number of units	$=$	total cost
9.96	$\times$	7	$=$	69.72

The total cost is $69.72.

SECTION 6.2

You Try It 1

Strategy   To find the percent increase:
• Find the amount of the increase.
• Solve the basic percent equation for *percent*.

Solution

New value	$-$	original value	$=$	amount of increase
1.83	$-$	1.46	$=$	0.37

Percent $\times$ base = amount
$n \quad \times 1.46 = \quad 0.37$
$n = 0.37 \div 1.46$
$n \approx 0.25 = 25\%$

The percent increase was 25%.

You Try It 2

Strategy   To find the new hourly wage:
• Solve the basic percent equation for *amount*.
• Add the amount of the increase to the original wage.

Solution   Percent $\times$ base $=$ amount
$0.14 \quad \times 12.50 = \quad n$
$1.75 = n$

$12.50 + 1.75 = 14.25$

The new hourly wage is $14.25.

You Try It 3

Strategy   To find the markup, solve the basic percent equation for *amount*.

Solution   Percent $\times$ base $=$ amount

Markup rate	$\times$	cost	$=$	markup
0.20	$\times$	8	$=$	n
			1.60	$= n$

The markup is $1.60.

You Try It 4

Strategy   To find the selling price:
• Find the markup by solving the basic percent equation for *amount*.
• Add the markup to the cost.

Solution   Percent $\times$ base $=$ amount

Markup rate	$\times$	cost	$=$	markup
0.55	$\times$	72	$=$	n
			39.60	$= n$

Cost	$+$	markup	$=$	selling price
72	$+$	39.60	$=$	111.60

The selling price is $111.60.

You Try It 5

Strategy   To find the percent decrease:
• Find the amount of the decrease.
• Solve the basic percent equation for *percent*.

Solution

Original value	$-$	new value	$=$	amount of decrease
261,000	$-$	215,000	$=$	46,000

Percent $\times$ base $=$ amount
$n \quad \times 261,000 = 46,000$
$n = 46,000 \div 261,000$
$n \approx 0.176$

The percent decrease is 17.6%.

Copyright © Houghton Mifflin Company. All rights reserved.

You Try It 6

Strategy To find the visibility:
- Find the amount of decrease by solving the basic percent equation for *amount*.
- Subtract the amount of decrease from the original visibility.

Solution Percent × base = amount
$$0.40 \times 5 = n$$
$$2 = n$$

$$5 - 2 = 3$$

The visibility was 3 miles.

You Try It 7

Strategy To find the discount rate:
- Find the discount.
- Solve the basic percent equation for *percent*.

Solution

Regular price	−	sale price	=	discount
12.50	−	10.99	=	1.51

Percent	×	base	=	amount

Discount rate	×	regular price	=	discount
n	×	12.50	=	1.51

$$n = 1.51 \div 12.50$$
$$n = 0.1208$$

The discount rate is 12.1%.

You Try It 8

Strategy To find the sale price:
- Find the discount by solving the basic percent equation for *amount*.
- Subtract to find the sale price.

Solution Percent × base = amount

Discount rate	×	regular price	=	discount
0.15	×	110	=	n

$$16.5 = n$$

Regular price	−	discount	=	sale price
110	−	16.5	=	93.5

The sale price is $93.50.

SECTION 6.3

You Try It 1

Strategy To find the simple interest due, multiply the principal (15,000) times the annual interest rate (8% = 0.08) times the time, in years (18 months = $\frac{18}{12}$ years = 1.5 years).

Solution

Principal	×	annual interest rate	×	time (in years)	=	interest
15,000	×	0.08	×	1.5	=	1800

The interest due is $1800.

You Try It 2

Strategy To find the maturity value:
- Use the simple interest formula to find the simple interest due.
- Find the maturity value by adding the principal and the interest.

Solution

Principal	×	annual interest rate	×	time (in years)	=	interest
3800	×	0.06	×	$\frac{90}{365}$	≈	56.22

Principal	+	interest	=	maturity value
3800	+	56.22	=	3856.22

The maturity value is $3856.22.

You Try It 3

Strategy To find the monthly payment:
- Find the maturity value by adding the principal and the interest.
- Divide the maturity value by the length of the loan in months (12).

Solution Principal + interest = maturity value
$$1900 + 152 = 2052$$

Maturity value ÷ length of the loan = payment
$$2052 \div 12 = 171$$

The monthly payment is $171.

You Try It 4

Strategy To find the finance charge, multiply the principal, or unpaid balance (1250), times the monthly interest rate (1.6%) times the number of months (1).

Copyright © Houghton Mifflin Company. All rights reserved.

Solution

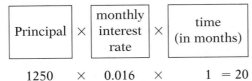

1250 × 0.016 × 1 = 20

The simple interest due is $20.

You Try It 5

Strategy To find the interest earned:
• Find the new principal by multiplying the original principal (1000) by the factor found in the Compound Interest Table (3.29066).
• Subtract the original principal from the new principal.

Solution 1000 × 3.29066 = 3290.66

The new principal is $3290.66.

3290.66 − 1000 = 2290.66

The interest earned is $2290.66.

SECTION 6.4

You Try It 1

Strategy To find the mortgage:
• Find the down payment by solving the basic percent equation for *amount*.
• Subtract the down payment from the purchase price.

Solution

Percent × base = amount

$$\boxed{\text{Percent}} \times \boxed{\begin{array}{c}\text{purchase}\\\text{price}\end{array}} = \boxed{\begin{array}{c}\text{down}\\\text{payment}\end{array}}$$

0.25 × 216,000 = n

54,000 = n

$$\boxed{\begin{array}{c}\text{Purchase}\\\text{price}\end{array}} - \boxed{\begin{array}{c}\text{down}\\\text{payment}\end{array}} = \boxed{\text{mortgage}}$$

216,000 − 54,000 = 162,000

The mortgage is $162,000.

You Try It 2

Strategy To find the loan origination fee, solve the basic percent equation for *amount*.

Solution

Percent × base = amount

$$\boxed{\text{Points}} \times \boxed{\text{mortgage}} = \boxed{\text{fee}}$$

0.045 × 80,000 = n

3600 = n

The loan origination fee was $3600.

You Try It 3

Strategy To find the monthly mortgage payment:
• Subtract the down payment from the purchase price to find the mortgage.
• Multiply the mortgage by the factor found in the Monthly Payment Table.

Solution

$$\boxed{\begin{array}{c}\text{Purchase}\\\text{price}\end{array}} - \boxed{\begin{array}{c}\text{down}\\\text{payment}\end{array}} = \boxed{\text{mortgage}}$$

75,000 − 15,000 = 60,000

60,000 × 0.0089973 = 539.838
             ↑
        From the table

The monthly mortgage payment is $539.84.

You Try It 4

Strategy To find the interest:
• Multiply the mortgage by the factor found in the Monthly Payment Table to find the monthly mortgage payment.
• Subtract the principal from the monthly mortgage payment.

Solution 125,000 × 0.0083920 = 1049
                    ↑   ↑
              From the   Monthly mortgage
                table   payment

$$\boxed{\begin{array}{c}\text{Monthly}\\\text{mortgage}\\\text{payment}\end{array}} - \boxed{\text{principal}} = \boxed{\text{interest}}$$

1049 − 492.65 = 556.35

The interest on the mortgage is $556.35.

You Try It 5

Strategy To find the monthly payment:
• Divide the annual property tax by 12 to find the monthly property tax.
• Add the monthly property tax to the monthly mortgage payment.

(Continued)

Copyright © Houghton Mifflin Company. All rights reserved.

(Continued)

Solution $744 \div 12 = 62$

The monthly property tax is $62.

$415.20 + 62 = 477.20$

The total monthly payment is $477.20.

SECTION 6.5

You Try It 1

Strategy   To find the amount financed:
- Find the down payment by solving the basic percent equation for *amount*.
- Subtract the down payment from the purchase price.

Solution

Percent $\times$ base $=$ amount

Percent	$\times$	purchase price	$=$	down payment

$$0.20 \quad \times \quad 19{,}200 \quad = \quad n$$
$$3840 = n$$

The down payment is $3840.

$19{,}200 - 3840 = 15{,}360$

The amount financed is $15,360.

You Try It 2

Strategy   To find the license fee, solve the basic percent equation for *amount*.

Solution

Percent $\times$ base $=$ amount

Percent	$\times$	purchase price	$=$	sales tax

$$0.015 \quad \times \quad 17{,}350 \quad = \quad n$$
$$260.25 = n$$

The license fee is $260.25.

You Try It 3

Strategy   To find the cost, multiply the cost per mile (0.31) by the number of miles driven (23,000).

Solution $23{,}000 \times 0.31 = 7130$

The cost is $7130.

You Try It 4

Strategy   To find the cost per mile for car insurance, divide the cost for insurance (360) by the number of miles driven (15,000).

Solution $360 \div 15{,}000 = 0.024$

The cost per mile for insurance is $.024.

You Try It 5

Strategy   To find the monthly payment:
- Subtract the down payment from the purchase price to find the amount financed.
- Multiply the amount financed by the factor found in the Monthly Payment Table.

Solution $25{,}900 - 6475 = 19{,}425$

$19{,}425 \times 0.0244129 \approx 474.22$

The monthly payment is $474.22.

SECTION 6.6

You Try It 1

Strategy   To find the worker's earnings:
- Find the worker's overtime wage by multiplying the hourly wage by 2.
- Multiply the number of overtime hours worked by the overtime wage.

Solution $18.50 \times 2 = 37$

The hourly wage for overtime is $37.

$37 \times 8 = 296$

The construction worker earns $296.

You Try It 2

Strategy   To find the salary per month, divide the annual salary by the number of months in a year (12).

Solution $48{,}228 \div 12 = 4019$

The contractor's monthly salary is $4019.

You Try It 3

Strategy   To find the total earnings:
- Find the sales over $50,000.
- Multiply the commission rate by sales over $50,000.
- Add the commission to the annual salary.

Solution $175{,}000 - 50{,}000 = 125{,}000$

Sales over $50,000 totaled $125,000.

$125{,}000 \times 0.095 = 11{,}875$

Copyright © Houghton Mifflin Company. All rights reserved.

Earnings from commissions totaled $11,875.

$27,000 + 11,875 = 38,875$

The insurance agent earned $38,875.

SECTION 6.7

You Try It 1

Strategy   To find the current balance:
- Subtract the amount of the check from the old balance.
- Add the amount of each deposit.

Solution

$$
\begin{array}{ll}
302.46 & \\
-\ \ 20.59 & \text{check} \\
\hline
281.87 & \\
176.86 & \text{first deposit} \\
+\ \ 94.73 & \text{second deposit} \\
\hline
553.46 &
\end{array}
$$

The current checking account balance is $553.46.

You Try It 2

Current checkbook balance: 623.41
Check: 237   + 78.73

702.14

Interest: +   2.11

704.25

Deposit: −523.84

180.41

Closing bank balance from bank statement: $180.41

Checkbook balance: $180.41

The bank statement and the checkbook balance.

Copyright © Houghton Mifflin Company. All rights reserved.

Answers to Chapter 1 Selected Exercises

PREP TEST

1. 8   **2.** 1 2 3 4 5 6 7 8 9 10   **3.** a and D; b and E; c and A; d and B; e and F; f and C

SECTION 1.1

1. [number line with point at 3, marks 0–12]   **3.** [number line with point at 9, marks 0–12]   **5.** $37 < 49$ **7.** $101 > 87$
9. $245 > 158$ **11.** $0 < 45$ **13.** $815 < 928$ **15.** Millions   **17.** Hundred-thousands   **19.** Three thousand seven hundred ninety   **21.** Fifty-eight thousand four hundred seventy-three   **23.** Four hundred ninety-eight thousand five hundred twelve   **25.** Six million eight hundred forty-two thousand seven hundred fifteen   **27.** 357   **29.** 63,780
31. 7,024,709   **33.** $6000 + 200 + 90 + 5$ **35.** $400,000 + 50,000 + 3000 + 900 + 20 + 1$
37. $300,000 + 1000 + 800 + 9$ **39.** $3,000,000 + 600 + 40 + 2$ **41.** 850   **43.** 4000   **45.** 53,000   **47.** 630,000
49. 250,000   **51.** 72,000,000   **53.** 999; 10,000

SECTION 1.2

1. 28   **3.** 125   **5.** 102   **7.** 154   **9.** 1489   **11.** 828   **13.** 1584   **15.** 1219   **17.** 102,317   **19.** 79,326
21. 1804   **23.** 1579   **25.** 19,740   **27.** 7420   **29.** 120,570   **31.** 207,453   **33.** 24,218   **35.** 11,974
37. 9323   **39.** 77,139   **41.** 14,383   **43.** 9473   **45.** 33,247   **47.** 5058   **49.** 1992   **51.** 68,263
53. Cal.: 17,754   **55.** Cal.: 2872   **57.** Cal.: 101,712   **59.** Cal.: 158,763   **61.** Cal.: 261,595   **63.** Cal.: 946,718
Est.: 17,700   Est.: 2900   Est.: 101,000   Est.: 158,000   Est.: 260,000   Est.: 940,000
65. Cal.: 32,691,621   **67.** Cal.: 34,420,922   **69.** There were 118,295 multiple births during the year.
Est.: 33,000,000   Est.: 34,000,000

71. The estimated income from the four *Star Wars* movies was $1,500,000,000.   **73. a.** The income from the two movies with the lowest box-office returns is $599,300,000.   **b.** Yes, this income exceeds the income from the 1977 *Star Wars* production.   **75. a.** During the three days, 1285 miles will be driven.   **b.** At the end of the trip, the odometer will read 69,977 miles.   **77.** Americans ages 16 to 34 have invested $1796 in these three investments.   **79.** The sum of the average amounts invested in home equity and retirement for all Americans is greater than the sum of all categories for Americans aged 16 and 34.   **81.** 11 different sums   **83.** No. $0 + 0 = 0$ **85.** 10 numbers

SECTION 1.3

1. 4   **3.** 4   **5.** 10   **7.** 4   **9.** 9   **11.** 22   **13.** 60   **15.** 66   **17.** 31   **19.** 901   **21.** 791   **23.** 1125
25. 3131   **27.** 47   **29.** 925   **31.** 4561   **33.** 3205   **35.** 1222   **37.** 3021   **39.** 3022   **41.** 3040
43. 212   **45.** 60,245   **47.** 65   **49.** 17   **51.** 23   **53.** 456   **55.** 57   **57.** 375   **59.** 3139   **61.** 3621
63. 738   **65.** 3545   **67.** 749   **69.** 5343   **71.** 66,463   **73.** 16,590   **75.** 52,404   **77.** 38,777   **79.** 4638
81. 3612   **83.** 2913   **85.** 2583   **87.** 5268   **89.** 71,767   **91.** 11,239   **93.** 8482   **95.** 625   **97.** 76,725
99. 23   **101.** 4648   **103.** Cal.: 29,837   **105.** Cal.: 36,668   **107.** Cal.: 101,998
Est.: 30,000   Est.: 40,000   Est.: 100,000
109. There was an increase of 75,621 complaints from 2001 to 2002.   **111.** The amount that remains to be paid is $10,275.   **113.** The Giant erupts 25 feet higher than Old Faithful.   **115.** The expected increase over 10 years is 106,000.   **117.** The amount remaining is $436.   **119. a.** True   **b.** False   **c.** False

SECTION 1.4

1. 6×2 or $6 \cdot 2$ **3.** 4×7 or $4 \cdot 7$ **5.** 12   **7.** 35   **9.** 25   **11.** 0   **13.** 72   **15.** 198   **17.** 335
19. 2492   **21.** 5463   **23.** 4200   **25.** 6327   **27.** 1896   **29.** 5056   **31.** 1685   **33.** 46,963   **35.** 59,976
37. 19,120   **39.** 19,790   **41.** 108   **43.** 3664   **45.** 20,036   **47.** 71,656   **49.** 432   **51.** 1944   **53.** 41,832
55. 43,620   **57.** 335,195   **59.** 594,625   **61.** 321,696   **63.** 342,171   **65.** 279,220   **67.** 191,800
69. 463,712   **71.** 180,621   **73.** 478,800   **75.** 158,422   **77.** 4,696,714   **79.** 5,542,452   **81.** 198,423
83. 18,834   **85.** 260,178   **87.** 315,109,895   **89.** Cal.: 440,076   **91.** Cal.: 6,491,166   **93.** Cal.: 18,728,744
Est.: 450,000   Est.: 6,300,000   Est.: 18,000,000

Copyright © Houghton Mifflin Company. All rights reserved.

95. Cal.: 57,691,192   **97.** The plane used 5190 gallons of fuel on a 6-hour flight.   **99.** The area is 256 square miles.
Est.: 54,000,000

101. Company A offers the lower total price.   **103.** The estimated cost of the electricians' labor is $5100.   **105.** The total cost is $2138.   **107.** There are 12 accidental deaths each hour; 288 deaths each day; and 105,120 deaths each year.

SECTION 1.5

1. 2   **3.** 6   **5.** 7   **7.** 16   **9.** 210   **11.** 44   **13.** 703   **15.** 910   **17.** 21,560   **19.** 3580   **21.** 482

23. 1075   **25.** 52   **27.** 34   **29.** 2 r1   **31.** 5 r2   **33.** 13 r1   **35.** 10 r3   **37.** 90 r2   **39.** 230 r1

41. 204 r3   **43.** 1347 r3   **45.** 1720 r2   **47.** 409 r2   **49.** 6214 r2   **51.** 8708 r2   **53.** 1080 r2   **55.** 4200

57. 19,600   **59.** 1 r38   **61.** 1 r26   **63.** 21 r21   **65.** 30 r22   **67.** 5 r40   **69.** 9 r17   **71.** 200 r21

73. 303 r1   **75.** 67 r13   **77.** 176 r13   **79.** 1086 r7   **81.** 403   **83.** 12 r456   **85.** 4 r160   **87.** 160 r27

89. 1669 r14   **91.** 7950   **93.** Cal.: 2225   **95.** Cal.: 11,016   **97.** Cal.: 26,656   **99.** Cal.: 504   **101.** Cal.: 541
Est.: 2000   Est.: 10,000   Est.: 30,000   Est.: 500   Est.: 500

103. Cal.: 20,621   **105.** The average monthly expense for housing is $976.   **107.** The average monthly claim for theft
Est.: 20,000

is $25,000.   **109.** The average number of hours worked by employees in Britain is 35 hours.   **111.** 380 pennies are in circulation for each person.   **113.** 198,400,000 cases of eggs are produced during the year.   **115.** The average U.S. household will spend $130 on gasoline each month.   **117.** The average starting salary of an accounting major is $13,092 greater than that of a psychology major.   **119.** The difference is $10,848.   **121.** The total pay is $491.   **123.** 212

SECTION 1.6

1. 2^3 **3.** $6^3 \cdot 7^4$ **5.** $2^3 \cdot 3^3$ **7.** $5 \cdot 7^5$ **9.** $3^3 \cdot 6^4$ **11.** $3^3 \cdot 5 \cdot 9^3$ **13.** 8   **15.** 400   **17.** 900   **19.** 972

21. 120   **23.** 360   **25.** 0   **27.** 90,000   **29.** 540   **31.** 4050   **33.** 11,025   **35.** 25,920   **37.** 4,320,000

39. 5   **41.** 10   **43.** 47   **45.** 8   **47.** 5   **49.** 8   **51.** 6   **53.** 53   **55.** 44   **57.** 19   **59.** 67   **61.** 168

63. 27   **65.** 14   **67.** 10   **69.** 9   **71.** 12   **73.** 32   **75.** 39   **77.** 1024

SECTION 1.7

1. 1, 2, 4   **3.** 1, 2, 5, 10   **5.** 1, 7   **7.** 1, 3, 9   **9.** 1, 13   **11.** 1, 2, 3, 6, 9, 18   **13.** 1, 2, 4, 7, 8, 14, 28, 56

15. 1, 3, 5, 9, 15, 45   **17.** 1, 29   **19.** 1, 2, 11, 22   **21.** 1, 2, 4, 13, 26, 52   **23.** 1, 2, 41, 82   **25.** 1, 3, 19, 57

27. 1, 2, 3, 4, 6, 8, 12, 16, 24, 48   **29.** 1, 5, 19, 95   **31.** 1, 2, 3, 6, 9, 18, 27, 54   **33.** 1, 2, 3, 6, 11, 22, 33, 66

35. 1, 2, 4, 5, 8, 10, 16, 20, 40, 80   **37.** 1, 2, 3, 4, 6, 8, 12, 16, 24, 32, 48, 96   **39.** 1, 2, 3, 5, 6, 9, 10, 15, 18, 30, 45, 90

41. $2 \cdot 3$ **43.** Prime   **45.** $2 \cdot 2 \cdot 2 \cdot 3$ **47.** $3 \cdot 3 \cdot 3$ **49.** $2 \cdot 2 \cdot 3 \cdot 3$ **51.** Prime   **53.** $2 \cdot 3 \cdot 3 \cdot 5$

55. $5 \cdot 23$ **57.** $2 \cdot 3 \cdot 3$ **59.** $2 \cdot 2 \cdot 7$ **61.** Prime   **63.** $2 \cdot 31$ **65.** $2 \cdot 11$ **67.** Prime   **69.** $2 \cdot 3 \cdot 11$

71. $2 \cdot 37$ **73.** Prime   **75.** $5 \cdot 11$ **77.** $2 \cdot 2 \cdot 2 \cdot 3 \cdot 5$ **79.** $2 \cdot 2 \cdot 2 \cdot 2 \cdot 2 \cdot 5$ **81.** $2 \cdot 2 \cdot 2 \cdot 3 \cdot 3 \cdot 3$

83. $5 \cdot 5 \cdot 5 \cdot 5$ **85.** 3, 5; 5, 7; 11, 13; 41, 43; 71, 73; and 17, 19 are all the twin primes less than 100.

CHAPTER 1 REVIEW EXERCISES*

1. 600 [1.6A]   **2.** 10,000 + 300 + 20 + 7 [1.1C]   **3.** 1, 2, 3, 6, 9, 18 [1.7A]   **4.** 12,493 [1.2A]

5. 1749 [1.3B]   **6.** 2135 [1.5A]   **7.** 101 > 87 [1.1A]   **8.** $5^2 \cdot 7^5$ [1.6A]   **9.** 619,833 [1.4B]

10. 5409 [1.3B]   **11.** 1081 [1.2A]   **12.** 2 [1.6B]   **13.** 45,700 [1.1D]   **14.** Two hundred seventy-six thousand fifty-seven [1.1B]   **15.** 1306 r59 [1.5C]   **16.** 2,011,044 [1.1B]   **17.** 488 r2 [1.5B]   **18.** 17 [1.6B]

19. 32 [1.6B]   **20.** $2 \cdot 2 \cdot 2 \cdot 3 \cdot 3$ [1.7B]   **21.** 2133 [1.3A]   **22.** 22,761 [1.4B]   **23.** The total pay for last week's work is $768. [1.4C]   **24.** He drove 27 miles per gallon of gasoline. [1.5D]   **25.** Each monthly car payment is $310. [1.5D]   **26.** The total income from commissions is $2567. [1.2B]   **27.** The total amount deposited is $301.

*Note: The numbers in brackets following the answers in the Chapter Review are a reference to the objective that corresponds to that problem. For example, the reference [1.2A] stands for Section 1.2, Objective A. This notation will be used for all Prep Tests, Chapter Reviews, Chapter Tests, and Cumulative Reviews throughout the text.

Copyright © Houghton Mifflin Company. All rights reserved.

The new checking account balance is $817.   [1.2B]   **28.** The total of the car payments is $2952.   [1.4C]
29. More males were involved in college sports in 2001 than in 1972.   [1.1A]   **30.** The difference between the numbers of male and female athletes in 1972 was 140,407 students.   [1.3C]   **31.** The number of female athletes increased by 120,939 students from 1972 to 2001.   [1.3C]   **32.** 159,421 more students were involved in athletics in 2001 than in 1972. [1.3C]

CHAPTER 1 TEST

1. 432   [1.6A]   **2.** Two hundred seven thousand sixty-eight   [1.1B]   **3.** 9333   [1.3B]   **4.** 1, 2, 4, 5, 10, 20   [1.7A]
5. 6,854,144   [1.4B]   **6.** 9   [1.6B]   **7.** $900,000 + 6000 + 300 + 70 + 8$ [1.1C]   **8.** 75,000   [1.1D]
9. 1121 r27   [1.5C]   **10.** $3^3 \cdot 7^2$ [1.6A]   **11.** 54,915   [1.2A]   **12.** $2 \cdot 2 \cdot 3 \cdot 7$ [1.7B]   **13.** 4   [1.6B]
14. 726,104   [1.4A]   **15.** 1,204,006   [1.1B]   **16.** 8710 r2   [1.5B]   **17.** $21 > 19$ [1.1A]   **18.** 703   [1.5A]
19. 96,798   [1.2A]   **20.** 19,922   [1.3B]   **21.** The difference in projected total enrollment between 2012 and 2009 is 154,000 students.   [1.3C]   **22.** The average enrollment in each of the grades 9 through 12 in 2012 is 3,858,500 students. [1.5D]   **23.** 3000 boxes were needed to pack the lemons.   [1.5D]   **24.** The investor receives $2844 over the 12-month period.   [1.4C]   **25. a.** 855 miles were driven during the 3 days.   [1.2B]   **b.** The odometer reading at the end of the 3 days is 48,481 miles.   [1.2B]

Answers to Chapter 2 Selected Exercises

PREP TEST

1. 20   [1.4A]   **2.** 120   [1.4A]   **3.** 9   [1.4A]   **4.** 10   [1.2A]   **5.** 7   [1.3A]   **6.** 2 r3   [1.5C]   **7.** 1, 2, 3, 4, 6, 12   [1.5B]   **8.** 59   [1.6B]   **9.** 7   [1.3A]   **10.** $44 < 48$ [1.1A]

SECTION 2.1

1. 40   **3.** 24   **5.** 30   **7.** 12   **9.** 24   **11.** 60   **13.** 56   **15.** 9   **17.** 32   **19.** 36   **21.** 660
23. 9384   **25.** 24   **27.** 30   **29.** 24   **31.** 576   **33.** 1680   **35.** 1   **37.** 3   **39.** 5   **41.** 25   **43.** 1
45. 4   **47.** 4   **49.** 6   **51.** 4   **53.** 1   **55.** 7   **57.** 5   **59.** 8   **61.** 1   **63.** 25   **65.** 7   **67.** 8
69. Two composite numbers are relatively prime if they do not have any common factor except the factor 1. Examples: 4 and 5, 8 and 9, and 16 and 21.

SECTION 2.2

1. Improper fraction   **3.** Proper fraction   **5.** $\frac{3}{4}$ **7.** $\frac{7}{8}$ **9.** $1\frac{1}{2}$ **11.** $2\frac{5}{8}$ **13.** $3\frac{3}{5}$ **15.** $\frac{5}{4}$ **17.** $\frac{8}{3}$

19. $\frac{28}{8}$ **21.**   **23.**   **25.**   **27.** $2\frac{3}{4}$ **29.** 5   **31.** $1\frac{1}{8}$ **33.** $2\frac{3}{10}$ **35.** 3

37. $1\frac{1}{7}$ **39.** $2\frac{1}{3}$ **41.** 16   **43.** $2\frac{1}{8}$ **45.** $2\frac{2}{5}$ **47.** 1   **49.** 9   **51.** $\frac{7}{3}$ **53.** $\frac{13}{2}$ **55.** $\frac{41}{6}$ **57.** $\frac{37}{4}$

59. $\frac{21}{2}$ **61.** $\frac{73}{9}$ **63.** $\frac{58}{11}$ **65.** $\frac{21}{8}$ **67.** $\frac{13}{8}$ **69.** $\frac{100}{9}$ **71.** $\frac{27}{8}$ **73.** $\frac{85}{13}$

SECTION 2.3

1. $\frac{5}{10}$ **3.** $\frac{9}{48}$ **5.** $\frac{12}{32}$ **7.** $\frac{9}{51}$ **9.** $\frac{12}{16}$ **11.** $\frac{27}{9}$ **13.** $\frac{20}{60}$ **15.** $\frac{44}{60}$ **17.** $\frac{12}{18}$ **19.** $\frac{35}{49}$ **21.** $\frac{10}{18}$

23. $\frac{21}{3}$ **25.** $\frac{35}{45}$ **27.** $\frac{60}{64}$ **29.** $\frac{21}{98}$ **31.** $\frac{30}{48}$ **33.** $\frac{15}{42}$ **35.** $\frac{102}{144}$ **37.** $\frac{153}{408}$ **39.** $\frac{340}{800}$ **41.** $\frac{1}{3}$ **43.** $\frac{1}{2}$

45. $\frac{1}{6}$ **47.** $1\frac{1}{9}$ **49.** 0   **51.** $\frac{9}{22}$ **53.** 3   **55.** $\frac{4}{21}$ **57.** $\frac{12}{35}$ **59.** $\frac{7}{11}$ **61.** $1\frac{1}{3}$ **63.** $\frac{3}{5}$ **65.** $\frac{1}{11}$

Copyright © Houghton Mifflin Company. All rights reserved.

67. 4   **69.** $\frac{1}{3}$ **71.** $\frac{3}{5}$ **73.** $2\frac{1}{4}$ **75.** $\frac{1}{5}$ **77.** $\frac{3}{1}, \frac{6}{2}, \frac{9}{3}, \frac{12}{4}, \frac{15}{5}$ are fractions that are equal to 3.   **79. a.** $\frac{4}{25}$

b. $\frac{4}{25}$

SECTION 2.4

1. $\frac{3}{7}$ **3.** 1   **5.** $1\frac{4}{11}$ **7.** $3\frac{2}{5}$ **9.** $2\frac{4}{5}$ **11.** $2\frac{1}{4}$ **13.** $1\frac{3}{8}$ **15.** $1\frac{7}{15}$ **17.** $\frac{15}{16}$ **19.** $1\frac{4}{11}$ **21.** 1

23. $1\frac{7}{8}$ **25.** $1\frac{1}{6}$ **27.** $\frac{13}{14}$ **29.** $\frac{53}{60}$ **31.** $1\frac{1}{56}$ **33.** $\frac{23}{60}$ **35.** $\frac{56}{57}$ **37.** $1\frac{17}{18}$ **39.** $1\frac{11}{48}$ **41.** $1\frac{9}{20}$

43. $1\frac{109}{180}$ **45.** $1\frac{91}{144}$ **47.** $2\frac{5}{72}$ **49.** $\frac{39}{40}$ **51.** $1\frac{19}{24}$ **53.** $1\frac{65}{72}$ **55.** $3\frac{2}{3}$ **57.** $10\frac{1}{12}$ **59.** $9\frac{2}{7}$ **61.** $6\frac{7}{40}$

63. $9\frac{47}{48}$ **65.** $8\frac{3}{13}$ **67.** $16\frac{29}{120}$ **69.** $24\frac{29}{40}$ **71.** $33\frac{7}{24}$ **73.** $10\frac{5}{36}$ **75.** $10\frac{5}{12}$ **77.** $14\frac{73}{90}$ **79.** $10\frac{13}{48}$

81. $42\frac{13}{54}$ **83.** $8\frac{1}{36}$ **85.** $14\frac{1}{12}$ **87.** $8\frac{61}{72}$ **89.** The length of the shaft is $1\frac{5}{16}$ inches.   **91.** The total thickness

is $1\frac{5}{16}$ inches.   **93. a.** A total of 20 hours was worked.   **b.** Your total salary for the week is $220.   **95.** The total

length of wood needed is $55\frac{3}{4}$ feet.   **97.** Those who changed homes outside the county were $\frac{1}{3}$ of the people who

changed homes.

SECTION 2.5

1. $\frac{2}{17}$ **3.** $\frac{1}{3}$ **5.** $\frac{1}{10}$ **7.** $\frac{5}{13}$ **9.** $\frac{1}{3}$ **11.** $\frac{4}{7}$ **13.** $\frac{1}{4}$ **15.** $\frac{9}{23}$ **17.** $\frac{1}{4}$ **19.** $\frac{1}{2}$ **21.** $\frac{19}{56}$ **23.** $\frac{1}{2}$

25. $\frac{11}{60}$ **27.** $\frac{15}{56}$ **29.** $\frac{37}{51}$ **31.** $\frac{17}{70}$ **33.** $\frac{49}{120}$ **35.** $\frac{8}{45}$ **37.** $\frac{11}{21}$ **39.** $\frac{23}{60}$ **41.** $\frac{1}{18}$ **43.** $5\frac{1}{5}$ **45.** $10\frac{9}{17}$

47. $4\frac{7}{8}$ **49.** $\frac{16}{21}$ **51.** $5\frac{1}{2}$ **53.** $5\frac{4}{7}$ **55.** $7\frac{5}{24}$ **57.** $48\frac{31}{70}$ **59.** $85\frac{2}{9}$ **61.** $9\frac{5}{13}$ **63.** $15\frac{11}{20}$ **65.** $4\frac{23}{24}$

67. $2\frac{11}{15}$ **69.** The missing dimension is $9\frac{1}{2}$ inches.   **71.** The difference between Meyfarth's distance and Coachman's

distance was $9\frac{3}{8}$ inches. The difference between Kostadinova's distance and Meyfarth's distance was $5\frac{1}{4}$ inches.

73. a. The distance from the starting point to the second checkpoint is $7\frac{17}{24}$ miles.   **b.** The distance from the second

checkpoint to the finish line is $4\frac{7}{24}$ miles.   **75. a.** Yes   **b.** The wrestler needs to lose $3\frac{1}{4}$ pounds to reach the desired

weight.   **77.** $2\frac{5}{6}$ **79.**

$\frac{3}{8}$	$\frac{3}{4}$	$\frac{3}{4}$
1	$\frac{5}{8}$	$\frac{1}{4}$
$\frac{1}{2}$	$\frac{1}{2}$	$\frac{7}{8}$

SECTION 2.6

1. $\frac{7}{12}$ **3.** $\frac{7}{48}$ **5.** $\frac{1}{48}$ **7.** $\frac{11}{14}$ **9.** $\frac{1}{7}$ **11.** $\frac{1}{8}$ **13.** 6   **15.** $\frac{5}{12}$ **17.** 6   **19.** $\frac{2}{3}$ **21.** $\frac{3}{16}$ **23.** $\frac{3}{80}$

25. 10   **27.** $\frac{1}{15}$ **29.** $\frac{2}{3}$ **31.** $\frac{7}{26}$ **33.** 4   **35.** $\frac{100}{357}$ **37.** $\frac{5}{24}$ **39.** $\frac{1}{12}$ **41.** $\frac{4}{15}$ **43.** $1\frac{1}{2}$ **45.** 4

47. $\frac{4}{9}$ **49.** $\frac{1}{2}$ **51.** $16\frac{1}{2}$ **53.** 10   **55.** $6\frac{3}{7}$ **57.** $18\frac{1}{3}$ **59.** $1\frac{5}{7}$ **61.** $3\frac{1}{2}$ **63.** $25\frac{5}{8}$ **65.** $1\frac{11}{16}$ **67.** 16

69. 3   **71.** $8\frac{37}{40}$ **73.** $6\frac{19}{28}$ **75.** $7\frac{2}{5}$ **77.** 16   **79.** $32\frac{4}{5}$ **81.** $28\frac{1}{2}$ **83.** 9   **85.** $\frac{5}{8}$ **87.** $3\frac{1}{40}$

89. The salmon costs $11.   **91. a.** No. $\frac{1}{3}$ of 9 feet is approximately 3 ft.   **b.** The length of the board cut off is $3\frac{1}{12}$ feet.

93. The area of the square is $27\frac{9}{16}$ square feet.   **95. a.** The amount budgeted for housing and utilities is $1680.

Copyright © Houghton Mifflin Company. All rights reserved.

b. The amount remaining for other than housing and utilities is $2520.   **97.** The total cost is $363.   **99.** The weight of the $12\frac{7}{12}$-foot steel rod is $54\frac{19}{36}$ pounds.   **101.** $\frac{3}{16}$ of the total portfolio is invested in corporate bonds.   **105.** *A*

SECTION 2.7

1. $\frac{5}{6}$ **3.** 1   **5.** 0   **7.** $\frac{1}{2}$ **9.** $2\frac{73}{256}$ **11.** $1\frac{1}{9}$ **13.** $\frac{1}{6}$ **15.** $\frac{7}{10}$ **17.** 2   **19.** 2   **21.** $\frac{1}{6}$ **23.** 6

25. $\frac{1}{15}$ **27.** 2   **29.** $2\frac{1}{2}$ **31.** 3   **33.** $1\frac{1}{6}$ **35.** $3\frac{1}{3}$ **37.** 6   **39.** $\frac{1}{2}$ **41.** $\frac{1}{30}$ **43.** $1\frac{4}{5}$ **45.** 13

47. $\frac{25}{288}$ **49.** 3   **51.** $\frac{1}{5}$ **53.** $\frac{11}{28}$ **55.** $\frac{5}{86}$ **57.** 120   **59.** $\frac{11}{40}$ **61.** $\frac{33}{40}$ **63.** $4\frac{4}{9}$ **65.** $\frac{13}{32}$ **67.** $10\frac{2}{3}$

69. $9\frac{39}{40}$ **71.** $\frac{12}{53}$ **73.** $4\frac{62}{191}$ **75.** 68   **77.** $8\frac{2}{7}$ **79.** $3\frac{13}{49}$ **81.** 4   **83.** $1\frac{3}{5}$ **85.** $\frac{9}{34}$ **87.** There are 12 servings in 16 ounces of cereal.   **89.** Each acre costs $24,000.   **91.** The nut will make 12 turns in moving $1\frac{7}{8}$ inches.

93. a. The total weight of the fat and bone is $1\frac{5}{12}$ pounds.   **b.** The chef can cut 28 servings from the roast.

95. The actual length of wall a is $12\frac{1}{2}$ feet. The actual length of wall b is 18 feet. The actual length of wall c is $15\frac{3}{4}$ feet.

97. $\frac{17}{50}$ of the money borrowed is spent on home improvement, cars and tuition.   **99.** The capacity of the music center is 1800 people.   **101.** $\frac{1}{6}$ of the puzzle is left to complete.   **103.** Your total earnings for last week's work are $111.

105. The average teenage boy consumes 3500 calories each week in soda.   **107.** 740 miles   **109. a.** $\frac{2}{3}$ **b.** $2\frac{5}{8}$

111. The quotient

SECTION 2.8

1. $\frac{11}{40} < \frac{19}{40}$ **3.** $\frac{2}{3} < \frac{5}{7}$ **5.** $\frac{5}{8} > \frac{7}{12}$ **7.** $\frac{7}{9} < \frac{11}{12}$ **9.** $\frac{13}{14} > \frac{19}{21}$ **11.** $\frac{7}{24} < \frac{11}{30}$ **13.** $\frac{9}{64}$ **15.** $\frac{8}{729}$ **17.** $\frac{1}{24}$

19. $\frac{8}{245}$ **21.** $\frac{1}{121}$ **23.** $\frac{81}{625}$ **25.** $\frac{7}{36}$ **27.** $\frac{27}{49}$ **29.** $\frac{5}{6}$ **31.** $1\frac{5}{12}$ **33.** $\frac{7}{48}$ **35.** $\frac{29}{36}$ **37.** $\frac{55}{72}$ **39.** $\frac{35}{54}$

41. 2   **43.** $\frac{9}{19}$ **45.** $\frac{7}{32}$ **47.** $\frac{64}{75}$

CHAPTER 2 REVIEW EXERCISES

1. $\frac{2}{3}$ [2.3B]   **2.** $\frac{5}{16}$ [2.8B]   **3.** $\frac{13}{4}$ [2.2A]   **4.** $1\frac{13}{18}$ [2.4B]   **5.** $\frac{11}{18} < \frac{17}{24}$ [2.8A]   **6.** $14\frac{19}{42}$ [2.5C]

7. $\frac{5}{36}$ [2.8C]   **8.** $9\frac{1}{24}$ [2.6B]   **9.** 2 [2.7B]   **10.** $\frac{25}{48}$ [2.5B]   **11.** $3\frac{1}{3}$ [2.7B]   **12.** 4 [2.1B]

13. $\frac{24}{36}$ [2.3A]   **14.** $\frac{3}{4}$ [2.7A]   **15.** $\frac{32}{44}$ [2.3A]   **16.** $16\frac{1}{2}$ [2.6B]   **17.** 36 [2.1A]   **18.** $\frac{4}{11}$ [2.3B]

19. $1\frac{1}{8}$ [2.4A]   **20.** $10\frac{1}{8}$ [2.5C]   **21.** $18\frac{13}{54}$ [2.4C]   **22.** 5 [2.1B]   **23.** $3\frac{2}{5}$ [2.2B]   **24.** $\frac{1}{15}$ [2.8C]

25. $5\frac{7}{8}$ [2.4C]   **26.** 54 [2.1A]   **27.** $\frac{1}{3}$ [2.5A]   **28.** $\frac{19}{7}$ [2.2B]   **29.** 2 [2.7A]   **30.** $\frac{1}{15}$ [2.6A]

31. $\frac{1}{8}$ [2.6A]   **32.** $1\frac{7}{8}$ [2.2A]   **33.** The total rainfall for the 3 months was $21\frac{7}{24}$ inches. [2.4D]   **34.** The cost per acre was $36,000. [2.7C]   **35.** The second checkpoint is $4\frac{3}{4}$ miles from the finish line. [2.5D]   **36.** The car can travel 243 miles. [2.6C]

CHAPTER 2 TEST

1. $\frac{4}{9}$ [2.6A]   **2.** 8 [2.1B]   **3.** $1\frac{3}{7}$ [2.7A]   **4.** $\frac{7}{24}$ [2.8C]   **5.** $\frac{49}{5}$ [2.2B]   **6.** 8 [2.6B]   **7.** $\frac{5}{8}$ [2.3B]

Copyright © Houghton Mifflin Company. All rights reserved.

8. $\frac{3}{8} < \frac{5}{12}$ [2.8A] **9.** $\frac{5}{6}$ [2.8C] **10.** 120 [2.1A] **11.** $\frac{1}{4}$ [2.5A] **12.** $3\frac{3}{5}$ [2.2B] **13.** $2\frac{2}{19}$ [2.7B]

14. $\frac{45}{72}$ [2.3A] **15.** $1\frac{61}{90}$ [2.4B] **16.** $13\frac{81}{88}$ [2.5C] **17.** $\frac{7}{48}$ [2.5B] **18.** $\frac{1}{6}$ [2.8B] **19.** $1\frac{11}{12}$ [2.4A]

20. $22\frac{4}{15}$ [2.4C] **21.** $\frac{11}{4}$ [2.2A] **22.** The electrician earns \$840. [2.6C] **23.** 11 lots were available for sale. [2.7C] **24.** The developer plans to build 30 houses on the property. [2.7C] **25.** The total rainfall for the 3-month period was $21\frac{11}{24}$ inches. [2.4D]

CUMULATIVE REVIEW EXERCISES

1. 290,000 [1.1D] **2.** 291,278 [1.3B] **3.** 73,154 [1.4B] **4.** 540 r12 [1.5C] **5.** 1 [1.6B]

6. $2 \cdot 2 \cdot 11$ [1.7B] **7.** 210 [2.1A] **8.** 20 [2.1B] **9.** $\frac{23}{3}$ [2.2B] **10.** $6\frac{1}{4}$ [2.2B] **11.** $\frac{15}{48}$ [2.3A]

12. $\frac{2}{5}$ [2.3B] **13.** $1\frac{7}{48}$ [2.4B] **14.** $14\frac{11}{48}$ [2.4C] **15.** $\frac{13}{24}$ [2.5B] **16.** $1\frac{7}{9}$ [2.5C] **17.** $\frac{7}{20}$ [2.6A]

18. $7\frac{1}{2}$ [2.6B] **19.** $1\frac{1}{20}$ [2.7A] **20.** $2\frac{5}{8}$ [2.7B] **21.** $\frac{1}{9}$ [2.8B] **22.** $5\frac{5}{24}$ [2.8C]

23. The amount in the checkbook account at the end of the week was \$862. [1.3C] **24.** The total income from the tickets was \$1410. [1.4C] **25.** The total weight is $12\frac{1}{24}$ pounds. [2.4D] **26.** The length of the remaining piece is $4\frac{17}{24}$ feet. [2.5D] **27.** The car travels 225 miles on $8\frac{1}{3}$ gallons of gas. [2.6C] **28.** 25 parcels can be sold from the remaining land. [2.7C]

Answers to Chapter 3 Selected Exercises

PREP TEST

1. $\frac{3}{10}$ [2.2A] **2.** 36,900 [1.1D] **3.** Four thousand seven hundred ninety-one [1.1B] **4.** 6842 [1.1B]

5. 9394 [1.2A] **6.** 1638 [1.3B] **7.** 76,804 [1.4B] **8.** 278 r18 [1.5C]

SECTION 3.1

1. Thousandths **3.** Ten-thousandths **5.** Hundredths **7.** 0.3 **9.** 0.21 **11.** 0.461 **13.** 0.093 **15.** $\frac{1}{10}$

17. $\frac{47}{100}$ **19.** $\frac{289}{1000}$ **21.** $\frac{9}{100}$ **23.** Thirty-seven hundredths **25.** Nine and four tenths **27.** Fifty-three ten-thousandths **29.** Forty-five thousandths **31.** Twenty-six and four hundredths **33.** 3.0806 **35.** 407.03 **37.** 246.024 **39.** 73.02684 **41.** 5.4 **43.** 30.0 **45.** 413.60 **47.** 6.062 **49.** 97 **51.** 5440 **53.** 0.0236 **55.** 0.18 ounce **57.** 26.2 miles **59.** For example, **a.** 0.15 **b.** 1.05 **c.** 0.001

SECTION 3.2

1. 150.1065 **3.** 95.8446 **5.** 69.644 **7.** 92.883 **9.** 113.205 **11.** 0.69 **13.** 16.305 **15.** 110.7666 **17.** 104.4959 **19.** Cal.: 234.192 Est.: 234 **21.** Cal.: 781.943 Est.: 782 **23.** The total length of the shaft is 5.65 inches. **25.** The amount in the checking account is \$3664.20. **27.** The combined populations of Asia and Africa in 2050 are expected to be 7.1 billion people. **29.** The number of self-employed people who earn more than \$5000 is 6.5 million. **31.** No, \$10 is not enough. **33.** No, a 4-foot rope cannot be wrapped around the box.

SECTION 3.3

1. 5.627 **3.** 113.6427 **5.** 6.7098 **7.** 215.697 **9.** 53.8776 **11.** 72.7091 **13.** 0.3142 **15.** 1.023 **17.** 261.166 **19.** 655.32 **21.** 342.9268 **23.** 8.628 **25.** 4.685 **27.** 10.383 **29.** Cal.: 2.74506 Est.: 3 **31.** Cal.: 7.14925 Est.: 7 **33.** The missing dimension is 7.55 inches. **35. a.** The total amount of

Copyright © Houghton Mifflin Company. All rights reserved.

the checks is $607.36.   **b.** Your new balance is $422.38.   **37.** The increase is 0.39 million births.   **39.** The growth in online shopping from 2000 to 2003 is 22.6 million households.   **41. a.** 0.1   **b.** 0.01   **c.** 0.001

SECTION 3.4

1. 0.36   **3.** 0.30   **5.** 0.25   **7.** 0.45   **9.** 6.93   **11.** 1.84   **13.** 4.32   **15.** 0.74   **17.** 39.5   **19.** 2.72
21. 0.603   **23.** 0.096   **25.** 13.50   **27.** 79.80   **29.** 4.316   **31.** 1.794   **33.** 0.06   **35.** 0.072   **37.** 0.1323
39. 0.03568   **41.** 0.0784   **43.** 0.076   **45.** 34.48   **47.** 580.5   **49.** 20.148   **51.** 0.04255   **53.** 0.17686
55. 0.19803   **57.** 14.8657   **59.** 0.0006608   **61.** 53.9961   **63.** 0.536335   **65.** 0.429   **67.** 2.116   **69.** 0.476
71. 1.022   **73.** 37.96   **75.** 2.318   **77.** 3.2   **79.** 6.5   **81.** 6285.6   **83.** 3200   **85.** 35,700   **87.** 6.3
89. 3.9   **91.** 49,000   **93.** 6.7   **95.** 0.012075   **97.** 0.0117796   **99.** 0.31004   **101.** 0.082845   **103.** 5.175
105. Cal.: 91.2   **107.** Cal.: 1.0472   **109.** Cal.: 3.897   **111.** Cal.: 11.2406   **113.** Cal.: 0.371096
Est.: 90   Est.: 0.8   Est.: 4.5   Est.: 12   Est.: 0.32
115. Cal.: 31.8528   **117.** Cal.: 1941.069459   **119.** Cal.: 0.00004351152   **121.** The motor costs $1.51 to operate.
Est.: 30   Est.: 2000   Est.: 0.00005
123. a. The estimated amount received is $25.00.   **b.** The amount received was $23.40.   **125.** The area is 23.625 square feet.   **127. a.** The amount of overtime pay is $650.25.   **b.** The nurse's total income is $1806.25.
129. The cost is $341.25.   **131. a.** The total cost for grade 1 is $21.12.   **b.** The total cost for grade 2 is $29.84.
c. The total cost for grade 3 is $201.66.   **d.** The total cost is $252.62.   **133.** Cars 2 and 5 would fail the test.
137. $1\dfrac{3}{10} \times 2\dfrac{31}{100} = \dfrac{13}{10} \times \dfrac{231}{100} = \dfrac{3003}{1000} = 3\dfrac{3}{1000} = 3.003$

SECTION 3.5

1. 0.82   **3.** 4.8   **5.** 89   **7.** 60   **9.** 84.3   **11.** 32.3   **13.** 5.06   **15.** 1.3   **17.** 0.11   **19.** 3.8   **21.** 6.3
23. 0.6   **25.** 2.5   **27.** 1.1   **29.** 130.6   **31.** 0.81   **33.** 0.09   **35.** 40.70   **37.** 0.46   **39.** 0.019
41. 0.087   **43.** 0.360   **45.** 0.103   **47.** 0.009   **49.** 1   **51.** 3   **53.** 1   **55.** 57   **57.** 0.407   **59.** 4.267
61. 0.01037   **63.** 0.008295   **65.** 0.82537   **67.** 0.032   **69.** 0.23627   **71.** 0.000053   **73.** 0.0018932
75. 18.42   **77.** 16.07   **79.** 0.0135   **81.** 0.023678   **83.** 0.112   **85.** Cal.: 11.1632   **87.** Cal.: 884.0909
Est.: 10   Est.: 1000
89. Cal.: 1.8269   **91.** Cal.: 58.8095   **93.** Cal.: 72.3053   **95.** Cal.: 0.0023   **97.** 6.23 yards are gained per carry.
Est.: 1.5   Est.: 50   Est.: 100   Est.: 0.0025
99. The cost per can is $.28.   **101.** The cost per mile is $.04.   **103.** The monthly payment is $58.65.
105. 2.9 million more women are expected to be attending institutions of higher learning in 2010.   **107.** The Army's budget was 4.2 times greater than the Navy's advertising budget.   **109.** The population of this segment is expected to be 2.1 times greater in 2030 than in 2000.   **111.** The cigarette consumption in 2000 was 2.5 times greater than in 1960.
113. A total of 9.7 million acres was burned.   **119.** ×   **121.** ×   **123.** ÷   **125.** 2.53

SECTION 3.6

1. 0.625   **3.** 0.667   **5.** 0.167   **7.** 0.417   **9.** 1.750   **11.** 1.500   **13.** 4.000   **15.** 0.003   **17.** 7.080
19. 37.500   **21.** 0.160   **23.** 8.400   **25.** $\dfrac{4}{5}$ **27.** $\dfrac{8}{25}$ **29.** $\dfrac{1}{8}$ **31.** $1\dfrac{1}{4}$ **33.** $16\dfrac{9}{10}$ **35.** $8\dfrac{2}{5}$ **37.** $8\dfrac{437}{1000}$
39. $2\dfrac{1}{4}$ **41.** $\dfrac{23}{150}$ **43.** $\dfrac{703}{800}$ **45.** $7\dfrac{19}{50}$ **47.** $\dfrac{57}{100}$ **49.** $\dfrac{2}{3}$ **51.** 0.6 > 0.45   **53.** 3.89 < 3.98
55. 0.025 < 0.105   **57.** $\dfrac{4}{5}$ < 0.802   **59.** 0.85 < $\dfrac{7}{8}$ **61.** $\dfrac{7}{12}$ > 0.58   **63.** $\dfrac{11}{12}$ < 0.92   **65.** 0.623 > 0.6023
67. 0.87 > 0.087   **69.** 0.033 < 0.3   **71.** The population ages 0 to 19 is more than $\dfrac{1}{4}$ of the total population.
73. Yes, the digits 538461 repeat.

CHAPTER 3 REVIEW EXERCISES

1. 54.5 [3.5A]   **2.** 833.958 [3.2A]   **3.** 0.055 < 0.1 [3.6C]   **4.** Twenty-two and ninety-two ten-thousandths [3.1A]
5. 0.05678 [3.1B]   **6.** 2.33 [3.6A]   **7.** $\dfrac{3}{8}$ [3.6B]   **8.** 36.714 [3.2A]   **9.** 34.025 [3.1A]   **10.** $\dfrac{5}{8}$ > 0.62 [3.6C]

Copyright © Houghton Mifflin Company. All rights reserved.

11. 0.778 [3.6A]   **12.** $\frac{2}{3}$ [3.6B]   **13.** 22.8635 [3.3A]   **14.** 7.94 [3.1B]   **15.** 8.932 [3.4A]

16. Three hundred forty-two and thirty-seven hundredths [3.1A]   **17.** 3.06753 [3.1A]   **18.** 25.7446 [3.4A]

19. 6.594 [3.5A]   **20.** 4.8785 [3.3A]   **21.** The new balance in your account is $661.51. [3.3B]   **22.** There are 53.466 million children in grades K–12. [3.2B]   **23.** There are 40.49 million more children in public school than in private school. [3.3B]   **24.** During a 5-day school week, 9.5 million gallons of milk are served. [3.4B]   **25.** The number who drove is 6.4 times greater than the number who flew. [3.5B]

CHAPTER 3 TEST

1. 0.66 < 0.666 [3.6C]   **2.** 4.087 [3.3A]   **3.** Forty-five and three hundred two ten-thousandths [3.1A]

4. 0.692 [3.6A]   **5.** $\frac{33}{40}$ [3.6B]   **6.** 0.0740 [3.1B]   **7.** 1.538 [3.5A]   **8.** 27.76626 [3.3A]   **9.** 7.095 [3.1B]

10. 232 [3.5A]   **11.** 458.581 [3.2A]   **12.** The missing dimension is 1.37 inches. [3.3B]   **13.** 0.00548 [3.4A]

14. 255.957 [3.2A]   **15.** 209.07086 [3.1A]   **16.** Each payment is $395.40. [3.5B]   **17.** Your total income is $3087.14. [3.2B]   **18.** The cost of the call is $4.63. [3.4B]   **19.** The yearly average computer use by a 10th-grade student is 348.4 hours. [3.4B]   **20.** On average a 2nd-grade student uses the computer 36.4 hours more per year than a 5th-grade student. [3.4B]

CUMULATIVE REVIEW EXERCISES

1. 235 r17 [1.5C]   **2.** 128 [1.6A]   **3.** 3 [1.6B]   **4.** 72 [2.1A]   **5.** $4\frac{2}{5}$ [2.2B]   **6.** $\frac{37}{8}$ [2.2B]

7. $\frac{25}{60}$ [2.3A]   **8.** $1\frac{17}{48}$ [2.4B]   **9.** $8\frac{35}{36}$ [2.4C]   **10.** $5\frac{23}{36}$ [2.5C]   **11.** $\frac{1}{12}$ [2.6A]   **12.** $9\frac{1}{8}$ [2.6B]

13. $1\frac{2}{9}$ [2.7A]   **14.** $\frac{19}{20}$ [2.7B]   **15.** $\frac{3}{16}$ [2.8B]   **16.** $2\frac{5}{18}$ [2.8C]   **17.** Sixty-five and three hundred nine ten-thousandths [3.1A]   **18.** 504.6991 [3.2A]   **19.** 21.0764 [3.3A]   **20.** 55.26066 [3.4A]   **21.** 2.154 [3.5A]

22. 0.733 [3.6A]   **23.** $\frac{1}{6}$ [3.6B]   **24.** $\frac{8}{9}$ < 0.98 [3.6C]   **25.** Sweden mandates 14 days more vacation than Germany. [1.3C]   **26.** The patient must lose $7\frac{3}{4}$ pounds the third month to achieve the goal. [2.5D]

27. Your checking account balance is $617.38. [3.3B]   **28.** The resulting thickness is 1.395 inches. [3.3B]   **29.** You paid $6008.80 in income tax last year. [3.4B]   **30.** The amount of each payment is $23.87. [3.5B]

Answers to Chapter 4 Selected Exercises

PREP TEST

1. $\frac{4}{5}$ [2.3B]   **2.** $\frac{1}{2}$ [2.3B]   **3.** 24.8 [3.6A]   **4.** 4 × 33 [1.4A]   **5.** 4 [1.5A]

SECTION 4.1

1. $\frac{1}{5}$ 1:5 1 to 5   **3.** $\frac{2}{1}$ 2:1 2 to 1   **5.** $\frac{3}{8}$ 3:8 3 to 8   **7.** $\frac{37}{24}$ 37:24 37 to 24   **9.** $\frac{1}{1}$ 1:1 1 to 1

11. $\frac{7}{10}$ 7:10 7 to 10   **13.** $\frac{1}{2}$ 1:2 1 to 2   **15.** $\frac{2}{1}$ 2:1 2 to 1   **17.** $\frac{3}{4}$ 3:4 3 to 4   **19.** $\frac{5}{3}$ 5:3 5 to 3

21. $\frac{2}{3}$ 2:3 2 to 3   **23.** $\frac{2}{1}$ 2:1 2 to 1   **25.** The ratio is $\frac{1}{3}$.   **27.** The ratio is $\frac{3}{8}$.   **29.** The ratio is $\frac{2}{77}$.

31. The ratio is $\frac{1}{12}$.   **33. a.** The amount of increase is $20,000.   **b.** The ratio is $\frac{2}{9}$.   **35.** The ratio is $\frac{11}{31}$.

37. No, the ratio = $\frac{830}{1389} \approx 0.5976$ is greater than $\frac{2}{5}$ (0.4).

Copyright © Houghton Mifflin Company. All rights reserved.

SECTION 4.2

1. $\dfrac{3 \text{ pounds}}{4 \text{ people}}$ **3.** $\dfrac{\$20}{3 \text{ boards}}$ **5.** $\dfrac{20 \text{ miles}}{1 \text{ gallon}}$ **7.** $\dfrac{5 \text{ children}}{2 \text{ families}}$ **9.** $\dfrac{8 \text{ gallons}}{1 \text{ hour}}$ **11.** 2.5 feet/second   **13.** $975/week

15. 110 trees/acre   **17.** $18.84/hour   **19.** 52.4 miles/hour   **21.** 28 miles/gallon   **23.** $1.65/pound   **25.** The gas mileage was 28.4 miles/gallon.   **27.** The rocket uses 213,600 gallons/minute.   **29. a.** 175 pounds of beef was packaged.   **b.** The beef cost $2.09/pound.   **31. a.** The camera goes through film at the rate of 336 feet/minute. **b.** The camera uses the film at a rate of 89 seconds/roll.   **33. a.** The price of the computer hardware would be 109,236 euros.   **b.** The cost of the car would be 3,978,000 yen.

SECTION 4.3

1. True   **3.** Not true   **5.** Not true   **7.** True   **9.** True   **11.** Not true   **13.** True   **15.** True   **17.** True   **19.** Not true   **21.** True   **23.** Not true   **25.** 3   **27.** 6   **29.** 9   **31.** 5.67   **33.** 4   **35.** 4.38   **37.** 88   **39.** 3.33   **41.** 26.25   **43.** 96   **45.** 9.78   **47.** 3.43   **49.** 1.34   **51.** 50.4   **53.** A 0.5-ounce serving contains 50 calories.   **55.** Ron used 70 pounds of fertilizer.   **57.** There were 375 wooden bats produced.   **59.** The distance is 16 miles.   **61.** 1.25 ounces are required.   **63.** 160,000 people would vote.   **65.** The monthly payment is $176.75.   **67.** You will own 400 shares.   **69.** A bowling ball would weigh 2.67 pounds on the moon.   **71.** The dividend would be $1071.

CHAPTER 4 REVIEW EXERCISES

1. True [4.3A]   **2.** $\dfrac{2}{5}$ 2:5 2 to 5 [4.1A]   **3.** 62.5 miles/hour [4.2B]   **4.** True [4.3A]   **5.** 68 [4.3B]

6. $7.50/hour [4.2B]   **7.** $1.75/pound [4.2B]   **8.** $\dfrac{2}{7}$ 2:7 2 to 7 [4.1A]   **9.** 36 [4.3B]   **10.** 19.44 [4.3B]

11. $\dfrac{2}{5}$ 2:5 2 to 5 [4.1A]   **12.** Not true [4.3A]   **13.** $\dfrac{\$15}{4 \text{ hours}}$ [4.2A]   **14.** 27.2 miles/gallon [4.2B]

15. $\dfrac{1}{1}$ 1:1 1 to 1 [4.1A]   **16.** True [4.3A]   **17.** 65.45 [4.3B]   **18.** $\dfrac{100 \text{ miles}}{3 \text{ hours}}$ [4.2A]   **19.** The ratio is $\dfrac{2}{5}$. [4.1B]   **20.** The property tax is $6400. [4.3C]   **21.** The ratio is $\dfrac{2}{1}$. [4.1B]   **22.** The cost per phone is $37.50. [4.2C]   **23.** 1344 blocks would be needed. [4.3C]   **24.** The ratio is $\dfrac{5}{2}$. [4.1B]   **25.** The turkey costs $.93/pound. [4.2C]   **26.** The average was 56.8 miles/hour. [4.2C]   **27.** The cost is $493.50. [4.3C]   **28.** The cost is $44.75/share. [4.2C]   **29.** 22.5 pounds of fertilizer will be used. [4.3C]   **30.** The ratio is $\dfrac{1}{2}$. [4.1B]

CHAPTER 4 TEST

1. $3836.40/month [4.2B]   **2.** $\dfrac{1}{6}$ 1:6 1 to 6 [4.1A]   **3.** $\dfrac{9 \text{ supports}}{4 \text{ feet}}$ [4.2A]   **4.** Not true [4.3A]

5. $\dfrac{3}{2}$ 3:2 3 to 2 [4.1A]   **6.** 144 [4.3B]   **7.** 30.5 miles/gallon [4.2B]   **8.** $\dfrac{1}{3}$ 1:3 1 to 3 [4.1A]

9. True [4.3A]   **10.** 40.5 [4.3B]   **11.** $\dfrac{\$27}{4 \text{ boards}}$ [4.2A]   **12.** $\dfrac{3}{5}$ 3:5 3 to 5 [4.1A]   **13.** The dividend is $625. [4.3C]   **14.** The ratio is $\dfrac{43}{56}$. [4.1B]   **15.** The plane's speed is 538 miles/hour. [4.2C]   **16.** The college student's body contains 132 pounds of water. [4.3C]   **17.** The cost of the lumber is $1.73/foot. [4.2C]   **18.** The amount of medication required is 0.875 ounce. [4.3C]   **19.** The ratio is $\dfrac{4}{5}$. [4.1B]   **20.** 36 defective hard drives are expected to be found in the production of 1200 hard drives. [4.3C]

CUMULATIVE REVIEW EXERCISES

1. 9158 [1.3B]   **2.** $2^4 \cdot 3^3$ [1.6A]   **3.** 3 [1.6B]   **4.** $2 \cdot 2 \cdot 2 \cdot 2 \cdot 2 \cdot 5$ [1.7B]   **5.** 36 [2.1A]   **6.** 14 [2.1B]

7. $\dfrac{5}{8}$ [2.3B]   **8.** $8\dfrac{3}{10}$ [2.4C]   **9.** $5\dfrac{11}{18}$ [2.5C]   **10.** $2\dfrac{5}{6}$ [2.6B]   **11.** $4\dfrac{2}{3}$ [2.7B]   **12.** $\dfrac{23}{30}$ [2.8C]

Copyright © Houghton Mifflin Company. All rights reserved.

13. Four and seven hundred nine ten-thousandths [3.1A] **14.** 2.10 [3.1B] **15.** 1.990 [3.5A] **16.** $\frac{1}{15}$ [3.6B]

17. $\frac{1}{8}$ [4.1A] **18.** $\frac{29¢}{2 \text{ pencils}}$ [4.2A] **19.** 33.4 miles/gallon [4.2B] **20.** 4.25 [4.3B] **21.** The car's speed is 57.2 miles/hour. [4.2C] **22.** 36 [4.3B] **23.** Your new balance is $744. [1.3C] **24.** The monthly payment is $570. [1.5D] **25.** 105 pages remain to be read. [2.6C] **26.** The cost per acre was $36,000. [2.7C] **27.** The change was $17.62. [3.3B] **28.** Your monthly salary is $3468.25. [3.5B] **29.** 25 inches will erode in 50 months. [4.3C] **30.** 1.6 ounces are required. [4.3C]

Answers to Chapter 5 Selected Exercises

PREP TEST

1. $\frac{19}{100}$ [2.6B] **2.** 0.23 [3.4A] **3.** 47 [3.4A] **4.** 2850 [3.4A] **5.** 4000 [3.5A] **6.** 32 [2.7B]

7. 62.5 [3.6A] **8.** $66\frac{2}{3}$ [2.2B] **9.** 1.75 [3.5A]

SECTION 5.1

1. $\frac{1}{4}$, 0.25 **3.** $1\frac{3}{10}$, 1.30 **5.** 1, 1.00 **7.** $\frac{73}{100}$, 0.73 **9.** $3\frac{83}{100}$, 3.83 **11.** $\frac{7}{10}$, 0.70 **13.** $\frac{22}{25}$, 0.88

15. $\frac{8}{25}$, 0.32 **17.** $\frac{2}{3}$ **19.** $\frac{5}{6}$ **21.** $\frac{1}{9}$ **23.** $\frac{5}{11}$ **25.** $\frac{3}{70}$ **27.** $\frac{1}{15}$ **29.** 0.065 **31.** 0.123 **33.** 0.0055

35. 0.0825 **37.** 0.0505 **39.** 0.02 **41.** 0.804 **43.** 0.049 **45.** 73% **47.** 1% **49.** 294% **51.** 0.6%

53. 310.6% **55.** 70% **57.** 37% **59.** 40% **61.** 12.5% **63.** 150% **65.** 166.7% **67.** 87.5% **69.** 48%

71. $33\frac{1}{3}$% **73.** $166\frac{2}{3}$% **75.** $87\frac{1}{2}$% **77.** 6% of those surveyed named something other than corn on the cob, cole slaw, corn bread, or fries. **79.** This represents $\frac{1}{2}$ off the regular price. **81. a.** False **b.** For example, 200% × 4 = 2 × 4 = 8

SECTION 5.2

1. 8 **3.** 10.8 **5.** 0.075 **7.** 80 **9.** 51.895 **11.** 7.5 **13.** 13 **15.** 3.75 **17.** 20 **19.** 210 **21.** 5% of 95 **23.** 79% of 16 **25.** 72% of 40 **27.** 25,805.0324 **29.** About 13.2 million people aged 18 to 24 do not have health insurance. **31.** 12,151 more faculty members described their political views as liberal than described their views as far left. **33. a.** The sales tax is $1770. **b.** The total cost of the car is $31,270. **35.** 6232 respondents did not answer yes to the question. **37.** Employees spend 48.2 hours with family and friends. **39.** There are approximately 3.4 hours difference between the actual and preferred amounts of time the employees spent on self.

SECTION 5.3

1. 32% **3.** $16\frac{2}{3}$% **5.** 200% **7.** 37.5% **9.** 18% **11.** 0.25% **13.** 20% **15.** 400% **17.** 2.5%

19. 37.5% **21.** 0.25% **23.** 2.4% **25.** 9.6% **27.** 70% of couples disagree about financial matters. **29.** Approximately 25.4% of the vegetables were wasted. **31.** The typical American household spends 6% of the total amount spent on energy utilities on lighting. **33.** 27% of the food produced in the United States is wasted. **35.** Approximately 27.4% of the total expenses is spent for food. **37.** 91.8% of the total is spent on all categories except training.

SECTION 5.4

1. 75 **3.** 50 **5.** 100 **7.** 85 **9.** 1200 **11.** 19.2 **13.** 7.5 **15.** 32 **17.** 200 **19.** 80 **21.** 9

Copyright © Houghton Mifflin Company. All rights reserved.

23. 504   **25.** 108   **27.** There were 15.8 million travelers who allowed their children to miss school to go along on a trip.   **29.** 3364 people responded to the survey.   **31. a.** 3000 boards were tested.   **b.** 2976 boards tested were not defective.   **33.** 200,935,308 people in the United States were age 20 and older in 2000.   **35.** The recommended daily allowance of copper for an adult is 2 milligrams.

SECTION 5.5

1. 65   **3.** 25%   **5.** 75   **7.** 12.5%   **9.** 400   **11.** 19.5   **13.** 14.8%   **15.** 62.62   **17.** 5   **19.** 45
21. 15   **23.** The total amount collected was $24,500.   **25.** The world's total land area is 58,750,000 square miles.
27. 22,577 hotels in the United States are located along highways.   **29.** 14,000,000 ounces of gold were mined in the United States that year.   **31.** 46.8% of the deaths were due to traffic accidents.   **33.** The 108th House of Representatives had the larger percent of Republicans.

CHAPTER 5 REVIEW EXERCISES

1. 60 [5.2A]   **2.** 20% [5.3A]   **3.** 175% [5.1B]   **4.** 75 [5.4A]   **5.** $\frac{3}{25}$ [5.1A]   **6.** 19.36 [5.2A]

7. 150% [5.3A]   **8.** 504 [5.4A]   **9.** 0.42 [5.1A]   **10.** 5.4 [5.2A]   **11.** 157.5 [5.4A]   **12.** 0.076 [5.1A]

13. 77.5 [5.2A]   **14.** $\frac{1}{6}$ [5.1A]   **15.** 160% [5.5A]   **16.** 75 [5.5A]   **17.** 38% [5.1B]   **18.** 10.9 [5.4A]

19. 7.3% [5.3A]   **20.** 613.3% [5.3A]   **21.** The student answered 85% of the questions correctly. [5.5B]
22. The company spent $4500 for newspaper advertising. [5.2B]   **23.** 31.7% of the cost is for electricity. [5.3B]
24. The total cost of the video camera is $1041.25. [5.2B]   **25.** Approximately 78.6% of the women wore sunscreen often. [5.3B]   **26.** The world's population in 2000 was approximately 6,100,000,000 people. [5.4B]   **27.** The cost of the computer 4 years ago was $3000. [5.5B]   **28.** The dollar value of all the online transactions made that year was $52 billion. [5.4B]

CHAPTER 5 TEST

1. 0.973 [5.1A]   **2.** $\frac{5}{6}$ [5.1A]   **3.** 30% [5.1B]   **4.** 163% [5.1B]   **5.** 150% [5.1B]   **6.** $66\frac{2}{3}$% [5.1B]

7. 50.05 [5.2A]   **8.** 61.36 [5.2A]   **9.** 76% of 13 [5.2A]   **10.** 212% of 12 [5.2A]   **11.** The amount spent for advertising is $4500. [5.2B]   **12.** 1170 pounds of vegetables were not spoiled. [5.2B]   **13.** 14.7% of the daily recommended amount of potassium is provided. [5.3B]   **14.** 9.1% of the daily recommended number of calories is provided. [5.3B]   **15.** 16% of the permanent employees are hired. [5.3B]   **16.** The student answered approximately 91.3% of the questions correctly. [5.3B]   **17.** 80 [5.4A]   **18.** 28.3 [5.4A]   **19.** 32,000 PDAs were tested. [5.4B]
20. The increase was 60% of the original price. [5.3B]   **21.** 143.0 [5.5A]   **22.** 1000% [5.5A]   **23.** The dollar increase is $1.74. [5.5B]   **24.** The population now is 220% of what it was 10 years ago. [5.5B]   **25.** The value of the car is $12,500. [5.5B]

CUMULATIVE REVIEW EXERCISES

1. 4 [1.6B]   **2.** 240 [2.1A]   **3.** $10\frac{11}{24}$ [2.4C]   **4.** $12\frac{41}{48}$ [2.5C]   **5.** $12\frac{4}{7}$ [2.6B]   **6.** $\frac{7}{24}$ [2.7B]

7. $\frac{1}{3}$ [2.8B]   **8.** $\frac{13}{36}$ [2.8C]   **9.** 3.08 [3.1B]   **10.** 1.1196 [3.3A]   **11.** 34.2813 [3.5A]   **12.** 3.625 [3.6A]

13. $1\frac{3}{4}$ [3.6B]   **14.** $\frac{3}{8} < 0.87$ [3.6C]   **15.** 53.3 [4.3B]   **16.** $9.60/hour [4.2B]   **17.** $\frac{11}{60}$ [5.1A]

18. $83\frac{1}{3}$% [5.1B]   **19.** 19.56 [5.2A/5.5A]   **20.** $133\frac{1}{3}$% [5.3A/5.5A]   **21.** 9.92 [5.4A/5.5A]

22. 342.9% [5.3A/5.5A]   **23.** Sergio's take-home pay is $592. [2.6C]   **24.** Each monthly payment is $204.25. [3.5B]
25. 420 gallons were used during the month. [3.5B]   **26.** The real estate tax is $5000. [4.3C]   **27.** The sales tax is 6% of the purchase price. [5.3B/5.5B]   **28.** 45% of the people did not favor the candidate. [5.3B/5.5B]   **29.** The approximate average number of hours spent watching TV in a week is 61.3 hours. [5.2B/5.5B]   **30.** 18% of the children tested had levels of lead that exceeded federal standards. [5.3B/5.5B]

Copyright © Houghton Mifflin Company. All rights reserved.

Answers to Chapter 6 Selected Exercises

PREP TEST

1. 0.75 [3.5A] **2.** 52.05 [3.4A] **3.** 504.51 [3.3A] **4.** 9750 [3.4A] **5.** 45 [3.4A] **6.** 1417.24 [3.2A]
7. 3.33 [3.5A] **8.** 0.605 [3.5A] **9.** 0.379 < 0.397 [3.6C]

SECTION 6.1

1. The unit cost is $.055 per ounce. **3.** The unit cost is $.374 per ounce. **5.** The unit cost is $.080 per tablet.
7. The unit price is $6.975 per clamp. **9.** The unit cost is $.199 per ounce. **11.** The unit cost is $.119 per screw.
13. The Sutter Home pasta sauce is the more economical purchase. **15.** 20 ounces is the more economical purchase.
17. 400 tablets is the more economical purchase. **19.** Land O' Lakes cheddar cheese is the more economical purchase.
21. Maxwell House coffee is the more economical purchase. **23.** Friskies Chef's Blend is the more economical purchase.
25. The total cost is $13.77. **27.** The total cost is $1.84. **29.** The total cost is $6.37. **31.** The total cost is $2.71.
33. The total cost is $5.96.

SECTION 6.2

1. The percent increase is 19.6%. **3.** The percent increase is 117.6%. **5.** The percent increase is 457.1%.
7. The population 8 years later was 157,861 people. **9.** The average age of American mothers giving birth in 2000 was
24.9 years. **11.** The markup is $35.70. **13.** The markup rate is 30%. **15. a.** The markup is $77.76. **b.** The
selling price is $239.76. **17. a.** The markup is $1.10. **b.** The selling price is $3.10. **19.** The selling price is $74.
21. The amount represents a decrease of 40%. **23.** There is a decrease of 540 employees. **25. a.** The amount of
decrease is 13 minutes. **b.** The amount represents a decrease of 25%. **27. a.** The amount of decrease is $.60.
b. The new dividend is $1.00. **29.** The amount represents a decrease of 6.6%. **31.** The discount rate is $33\frac{1}{3}$%.
33. The discount is $68. **35.** The discount rate is 30%. **37. a.** The discount is $.25 per pound. **b.** The sale price
is $1.00 per pound. **39. a.** The amount of the discount is $4. **b.** The discount rate is 20%. **41.** Yes

SECTION 6.3

1. a. $10,000 **b.** $850 **c.** 4.25% **d.** 2 years **3.** The simple interest owed is $960. **5.** The simple interest
due is $3375. **7.** The simple interest due is $1320. **9.** The simple interest due is $92.47. **11.** The simple interest
due is $84.76. **13.** The maturity value is $5120. **15.** The maturity value is $164,250. **17.** The total amount due
on the loan is $12,875. **19.** The maturity value is $14,543.70. **21.** The monthly payment is $6187.50.
23. a. The interest charged is $1080. **b.** The monthly payment is $545. **25.** The monthly payment is $1332.50.
27. The finance charge is $1.48. **29.** The finance charge is $185.53. **31.** The difference in finance charges is $10.07.
33. The value of the investment after 1 year is $780.60. **35.** The value of the investment after 15 years is $7281.78.
37. a. The value of the investment in 5 years will be $111,446.25. **b.** The amount of interest earned will be $36,446.25.
39. The amount of interest earned is $5799.48. **41.** The interest owed is $16.

SECTION 6.4

1. The mortgage is $82,450. **3.** The down payment is $7500. **5.** The down payment is $212,500. **7.** The loan
origination fee is $3750. **9. a.** The down payment is $7500. **b.** The mortgage is $142,500. **11.** The mortgage is
$189,000. **13.** The monthly mortgage payment is $1157.73. **15.** No, the couple cannot afford to buy the house.
17. The monthly property tax is $112.35. **19. a.** The monthly mortgage payment is $1678.40. **b.** The interest
payment is $736.68. **21.** The total monthly payment for the mortgage and property tax is $982.70. **23.** The monthly
mortgage payment is $1645.53. **25.** By using the 20-year loan, the couple will save $63,408.

Copyright © Houghton Mifflin Company. All rights reserved.

SECTION 6.5

1. No, Amanda does not have enough money for the down payment.   **3.** The sales tax is $1192.50.   **5.** The license fee is $450.   **7. a.** The sales tax is $1120.   **b.** The total cost of the sales tax and the license fee is $1395.
9. a. The down payment is $4050.   **b.** The amount financed is $12,150.   **11.** The amount financed is $28,000.
13. The monthly truck payment is $552.70.   **15.** The cost is $5120.   **17.** The cost is about $.11 per mile.
19. The amount of interest is $74.75.   **21. a.** The amount financed is $153,200.   **b.** The monthly payment is $2961.78.   **23.** The monthly payment is $649.63.   **25.** The 7% loan has the lesser loan cost.

SECTION 6.6

1. Lewis earns $380.   **3.** The real estate agent's commission is $3930.   **5.** The stockbroker's commission is $84.
7. The teacher's monthly salary is $3244.   **9.** The overtime wage is $51.60/hour.   **11.** The golf pro's commission was $112.50.   **13.** The typist earns $618.75.   **15.** Maxine's hourly wage is $85.   **17. a.** Mark's hourly wage on Saturday is $23.85.   **b.** Mark earns $190.80.   **19. a.** The nurse's increase in hourly pay is $2.15.   **b.** The nurse's hourly pay is $23.65.   **21.** Nicole's earnings were $475.   **23.** The starting salary for a chemical engineer increased $922 over that of the previous year.   **25.** The starting salary for a political science major decreased $4115 from that of the previous year.

SECTION 6.7

1. Your current checking account balance is $486.32.   **3.** The real estate firm's current checking account balance is $1222.47.   **5.** The nutritionist's current balance is $825.27.   **7.** The current checking account balance is $3000.82.
9. Yes, there is enough money in the carpenter's account to purchase the refrigerator.   **11.** Yes, there is enough money in the account to make the two purchases.   **13.** The bank statement and checkbook balance.   **15.** The bank statement and checkbook balance.   **17.** added   **19.** subtract

CHAPTER 6 REVIEW EXERCISES

1. The unit cost is $.195 per ounce or 19.5¢ per ounce. [6.1A]   **2.** The cost is $.164 or 16.4¢ per mile. [6.5B]
3. The percent increase is 30.4%. [6.2A]   **4.** The markup is $72. [6.2B]   **5.** The simple interest due is $3000. [6.3A]   **6.** The value of the investment after 10 years is $45,550.75. [6.3C]   **7.** The percent increase is 15%. [6.2A]
8. The total monthly payment for the mortgage and property tax is $578.53. [6.4B]   **9.** The monthly payment is $518.02. [6.5B]   **10.** The value of the investment will be $53,593. [6.3C]   **11.** The down payment is $18,750. [6.4A]   **12.** The total cost of the sales tax and license fee is $1471.25. [6.5A]   **13.** The selling price is $2079. [6.2B]
14. The interest paid is $97.33. [6.5B]   **15.** The commission was $3240. [6.6A]   **16.** The sale price is $141. [6.2D]
17. The current checkbook balance is $943.68. [6.7A]   **18.** The maturity value is $31,200. [6.3A]
19. The origination fee is $1875. [6.4A]   **20.** The more economical purchase is 33 ounces for $6.99. [6.1B]
21. The monthly mortgage payment is $934.08. [6.4B]   **22.** The total income was $655.20. [6.6A]   **23.** The donut shop's checkbook balance is $8866.58. [6.7A]   **24.** The monthly payment is $14,093.75. [6.3A]   **25.** The finance charge is $7.20. [6.3B]

CHAPTER 6 TEST

1. The cost per foot is $6.92. [6.1A]   **2.** The more economical purchase is 3 pounds for $7.49. [6.1B]
3. The total cost is $14.53. [6.1C]   **4.** The percent increase in the cost of the exercise bicycle is 20%. [6.2A]
5. The selling price of the compact disc player is $301. [6.2B]   **6.** The percent decrease is 7.7%. [6.2C]
7. The percent decrease is 20%. [6.2C]   **8.** The sale price of the corner hutch is $209.30. [6.2D]   **9.** The discount rate is 40%. [6.2D]   **10.** The simple interest due is $2000. [6.3A]   **11.** The maturity value is $26,725. [6.3A]
12. The finance charge is $4.50. [6.3B]   **13.** The amount of interest earned in 10 years will be $24,420.60. [6.3C]
14. The loan origination fee is $3350. [6.4A]   **15.** The monthly mortgage payment is $1713.44. [6.4B]
16. The amount financed is $19,000. [6.5A]   **17.** The monthly truck payment is $482.68. [6.5B]   **18.** Shaney earns $1071. [6.6A]   **19.** The current checkbook balance is $6612.25. [6.7A]   **20.** The bank statement and the checkbook balance. [6.7B]

Copyright © Houghton Mifflin Company. All rights reserved.

CUMULATIVE REVIEW EXERCISES

1. 13 [1.6B]   **2.** $8\frac{13}{24}$ [2.4C]   **3.** $2\frac{37}{48}$ [2.5C]   **4.** 9 [2.6B]   **5.** 2 [2.7B]   **6.** 5 [2.8C]   **7.** 52.2 [3.5A]

8. 1.417 [3.6A]   **9.** $51.25/hour [4.2B]   **10.** 10.94 [4.3B]   **11.** 62.5% [5.1B]   **12.** 27.3 [5.2A]

13. 0.182 [5.1A]   **14.** 42% [5.3A]   **15.** 250 [5.4A]   **16.** 154.76 [5.4A/5.5A]   **17.** The total rainfall is $13\frac{11}{12}$

inches. [2.4D]   **18.** The amount paid in taxes is $970. [2.6C]   **19.** The ratio is $\frac{3}{5}$. [4.1B]   **20.** 33.4 miles are

driven per gallon. [4.2C]   **21.** The cost is $.93 per pound. [4.2C]   **22.** The dividend is $280. [4.3C]   **23.** The

sale price is $720. [6.2D]   **24.** The selling price of the disc player is $119. [6.2B]   **25.** The percent increase in Sook

Kim's salary is 8%. [6.2A]   **26.** The simple interest due is $2700. [6.3A]   **27.** The monthly car payment is $791.81.

[6.5B]   **28.** The family's new checking account balance is $2243.77. [6.7A]   **29.** The cost per mile is $.20. [6.5B]

30. The monthly mortgage payment is $1232.26. [6.4B]

Copyright © Houghton Mifflin Company. All rights reserved.

Glossary

addend In addition, one of the numbers added. [1.2]

addition The process of finding the total of two numbers. [1.2]

Addition Property of Zero Zero added to a number does not change the number. [1.2]

approximation An estimated value obtained by rounding an exact value. [1.1]

Associative Property of Addition Numbers to be added can be grouped (with parentheses, for example) in any order; the sum will be the same. [1.2]

Associative Property of Multiplication Numbers to be multiplied can be grouped (with parentheses, for example) in any order; the product will be the same. [1.4]

average The sum of all the numbers divided by the number of those numbers. [1.5]

balancing a checkbook Determining whether the checking account balance is accurate. [6.7]

bank statement A document showing all the transactions in a bank account during the month. [6.7]

basic percent equation Percent times base equals amount. [5.2]

borrowing In subtraction, taking a unit from the next larger place value in the minuend and adding it to the number in the given place value in order to make that number larger than the number to be subtracted from it. [1.3]

carrying In addition, transferring a number to another column. [1.2]

check A printed form that, when filled out and signed, instructs a bank to pay a specified sum of money to the person named on it. [6.7]

checking account A bank account that enables you to withdraw money or make payments to other people, using checks. [6.7]

commission That part of the pay earned by a salesperson that is calculated as a percent of the salesperson's sales. [6.6]

common factor A number that is a factor of two or more numbers is a common factor of those numbers. [2.1]

common multiple A number that is a multiple of two or more numbers is a common multiple of those numbers. [2.1]

Commutative Property of Addition Two numbers can be added in either order; the sum will be the same. [1.2]

Commutative Property of Multiplication Two numbers can be multiplied in either order; the product will be the same. [1.4]

composite number A number that has whole-number factors besides 1 and itself. For instance, 18 is a composite number. [1.7]

compound interest Interest computed not only on the original principal but also on interest already earned. [6.3]

cost The price that a business pays for a product. [6.2]

cross product In a proportion, the product of the numerator on the left side of the proportion times the denominator on the right, and the product of the denominator on the left side of the proportion times the numerator on the right. [4.3]

decimal A number written in decimal notation. [3.1]

decimal notation Notation in which a number consists of a whole-number part, a decimal point, and a decimal part. [3.1]

decimal part In decimal notation, that part of the number that appears to the right of the decimal point. [3.1]

decimal point In decimal notation, the point that separates the whole-number part from the decimal part. [3.1]

denominator The part of a fraction that appears below the fraction bar. [2.2]

deposit slip A form for depositing money in a checking account. [6.7]

difference In subtraction, the result of subtracting two numbers. [1.3]

discount The difference between the regular price and the sale price. [6.2]

discount rate The percent of a product's regular price that is represented by the discount. [6.2]

dividend In division, the number into which the divisor is divided to yield the quotient. [1.5]

division The process of finding the quotient of two numbers. [1.5]

divisor In division, the number that is divided into the dividend to yield the quotient. [1.5]

down payment The percent of a home's purchase price that the bank, when issuing a mortgage, requires the borrower to provide. [6.4]

equivalent fractions Equal fractions with different denominators. [2.3]

expanded form The number 46,208 can be written in expanded form as $40,000 + 6000 + 200 + 0 + 8$. [1.1]

exponent In exponential notation, the raised number that indicates how many times the number to which it is attached is taken as a factor. [1.6]

exponential notation The expression of a number to some power, indicated by an exponent. [1.6]

factors In multiplication, the numbers that are multiplied. [1.4]

factors of a number The whole-number factors of a number divide that number evenly. [1.7]

finance charges Interest charges on purchases made with a credit card. [6.3]

fixed-rate mortgage A mortgage in which the monthly payment remains the same for the life of the loan. [6.4]

fraction The notation used to represent the number of equal parts of a whole. [2.2]

fraction bar The bar that separates the numerator of a fraction from the denominator. [2.2]

graph of a whole number A heavy dot placed directly above that number on the number line. [1.1]

greater than A number that appears to the right of a given number on the number line is greater than the given number. [1.1]

greatest common factor (GCF) The largest common factor of two or more numbers. [2.1]

hourly wage Pay calculated on the basis of a certain amount for each hour worked. [6.6]

improper fraction A fraction greater than or equal to 1. [2.2]

interest Money paid for the privilege of using someone else's money. [6.3]

interest rate The percent used to determine the amount of interest. [6.3]

inverting a fraction Interchanging the numerator and denominator. [2.7]

least common denominator (LCD) The least common multiple of denominators. [2.4]

Copyright © Houghton Mifflin Company. All rights reserved.

least common multiple (LCM) The smallest common multiple of two or more numbers. [2.1]

less than A number that appears to the left of a given number on the number line is less than the given number. [1.1]

license fees Fees charged for authorization to operate a vehicle. [6.5]

loan origination fee The fee a bank charges for processing mortgage papers. [6.4]

markup The difference between selling price and cost. [6.2]

markup rate The percent of a product's cost that is represented by the markup. [6.2]

maturity value of a loan The principal of a loan plus the interest owed on it. [6.3]

minuend In subtraction, the number from which another number (the subtrahend) is subtracted. [1.3]

mixed number A number greater than 1 that has a whole-number part and a fractional part. [2.2]

monthly mortgage payment One of 12 payments due each year to the lender of money to buy real estate. [6.4]

mortgage The amount borrowed to buy real estate. [6.4]

multiples of a number The products of that number and the numbers 1, 2, 3, ... [2.1]

multiplication The process of finding the product of two numbers. [1.4]

Multiplication Property of One The product of a number and one is the number. [1.4]

Multiplication Property of Zero The product of a number and zero is zero. [1.4]

number line A line on which a number can be graphed. [1.1]

numerator The part of a fraction that appears above the fraction bar. [2.2]

Order of Operations Agreement A set of rules that tells us in what order to perform the operations that occur in a numerical expression. [1.6]

percent Parts per hundred. [5.1]

percent decrease A decrease of a quantity, expressed as a percent of its original value. [6.2]

percent increase An increase of a quantity, expressed as a percent of its original value. [6.2]

period In a number written in standard form, each group of digits separated from other digits by a comma or commas. [1.1]

place value The position of each digit in a number written in standard form determines that digit's place value. [1.1]

place-value chart A chart that indicates the place value of every digit in a number. [1.1]

points A term banks use to mean percent of a mortgage; used to express the loan origination fee. [6.4]

prime factorization The expression of a number as the product of its prime factors. [1.7]

prime number A number whose only whole-number factors are 1 and itself. For instance, 13 is a prime number. [1.7]

principal The amount of money originally deposited or borrowed. [6.3]

product In multiplication, the result of multiplying two numbers. [1.4]

proper fraction A fraction less than 1. [2.2]

property tax A tax based on the value of real estate. [6.4]

proportion An expression of the equality of two ratios or rates. [4.3]

quotient In division, the result of dividing the divisor into the dividend. [1.5]

rate A comparison of two quantities that have different units. [4.2]

ratio A comparison of two quantities that have the same units. [4.1]

reciprocal of a fraction The fraction with the numerator and denominator interchanged. [2.7]

remainder In division, the quantity left over when it is not possible to separate objects or numbers into a whole number of equal groups. [1.5]

rounding Giving an approximate value of an exact number. [1.1]

salary Pay based on a weekly, biweekly, monthly, or annual time schedule. [6.6]

sale price The reduced price. [6.2]

sales tax A tax levied by a state or municipality on purchases. [6.5]

selling price The price for which a business sells a product to a customer. [6.2]

service charge A sum of money charged by a bank for handling a transaction. [6.7]

simple interest Interest computed on the original principal. [6.3]

simplest form of a fraction A fraction is in simplest form when there are no common factors in the numerator and denominator. [2.3]

simplest form of a rate A rate is in simplest form when the numbers that make up the rate have no common factor. [4.2]

simplest form of a ratio A ratio is in simplest form when the two numbers do not have a common factor. [4.1]

standard form A whole number is in standard form when it is written using the digits 0, 1, 2, . . . , 9. An example is 46,208. [1.1]

subtraction The process of finding the difference between two numbers. [1.3]

subtrahend In subtraction, the number that is subtracted from another number (the minuend). [1.3]

sum In addition, the total of the numbers added. [1.2]

total cost The unit cost multiplied by the number of units purchased. [6.1]

true proportion A proportion in which the fractions are equal. [4.3]

unit cost The cost of one item. [6.1]

unit rate A rate in which the number in the denominator is 1. [4.2]

whole numbers The whole numbers are 0, 1, 2, 3, ... [1.1]

whole-number part In decimal notation, that part of the number that appears to the left of the decimal point. [3.1]

Copyright © Houghton Mifflin Company. All rights reserved.

Index

Addends, 9
Addition, 9
  applications of, 12, 80, 134
  Associative Property of, 9
  calculator for, 10, 11
  carrying in, 10
  Commutative Property of, 9
  of decimals, 133–134
  estimating the sum, 11, 133
  of fractions, 77–80
  of mixed numbers, 78–80
  on the number line, 9
  Order of Operations Agreement and, 47
  properties of, 9
  related to subtraction, 17
  verbal phrases for, 10
  of whole numbers, 9–12
Addition Property of Zero, 9
Amount, in percent problems, 207, 211, 215, 219
Application problems, 12
Approximately equal to (≈), 11
Approximation, *see* Estimation; Rounding
Associative Property of Addition, 9
Associative Property of Multiplication, 25
Average, 39

Balancing a checkbook, 274–277
Bank statements, 274–277
Bars, musical, 115
Base, in percent problems, 207, 211, 215, 219
Basic percent equation, 207, 211, 215
  car expenses and, 265
  commission and, 269
  percent decrease and, 242
  percent increase and, 239
Borrowing in subtraction
  with mixed numbers, 86
  with whole numbers, 18

Calculator
  addition, 10, 11
  decimal places in display, 152
  decimal point, 141
  division, 10
  division by zero and, 33
  estimation of answer, 11
  exponential expressions, 46
  fraction key, 77
  multiplication, 10
  Order of Operations Agreement and, 55
  percent key, 211
  problem solving with, 223
  proportions, 219
  rounding to nearest cent, 260
  square of a number, 47
  subtraction, 10
Car expenses, 265–266, 284
Carrying
  in addition, 10
  in multiplication, 26

Chapter Review Exercises, 59, 119, 167, 195, 227, 287
Chapter Summary, 56, 116, 165, 194, 226, 285
Chapter Test, 61, 121, 169, 197, 229, 289
Check, 273
Checking account, 273
  balancing checkbook for, 274–277
Commissions, 269, 270
Common denominator
  in adding fractions, 77
  in subtracting fractions, 85
Common factor, 66
Common multiple, 65
Commutative Property of Addition, 9
Commutative Property of Multiplication, 25
Composite number, 51
Compound interest, 252–254
Construction
  floor plans for, 193
  of stairway, 115
Consumer price index (CPI), 225–226
Conversion
  between decimals and fractions, 160–161
  between improper fractions and mixed numbers, 70
  between percent and fraction or decimal, 203–204
Cost, 240
  total, 236
  unit, 235–236
Counterexample, 283
Credit card finance charges, 251–252
Cross products, 183
Cube of a number, 46
Cumulative Review Exercises, 123, 171, 199, 231, 291

Decimal notation, 127
Decimal part, 127
Decimal point, 127
  with calculator, 141
Decimals, 127
  addition of, 133–134
  applications of, 134, 138, 143–144, 154
  converting to/from fractions, 160–161
  converting to/from percents, 203–204
  dividing by powers of ten, 152
  division of, 151–154
  estimation of, 133, 138, 143, 153
  multiplication of, 141–144
  multiplying by powers of ten, 141–142
  on number line, 161
  order relations of, 161
  relationship to fractions, 127
  repeating, 165
  rounding of, 129–130
  subtraction of, 137–138
  terminating, 165
  word form of, 127–128

Decrease
  percent, 242–244
  as subtraction, 18
Denominator, 69
  common, 77, 85
Deposit slip, 273
Difference, 17, 18
  estimating, 19, 138
Discount, 243–244
Discount rate, 243–244
Dividend, 33
Divisibility rules, 50
Division, 33
  applications of, 39–40, 54, 103–104, 154
  of decimals, 151–154
  estimating the quotient, 39, 153
  factors of a number and, 50
  fraction as, 34
  of fractions, 101–104
  of mixed numbers, 102–104
  one in, 33
  Order of Operations Agreement and, 47
  by powers of ten, 152
  remainder in, 35–36
  verbal phrases for, 37
  of whole numbers, 33–40
  zero in, 33
Divisor, 33
Down payment
  on car, 265
  on house, 259

Economical purchase, 235–236
Equations, percent, 207–208, 211–212, 215–216
Equivalent fractions, 73
Estimation
  of decimals, 133, 138, 143, 153
  of percents, 224
  in problem solving, 54, 224
  in using calculator, 11
  of whole numbers, 11, 19, 27, 39
  see also Rounding
Euler, Leonhard, 283
Expanded form, of a whole number, 4–5
Exponent(s), 46
  with fractions, 110–111
  one as, 46
  powers of ten, 46, 141–142, 152
Exponential expressions, simplifying, 46, 47, 110–111
Exponential notation, 46

Factor(s), 50–51
  common, 66
  greatest common, 66, 93
  in multiplication, 25
Factorization, prime, 51, 65, 66
Fermat, Pierre de, 283
Finance charges, 251–252, 284

Copyright © Houghton Mifflin Company. All rights reserved.

Copyright © Houghton Mifflin Company. All rights reserved.

Copyright © Houghton Mifflin Company. All rights reserved.

Index of Applications

TI-83 Plus/84 Plus*

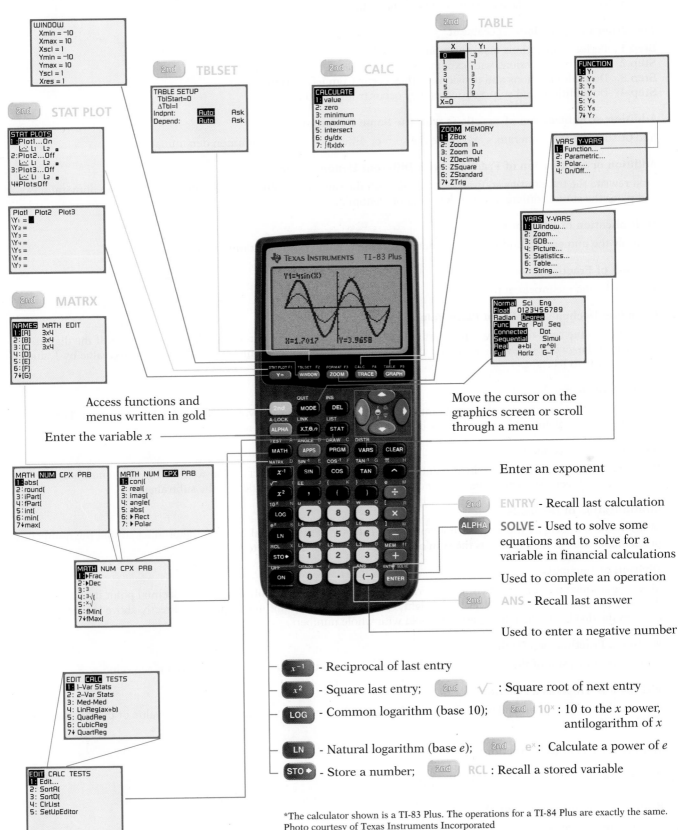

WINDOW
Xmin = -10
Xmax = 10
Xscl = 1
Ymin = -10
Ymax = 10
Yscl = 1
Xres = 1

2nd TBLSET

TABLE SETUP
TblStart=0
ΔTbl=1
Indpnt: Auto   Ask
Depend: Auto   Ask

2nd CALC

CALCULATE
1: value
2: zero
3: minimum
4: maximum
5: intersect
6: dy/dx
7: ∫f(x)dx

2nd TABLE

X	Y₁
0	-3
1	-1
2	1
3	3
4	5
5	7
6	9

X=0

FUNCTION
1: Y₁
2: Y₂
3: Y₃
4: Y₄
5: Y₅
6: Y₆
7↓ Y₇

ZOOM MEMORY
1: ZBox
2: Zoom In
3: Zoom Out
4: ZDecimal
5: ZSquare
6: ZStandard
7↓ ZTrig

VARS Y-VARS
1: Function...
2: Parametric...
3: Polar...
4: On/Off...

2nd STAT PLOT

STAT PLOTS
1: Plot1...On
   ⌐ L₁ L₂ ▪
2: Plot2...Off
   ⌐ L₁ L₂ ▪
3: Plot3...Off
   ⌐ L₁ L₂ ▪
4↓ PlotsOff

Plot1 Plot2 Plot3
\Y₁ = ▪
\Y₂ =
\Y₃ =
\Y₄ =
\Y₅ =
\Y₆ =
\Y₇ =

VARS Y-VARS
1: Window...
2: Zoom...
3: GDB...
4: Picture...
5: Statistics...
6: Table...
7: String...

Normal Sci Eng
Float 0123456789
Radian Degree
Func Par Pol Seq
Connected   Dot
Sequential   Simul
Real a+bi re^θi
Full Horiz G-T

2nd MATRX

NAMES MATH EDIT
1: [A]   3x4
2: [B]   3x4
3: [C]   3x4
4: [D]
5: [E]
6: [F]
7↓ [G]

Access functions and menus written in gold

Enter the variable *x*

Move the cursor on the graphics screen or scroll through a menu

MATH NUM CPX PRB
1: abs(
2: round(
3: iPart(
4: fPart(
5: int(
6: min(
7↓ max(

MATH NUM CPX PRB
1: conj(
2: real(
3: imag(
4: angle(
5: abs(
6: ▶Rect
7: ▶Polar

Enter an exponent

2nd   ENTRY - Recall last calculation

ALPHA   SOLVE - Used to solve some equations and to solve for a variable in financial calculations

Used to complete an operation

MATH NUM CPX PRB
1: ▶Frac
2: ▶Dec
3: ³
4: ³√(
5: ˣ√(
6: fMin(
7↓ fMax(

2nd   ANS - Recall last answer

Used to enter a negative number

x^{-1} - Reciprocal of last entry

x^2 - Square last entry;   2nd √ : Square root of next entry

LOG - Common logarithm (base 10);   2nd 10ˣ : 10 to the *x* power, antilogarithm of *x*

EDIT CALC TESTS
1: 1-Var Stats
2: 2-Var Stats
3: Med-Med
4: LinReg(ax+b)
5: QuadReg
6: CubicReg
7↓ QuartReg

LN - Natural logarithm (base *e*);   2nd eˣ : Calculate a power of *e*

STO - Store a number;   2nd RCL : Recall a stored variable

EDIT CALC TESTS
1: Edit...
2: SortA(
3: SortD(
4: ClrList
5: SetUpEditor

*The calculator shown is a TI-83 Plus. The operations for a TI-84 Plus are exactly the same.
Photo courtesy of Texas Instruments Incorporated

Essential Rules

Rounding Whole Numbers to a Given Place Value

If the digit to the right of the given place value is less than 5, that digit and all digits to the right are replaced by zeros. If the digit to the right of the given place value is greater than or equal to 5, increase the digit in the given place value by 1, and replace all other digits to the right by zeros.

The Order of Operations Agreement

Step 1 Perform operations inside grouping symbols.
Step 2 Simplify expressions with exponents.
Step 3 Do multiplications and divisions as they occur from left to right.
Step 4 Do additions and subtractions as they occur from left to right.

Addition or Subtraction of Fractions with the Same Denominator

Add or subtract the numerators and place the sum or difference over the common denominator.

Addition or Subtraction of Fractions with Different Denominators

First rewrite the fractions as equivalent fractions with the same denominator. Then add or subtract the numerators and place the sum or difference over the common denominator.

Multiplication of Fractions

Multiply the numerators and place the product over the product of the denominators.

Division of Fractions

Multiply by the reciprocal of the divisor.

Rounding Decimals to a Given Place Value

If the digit to the right of the given place value is less than 5, drop that digit and all digits to the right. If the digit to the right of the given place value is greater than or equal to 5, increase the number in the given place value by 1, and drop all digits to its right.

Addition of Decimals

Write the numbers so that the decimal points are on a vertical line. Add as you would with whole numbers. Then write the decimal point in the sum directly below the decimal points in the addends.

Subtraction of Decimals

Write the numbers so that the decimal points are on a vertical line. Subtract as you would with whole numbers. Then write the decimal point in the difference directly below the decimal point in the subtrahend.

Multiplication of Decimals

Multiply the numbers as you would whole numbers. Then write the decimal point in the product so that the number of decimal places in the product is the sum of the decimal places in the factors.

Division of Decimals

Move the decimal point in the divisor to the right so that it is a whole number. Move the decimal point in the dividend the same number of places to the right. Place the decimal point in the quotient directly above the decimal point in the dividend. Then divide as you would with whole numbers.

Writing a Fraction as a Decimal

Divide the numerator of the fraction by the denominator. Round the quotient to the desired place value.

Writing a Decimal as a Fraction

Remove the decimal point and place the decimal part over a denominator equal to the place value of the last digit in the decimal.

Writing a Percent as a Fraction

Remove the percent sign and multiply by $\frac{1}{100}$.

Writing a Percent as a Decimal

Remove the percent sign and multiply by 0.01.

Writing a Fraction or a Decimal as a Percent

Multiply by 100%.

Copyright © Houghton Mifflin Company. All rights reserved.